A SEA MONSTER'S TALE
In search of the Basking Shark

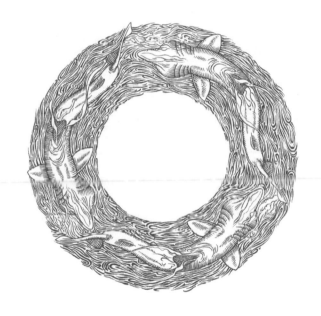

Colin Speedie

WILD
NATURE
PRESS

Published in 2017 by
Wild Nature Press
Winson House, Church Road
Plympton St. Maurice, Plymouth PL7 1NH

A catalogue record for this book is available from the British Library.

ISBN 978-0-9573946-8-1

Printed and bound in Malta on behalf of Latitude Press

10 9 8 7 6 5 4 3 2 1

www.wildnaturepress.com

Contents

Foreword

Sailing and looking for sharks are two of my favourite activities. They are trumped only by actually watching Basking Sharks, from a yacht, in the staggeringly beautiful waters of west Scotland. It is therefore a great privilege to be invited to write the foreword to a book that is a wonderful reminder of superb days at sea in the company of sharks and an expert on one of the world's largest fish species.

I have been fortunate enough to take part in several of Colin Speedie's expeditions to survey Basking Sharks in southwest England and the Western Isles of Scotland. Even if no sharks appear, Colin's crew invariably enjoy discussing the history of Basking Shark fisheries in Scotland and elsewhere, and debating the mysteries of their long-distance migrations and the changing patterns (in time and place) of their seasonal feeding and breeding aggregations. The regular appearance of Basking Sharks has attracted the attention of hunters for hundreds of years and these huge animals were being fished off the Scottish coast less than 50 years ago. The mouldering remains of this industry can still be found, sometimes more reliably than the surviving sharks themselves. However, when they do appear… Wow! Every sight of a gigantic shark is amazing. It's also a stark reminder of the bravery, foolishness or desperation of the fishers who used to catch these huge sea monsters, often from tiny boats, as described so grippingly here.

The author tracks the gradual change in public attitude, from bloodthirsty engagement in early shark eradication programmes, to participation in citizen-science research projects, to keen advocacy for shark conservation. Several decades after the last shark harpoon gun was fired in British waters, our Basking Shark population is showing signs of recovery. It now supports a sustainable, non-consumptive industry: shark watching. So, this book is very timely for the rising numbers of people able to go out to watch Basking Sharks. But it's far more than that: Colin is a superb raconteur, and this book demonstrates that he's also a fantastic writer who has done a huge amount of historical research. I used to think that I know quite a lot about Basking Sharks but I know far more now.

This book is also important because it demonstrates that shark conservation, management and the rebuilding of depleted populations is possible, particularly with the mobilisation of public support. Admittedly, it has taken a long time to get to this point. As described here, marine conservationists worked for decades to raise public awareness and to persuade decision-makers to end target fisheries and introduce wildlife protection or zero fisheries quotas for Basking Sharks in the UK and Europe, then to list the species in international wildlife conventions. These efforts led in due course to stronger conservation and management efforts in other countries, and have since been followed by similar successful campaigns for the conservation of other shark and ray species.

The Basking Shark was a path-breaker; an essential charismatic flagship species that helped to change public opinion and forge the understanding that sharks are wildlife too. Its story demonstrates that sharks can be commercially important for recreation and tourism as well as for fisheries exploitation. As the road to shark conservation and management broadens in the Basking Shark's wake, it has become slightly easier to achieve the protection of forgotten species, long since depleted by target and bycatch fisheries, and even to lever more sustainable management of sharks and rays before they become threatened. However, we can only continue to make such progress if there is overt public support for shark and ray conservation.

Dear reader, if you enjoy this book, you are a shark conservationist too. Please support organisations that campaign to ensure that Basking Shark conservation continues to address new threats, and other shark and ray species benefit from the efforts described here.

Sarah Fowler

Shark conservation and management policy advisor

For Louise, my partner in life, love and adventure

And for Guy and Frances who so kindly tolerated my extended absences

Introduction

In a still sea on a calm Hebridean morning a giant shark is following our research yacht. Scarcely making any forward progress, this nine-metre-long monster has taken up station off our stern, and is slowly wagging its head from side to side, as if sniffing for some scent that might identify the yacht as friend or foe. Beside me, one of our volunteer crew is jabbering like a game show host, almost overwhelmed by the experience of his first encounter with *Cetorhinus maximus*, the mighty Basking Shark.

As well he might be. It's not every day that you find yourself being approached intimately by such a behemoth. Capable of reaching up to 11m in length, this fish is longer than a London bus, but on first acquaintance it is not just the length but the colossal girth that astonishes. With its slow, sinuous movements, the Basking Shark is reminiscent of some creature from a long-forgotten age, and indeed it can trace its lineage back for many millions of years. Out here in this wild, intoxicating mix of land, sea and sky, it seems so perfectly at home, an essential ingredient of this timeless place, that we're almost unsurprised to encounter it.

Yet in some ways it's remarkable that we are able to experience this moment at all, as man has conducted a war of attrition on this harmless, plankton feeding fish for centuries, not least here in the Sea of the Hebrides off the West coast of Scotland.

Meet the beast

The Basking Shark belongs in the class of the Chondrichthyans, the cartilaginous fishes, in the sub-class Elasmobranchii, the sharks. The first shark is believed to have existed nearly 400 million years ago. Since then, sharks have diversified to inhabit a highly diverse range of habitats and a wide variety of lifestyles. Some evolved to become filter feeders and achieve great size, such as the Basking Shark, the Megamouth Shark (*Megachasma pelagios*) and the Whale Shark (*Rhincodon typus*). Of these three, only the Basking Shark feeds solely on plankton, and belongs in its own family, Cetorhinidae, of which it is the only living species. Fossil records suggest that sharks of the genus *Cetorhinus* that closely resemble today's animal have existed since the middle Eocene, making the family at least 38 million years old.

Out on the water, the Basking Shark is a creature of such massive scale and obvious power that at first glance it can be hard to believe it is a harmless filter feeder. That first impression is heightened by the way that it closely resembles some of its relatives. The Basking Shark belongs in the order Lamniformes, along with many of the predatory oceanic mackerel sharks such as the White Shark (*Carcharodon carcharias*), the Shortfin Mako (*Isurus oxyrinchus*) and the Porbeagle (*Lamna nasus*), and shares many external features with them, including horizontal keels in the area of the caudal peduncle and a second anal fin. No doubt it is this resemblance that is one of the main reasons for the annual 'silly season' sightings of White Sharks that form a perennial staple amongst some of the more colourful British newspapers.

The Basking Shark is widely distributed around the world, mainly in warm temperate and boreal waters in the North Atlantic and North Pacific oceans, in the Mediterranean and in the southern hemisphere between South America, South Africa, Australia and New Zealand. Basking Sharks are generally sighted at the surface during spring and summer in the North Atlantic, along the continental shelf edge and coastal waters. Efficient foragers, they travel constantly to identify and exploit the best patches of plankton, and satellite tagging studies have shown that the species will migrate over thousands of kilometres and dive to depths in excess of 1,000m to seek out their prey.

Appearance

The massive body is cylindrical with huge pectoral fins and a large crescent-shaped caudal (tail) fin. There are two dorsal fins, the first being the huge 'sail' that is the first visible sign that gives away the presence of a Basking Shark at the surface, which can reach over one metre tall in the biggest fish.

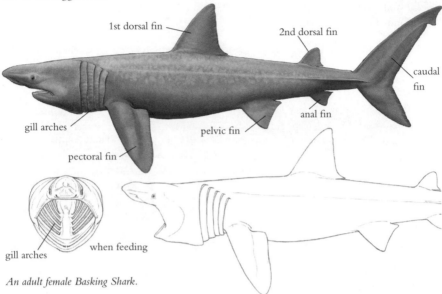

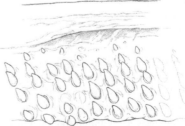

An adult female Basking Shark.

The head is the most obvious and impressive element in adult fish, especially when they are feeding. When swimming with the mouth partly open, the head is slender with a conical, pointed rostrum (snout), and the eyes well forward. Once the animal opens its mouth to feed, the appearance of the head is completely altered, not only as the gape is around one metre (three feet) in a seven-metre shark, but also because the whole of the branchial region enlarges, looking rather like an approaching jet engine when viewed from ahead.

Five pairs of gill arches almost completely encircle the body, and fan open as the shark swims forward, allowing the water entering the mouth to pass through the filamentous gill rakers in the arches, filtering out the shark's favoured planktonic prey, calanoid copepods, and allowing respiration. The inside of the mouth is bright white, with occasional small, darker blotches, a form of coloration that some researchers believe may attract the copepods. The jaws are rimmed by a pink strip of tissue with rows of minute, hooked teeth: six rows on the upper jaw and nine on the lower.

The Basking Shark possesses hundreds of tiny, hooked teeth, despite the fact that it only feeds on plankton.

Size

With a maximum total length in excess of ten metres (33 feet) the Basking Shark is the second largest fish in the world, exceeded only by another filter feeder, the Whale Shark of tropical seas, that can achieve lengths of up to 21 metres. It has been suggested that the maximum weight of a nine-metre (30ft) individual would be 6.5–7 tonnes, whilst a more recent study using an isometric length–weight relationship proposes a maximum weight for a 10m individual of 7.5 tonnes.

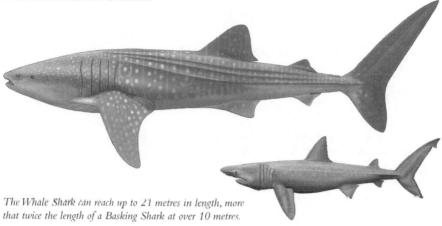

The Whale Shark can reach up to 21 metres in length, more that twice the length of a Basking Shark at over 10 metres.

Basking Sharks have often been described by researchers as being 'dark-grey, almost black', but this is probably due to the small number of sharks examined and the fact that they were dead animals that had been allowed to become dry. After death the colour and markings can fade and change, giving an erroneous impression of the true colour. In living animals, colour varies considerably, and the shark has attractive python-like markings that run from the top of the head down the length of the body.

The skin is deeply fissured and is covered in minute sharp denticles that are highly abrasive to the touch. The skin is coated in a dark, acrid-smelling slime that may have anti-fouling properties, to ward off parasitic copepods such as *Dinematura producta* and the Sea Lamprey (*Petromyzon marinus*) that attach themselves to the skin of the shark. Many sharks seem to carry these creatures, particularly in the ventral area around the cloaca, although the scars left by their attachment can often be seen even on the fins.

Life history and reproduction

Basking Sharks are believed to be long-lived, perhaps reaching 50 years of age. The females are believed to reach sexual maturity at around 18 years, with males reaching that life stage slightly younger, at around 17 years. Mature males are equipped with two external sexual organs called claspers, that are around one metre long, and form a claw to stay in place during copulation. The gestation period is believed to be one of the longest of all species, at up to 2.6 years, when five to six live pups of 1.5–1.7m in length are born. There may be a two to four-year gap between pregnancies. Due to a lack of natural predators, the rate of natural mortality may be as low as 0.091. If these figures are correct, the species may have among the lowest natural mortality and productivity levels calculated for a commercially fished marine species.

Senses

Basking Sharks are equipped with the same sensory faculties as other sharks, notably a powerful sense of smell that may help them locate dense patches of zooplankton, as well as electroreception via the ampullae of Lorenzini, pit organs concentrated on the snout that are sensitive to electrical fields. As dense aggregations of zooplankton will almost certainly produce detectable electrical activity, this may be a further cue that the shark can employ to forage actively for the best patches of plankton. Eyesight is not especially developed, but is acute enough to identify potential threats such as small craft or swimmers at close range.

Basking at the surface

Surface swimming is really the tip of the iceberg. At any one time that a shark is sighted at the surface, there are many, many more out of sight in the depths. Surface swimming is another form of foraging behaviour, when dense patches of zooplankton are high in the water column, and the sharks track those patches to exploit them. The Basking Shark, like other sharks, possesses no swim bladder to provide buoyancy, but instead has a massive bi-lobal liver filled with oil. This acts as an energy reserve when plankton is less dense or when the shark is migrating, and as a hydrostatic organ, a 'hepatic float' capable of maintaining the shark at almost neutral buoyancy. The liver oil comprises almost 40% squalene, a terpenoid hydrocarbon that has a low specific gravity, and it is this compound that is mainly responsible for buoyancy. And it was this oil that was the main driver for the large-scale pursuit of the Basking Shark.

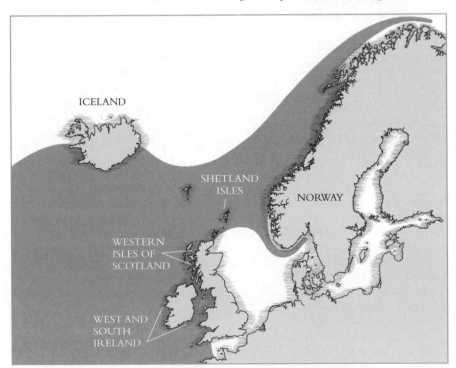

Basking Shark distribution in the North-east Atlantic, including key historic hunting sites.

Around the world, from Norway to the United States and from Ireland to Canada, many tens of thousands of these creatures were killed for that valuable liver oil, to light city streets and remote coastal dwellings. In the early years the hunt was on a relatively small scale, with Basking Sharks taken only on an opportunistic basis. But as demand increased to light city streets, so the hunt was intensified and became a significant threat to the species before the arrival of paraffin on the market offered a cheaper alternative.

At times considered a pest that damaged fishing nets and gear, fisheries protection vessels were sometimes employed to slaughter Basking Sharks in significant numbers. In the aftermath of World War II, a global shortage of high-grade oils meant that prices for the liver oil rocketed, prompting commercial fisheries for the species to become established and expand rapidly. Latterly, demand from the insatiable shark fin trade in Asia put a further bounty on the heads of Basking Sharks. These fisheries were on a hitherto unprecedented scale and were highly mechanised compared with previous exploitation efforts. As a result of these changes, the take of sharks increased exponentially, and it is likely that over 100,000 Basking Sharks were killed in the North-east Atlantic alone in the last 55 years of the twentieth century.

These pressures, over at least two centuries, combined to bring the Basking Shark to the brink of localised extinction in several places. Even today, at many sites where the Basking Shark was once considered seasonally abundant, sightings of animals remain drastically reduced. Perhaps now, with the collapse of those fisheries and the development of conservation measures to protect the Basking Shark, it may be hoped that populations can make a recovery and return to grace their old haunts once more.

But how to know whether such a recovery is underway, when the creature you wish to study spends more than 99% of its life out of sight, below the surface of the water? For decades the scientific record held little more than informed speculation about the ecology of the Basking Shark, combined with a broad understanding of the biology of the species gleaned from stranded or harpooned specimens.

Fortunately for the shark, at the same time that the nascent conservation movement identified the Basking Shark as a key species in need of protection, science was making giant steps in its ability to track animals wherever they might travel, and through the use of new technologies such as satellite tracking, many of the enduring mysteries concerning the shark and its whereabouts when below the surface have now been comprehensively laid to rest.

Hope for the future?

Over the last 70 years, attitudes towards the Basking Shark have swung dramatically from seeing the species as a valuable commercial prey or a 'plague on fisheries' to a 'gentle giant' in need of protection. For the hunters in Scotland after the cataclysm of World War II there must have seemed nothing ignoble about hunting the species to provide valuable liver oil in a time of necessity. To the fishermen in Canada who saw their nets destroyed by Basking Sharks, it must have seemed entirely reasonable to demand action to rid their waters of these 'pests'. But that was then and this is now, and the drastic reduction in numbers of these extraordinary creatures as a result of those historic actions demands that we do all we can to safeguard the remaining populations from further harm.

In the aftermath of the decimation of the species through exploitation, some questions remain unanswered. Where have the sharks gone from many of their favourite haunts, and is that absence entirely the fault of man? Fortunately, a road map of sorts exists for some areas, notably in Scotland. Some of the local hunters there during the last era of hunting left highly

11

readable accounts of their travails, allowing for insights to be made into the distribution and abundance of the Basking Shark, at least in some small areas of the sharks' former range.

Which is what has brought us here at this moment, as part of a project to establish whether there are still shark 'hotspots' in British waters, post-hunting, where these creatures can be seen in abundance. Our aim is to try and identify any such sites – if they still exist – to record Basking Shark numbers and social behaviour, with a view to securing further protection for those sites. This slow, painstaking work, in the often-difficult weather that plagues the wild waters of Britain's western seaboard, leaves us with a keen insight into the difficulties faced by the shark hunters, and a far greater respect for them as a result.

In time, our work will also teach us much about the local and global politics that dictate the terms on which creatures retain the chance to survive through protection – or not. It is, after all, one thing to build a case for a 'charismatic megafauna' beast like the Basking Shark, whilst many smaller, less obvious but no less important creatures never get noticed at all. But if you can't achieve protection for an iconic creature like the Basking Shark, then what hope is there for any species?

None of that matters right now, though. We're simply awestruck to find ourselves in this wonderful place with this extraordinary creature on our trail, and filled with excitement in anticipation of the journey of discovery ahead of us.

Chapter 1
The Early History

Given that whaling is believed to have started sometime in the eleventh century in northern Europe, it seems surprising that a huge fish that seasonally inhabited the same waters as whales might have gone unrecorded for so long. Yet it wasn't until over 500 years later that the Basking Shark first makes an appearance in the world of science.

The Basking Shark was first described, illustrated and given a scientific name in 1765 by Johan Ernst Gunnerus, bishop of Trondheim, in an article published in *Der Trondheimske Selskabs Skrifter* (Transactions of the Trondheim Society). As a travelling bishop with a large diocese, Gunnerus was able to indulge his enthusiasm for the natural world as he visited his parishes, making many worthwhile contributions to the understanding of the flora and fauna of Norway.

Having heard of a hunt developing locally for the Basking Shark (which he called the 'Brugde') in Norway, he made it his business to learn more about the creature, and during one of his regular diocesan journeys to Nordland, a region north of Trondheim, he met a

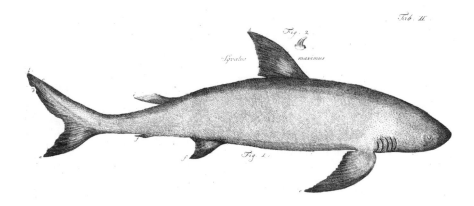

The first recorded illustration of a Basking Shark, as described by Johan Ernst Gunnerus.

A stranded male Basking Shark on a beach entertains sightseers. Note that the male sexual organs (claspers) have been interpreted as legs and feet! From Harpers Weekly 24th October 1868.

man involved in the nascent fishery. This landowner arranged for his foreman to send him an account of the fish, and in due time Gunnerus received a drawing, a piece of skin and a carved wooden model of the shark. Later, a stuffed skin nine metres long was sent to him from the island of Smolen, and upon this evidently well-preserved specimen he based his scientific description. Having examined the specimen, it was clear to him that this was not any previously described shark, but was in fact a new species to science.

A regular correspondent with the renowned Swedish systematist, Carl von Linne (Linnaeus), Gunnerus adopted Linnaeus's binomial system to name this new shark *Squalus maximus* (the biggest shark). Linnaeus incorporated this name into the 12th edition of his *Systema Naturae* of 1776, establishing once and for all that Gunnerus was the first person to make a scientific description of the shark. The scientific name remained in place until 1816 when Henri-Marie de Blainville of the Paris Natural History Museum created a new sub-genus *Cetorhinus* for the Basking Shark, and the species remains *Cetorhinus maximus* (Gunnerus 1765) to this day.

Around the same time another correspondent with Linnaeus, the Welsh naturalist Thomas Pennant, made the first report of the Basking Shark in his *British Zoology* of 1769. Working from reports sent to him by two Anglesey rectors well acquainted with a local fishery for the creature, and from samples of skin, jaws and gill rakers, Pennant proposed that the creature was indeed a shark, as opposed to a whale, especially due to its vertical tail. Despite containing a number of unavoidable errors (he had not yet examined a full specimen), his treatise contained one contribution that has lasted to this day. He proposed to name his shark the Basking Shark, after its habit of 'lying as if to sun itself on the surface of the water'.

He made this change from the name commonly used in Ireland and Wales, the sun-fish, largely to avoid confusion with the huge Ocean Sunfish (*Mola mola*) that was an irregular visitor to Britain's shores. However, he retained the imagery of sun-fish, that of a creature that rises to the surface to 'bask' in the sunlight, when the truth is far more prosaic, in that the shark only spends time at the surface when its planktonic prey are there. But the name has stayed and it is a popular name that works well.

Pennant travelled to Scotland some years later on a fact-gathering expedition to produce an enlarged edition of his book, *A Tour in Scotland*. On this, his second voyage to the region, he expended his itinerary to take in the Hebrides, landing at Loch Ranza on Arran in the Firth of Clyde in late June 1772, just in time to make his first encounter with a Basking Shark. An 8.3m shark had been captured and was lying on a nearby beach, giving him a perfect opportunity to inspect the carcass. The information he gleaned from the fishermen and the data that he gathered from the stranded shark greatly added to the text of *A Tour in Scotland and Voyage to the Hebrides 1772* and included a fine drawing of the hunt he had witnessed.

His later edition of *British Zoology* included an enhanced description of the Basking Shark derived from the animal he inspected at Loch Ranza. It also included one of the earliest sketches of a Basking Shark which, although far from exact, is certainly no more bizarre than many of the others that came later, all of which likely suffered from being depictions of dead animals that had collapse of tissue or other early forms of decomposition. It is worth considering just how different a Basking Shark looks alive in its natural habitat, than decomposing on a beach. Such a soft, cartilaginous creature adopts some really strange contours when removed from its buoyant environment. Coloration and body markings change or fade, and in no time at all, a stranded Basking Shark looks very altered from the living creature.

There is no actual record of who drew the shark, but Denis Fairfax, in his excellent book *The Basking Shark in Scotland*, suggests two plausible alternatives. The first is the more obvious, that Pennant instructed the artist who accompanied him on his second tour of Scotland, Moses Griffith, to draw the shark he examined at Loch Ranza. As Moses Griffith drew the harpoon hunting scene in the same book in which the shark depiction is published, this seems a reasonable hypothesis. But Fairfax also suggests an intriguing alternative, that one of Pennant's ecclesiastical correspondents, the Reverend George Low of Birsay in Orkney, also sent him a sketch of a Basking Shark that had been harpooned close to his home. Once again the sketch would have been from a dead specimen, with all of the usual difficulties inherent.

The same problem affected many of the speculative (and some downright wild) identifications that were to be subsequently derived, especially those that came from well-intentioned but amateur naturalists around Britain. Foremost amongst these was Dr Jonathan Couch, a physician from the tiny fishing port of Polperro on the south coast of Cornwall. Seldom travelling from his home town, Couch devoted his life to the well-being of his fishermen patients. In return he was able to indulge his fascination with fish, examining and drawing a wide variety of specimens that his patients brought to him, some of which were later published in scientific journals. He eventually produced a four-volume series entitled *A History of the Fishes of the British Islands*, that included not only the Basking Shark but also two 'new' species, based on drawings that had been sent to him by correspondents. He named these two creatures the Rashleigh Shark (*Polyprosopus rashleighanus*) and the Broad-headed Gazer (*Polyprosopus macer*). To be fair, the drawings suffered from the usual problems of drying out, shrinkage and decomposition of the head area, and the wonderfully weird looking Broad-headed Gazer, with its very pointed snout, looks to have been drawn from a very young animal where the rostral area is more elongated than in an adult. But there's no doubt that they are both Basking Sharks, even though the drawings do their best to convince you otherwise.

BASKING SHARK.

XIV

Early attempts to capture the living animal in its natural habitat were seldom very accurate, having been drawn from second-hand reports or stranded animals. From Couch, A History of the Fishes of the British Islands 1860–65.

Undoubtedly it is the way that a dead Basking Shark appears in the more advanced stages of decomposition that has led to some of the more eccentric attributions to the species, in the form of sea monsters. The first thing to disintegrate is usually the gill arch and jaw area, leaving the small skull attached to the long spinal column, resulting in an almost dinosaur-like appearance. The more solid cartilaginous attachments of the pectoral and anal fins, and the claspers (if a male) are much slower to break down, and the remnants of their attachments were sometimes attributed to being part of three apparent pairs of legs. The tail is another early departure, leaving a long, slender spine, and the whole resembles nothing more than a plesiosaur, albeit with legs instead of fins. So on many occasions when a badly decomposed Basking Shark comes ashore or is trawled up from the seabed, sensation, dispute and disappointment have soon appeared on the scene.

Perhaps the best documented of these creatures comes from Britain: the so-called Stronsay Monster. Denis Fairfax describes how in 1808 some fishermen found the decomposing carcass of what they thought to be a whale floating off the island of Stronsay in the Orkney Islands. When it washed ashore later the news soon spread, and it wasn't long before speculation as to its origin began. At a meeting of the learned Wernerian Natural History Society in Edinburgh, members heard that the remains most resembled the 'Great Sea Snake of the Northern Ocean' as described by Hans Egede, Bishop of Greenland in1734 and Erik Pontoppidan, Bishop of Bergen in 1755.

The scene having been set, nature lent a helping hand when a storm all but wiped out the remains, leaving only the skull, a few attached vertebrae, a fin and some gill cartilage. Fairfax reports that these remnants were sent to Edward Home, a celebrated surgeon and anatomist, and a regular contributor to scientific journals. Home gathered the affidavits taken from the men who had seen the creature, together with a drawing made some six weeks afterwards. Home didn't take long to come to his judgment; these were the remains of a Basking Shark. And whilst he didn't doubt the sincerity of the men who had provided the information that led to the drawing, he was very much of the opinion that the men had read that the creature might be Pontoppidan's maned sea serpent, which had clouded their judgment.

Not everybody shared his view. In January 1809 Dr John Barclay, anatomist and founder member of the Wernerian Society, made a presentation to the members of that Society on the 'Great Sea Snake'. At the same meeting one Patrick O'Neill proposed that the animal be from a new genus *Halsydrus*, with the proposed addition of *pontoppidani* to underpin his belief that the creature was the one described by Pontoppidan. A monster was born.

However, Home was a very learned man, and had an existing interest in the Basking Shark. As a surgeon and anatomist, his main interest lay in anatomy and physiology, not in sea monsters, and having made a careful examination of the remaining physical parts of the creature, he was well placed to pronounce on the matter. When he presented his finding to the Royal Society in London in May 1809, it might be easy to imagine that his word alone would have carried the day. But not a bit of it, and Barclay hit back with rebuttals of Home's arguments in 1810 and 1811. The dispute grumbled along for more than 40 years, with Orcadian naturalist T.S. Traill still insisting that the Stronsay monster was no known shark. There are probably people who would agree with him today, for there are still periodic 'discoveries' of sea monsters, that have all eventually been proven to be Basking Sharks, ranging from British Columbia (several), Orkney again (*Scapasaurus*) New Zealand, and most recently the Philippines.

Dr Jonathan Couch identified two 'new' species that were undoubtedly both Basking Sharks. Above: the Rashleigh Shark (Polyprospus rashleaighanus). Below: the Broad-headed Gazer (Polyprosopus macer). From Couch, A History of the Fishes of the British Islands 1860–65.

Another possible way in which the Basking Shark has entered the world of cryptozoology is via its behaviour observed at the sea surface. Groups of sharks feeding at the surface, and especially those engaged in what is believed to be courtship behaviour, often swim in long lines, closely following each other with noses, dorsal fins and caudal fins showing. When seen from a distance they look like nothing other than a sea snake or plesiosaur.

A famous example of this may be *Cadborosaurus willsi* ('Caddy'), named from an original sighting in Cadboro Bay, Victoria, British Columbia. Between 1881 and 1991 there were 181 reports of 'Caddy' from around that area, at least some of which proved to be Basking Sharks. A book on the mythical creature, *Cadborosaurus: Survivor from the Deep*, pointed out that there was a significant rise in the number of 'Caddy' sightings between 1930 and 1960, which Scott Wallace and Brian Gisborne identify in their fascinating book, *The Basking Shark, the slaughter of BC's gentle giants*, as a period coincidental with Basking Sharks being abundant in those waters.

Even closer to my home these reports still occur. Many years ago, on a beautiful calm evening, I stood watching Basking Sharks from the extremity of Black Head, a promontory on the eastern side of the Lizard peninsula in Cornwall. I was just about to pack up and go home, when I was joined by an elderly but very hale and hearty lady who asked me if I was watching 'Morgawr'. I replied no, I was watching Basking Sharks, and pointed out a small group well offshore. Emphatically, she said, 'No, that's Morgawr, Cornwall's sea serpent. I watch him all of the time'. To this day I have no idea if she was pulling my leg, but her scornful tone did suggest that I was obviously simple and should have known better.

Chapter 2
The First Hunters

The first descriptions of the Basking Shark came about as a by-product of commercial fisheries in 1765 (Gunnerus, from Norway) and 1769 (Pennant, from Wales). The close relationship between those dates suggests that the hunting of the Basking Shark began almost simultaneously across a wide range. It soon gathered momentum, developing into a sustained fishery that would be actively prosecuted over a wide front in the North-east Atlantic, from Ireland through Scotland, Iceland and Norway.

Fish, whale and seal oil were already established commodities in those places, mainly for local use. It would have been logical for the inhabitants of those coasts to take advantage of any other sizeable animals (including Basking Sharks) that came ashore, and they would surely have soon recognised the value of the sharks' livers to furnish oil. Basking Sharks do strand or are washed ashore from time to time, often as the result of accidental capture in fisheries for other species, and local people would undoubtedly have eyed such substantial prizes for their potential benefits such as the skin, which was occasionally used as a crude form of sandpaper, as well as their liver oil. Denis Fairfax cites historical reports from Iceland that describe liver oil being extracted from stranded Basking Sharks, and there is no reason to assume that fishermen in Ireland or Scotland would have been any less ready to take advantage of such an opportunity.

Drive fisheries, where boats trapped small whales in bays and then chased them ashore, were also widely practised all around the Highlands and Islands of Scotland (and still are in the Faeroe Islands) and there is some evidence that similar tactics were employed for the Basking Shark, at least in parts of Loch Fyne, Scotland:

> The sun or sail-fish occasionally visits us; this sluggish fish sometimes swims into the salmon nets, and suffers itself to be drawn towards the shore, without any resistance, till it gets near the land, that for want of a sufficient body of water, it cannot exert its strength, in disentangling itself from the net, the fishers in the meantime take advantage of its situation, and attack it with sticks and stones, till they have it secure.

Fairfax quotes another report from the Orkney Islands telling of local people catching a shark by 'throwing a noose of rope over his pectoral fins, and playing him', so it seems likely that opportunistic hunting of the Basking Shark to supply local needs may have taken place over a wide area for some considerable time before the development of a truly commercial hunt.

The first actual record of the process of hunting the Basking Shark comes from Ireland in 1739, when William Henry described a hunt in Donegal Bay:

> The fishermen, making up to them, strike them with their harpoon irons.
> Whereupon they dart to the bottom and rolling on the ground, work the harpoon into the wound. Then being irritated, they rise again to the surface and shoot away with an incredible velocity dragging the boat after them, and they bear way sometimes for leagues; till at last dying, they float on the surface till the fishermen come along their side, and cut out the liver, which affords several barrels of oyl. In this dangerous war with these smaller leviathans, it is necessary to have 100 fathoms of small cord fixt to the end of the harpoon, to give it play: and for a man to stand by the gunnel of the boat, with a hatchet, to cut the rope in case of any stop in its running off, or the fish's emerging too suddenly; either of which accidents might overset the boat.

This vivid account indicates that it was already a well-organised fishery.

In 1740 an invitation to tender to supply sun-fish oil appeared in the *Dublin News Letter*. As the tender also requested supplies of rapeseed oil, Kenneth McNally in his fascinating book, *The Sun-fish Hunt*, suggests that it seems likely that both requirements were intended for street-lighting purposes. This argument is further supported by a reference he quotes from 1742 concerning street lighting in Galway and Waterford that specified 'sufficient lights to be lighted up and continue burning with double wick and a sufficient quantity of rape oil of the produce of this Kingdom, or of sun-fish oil made in this Kingdom', which suggests that a local sun-fish hunt was already an established feature of Irish coastal life, at least on the west coast.

It appears that the fishery might have been generated by a combination of increased abundance in correlation with the development and expansion of whaling. In 1744 it was reported that 'the coasts of Ireland, especially those in the west, have of late years been much frequented by whales and sun-fish, which come in March or April and stay till November'. In the years just prior to that report there had been a popular campaign to encourage the establishment of whaling activity on the north-west coast of Ireland, encapsulated in the short poem that opens McNally's book:

> Equip your boats with sharp Harpoon and Lance,
> Let's strive our publick Treasure to Advance;
> So shall returning Gold reward our toil,
> When London Lamps shall glow with Irish Oil.

Indeed, a Lieutenant Chaplain had taken up the challenge in 1737, and had established a short-lived whaling venture based around Killybegs on the northern shore of Donegal Bay. Unsuccessful though it may have been, it would have at least offered some training to local men in the art of harpoon fisheries, and the rendering down of body material to obtain oil, which may well have proved useful in the prosecution of a parallel Basking Shark fishery.

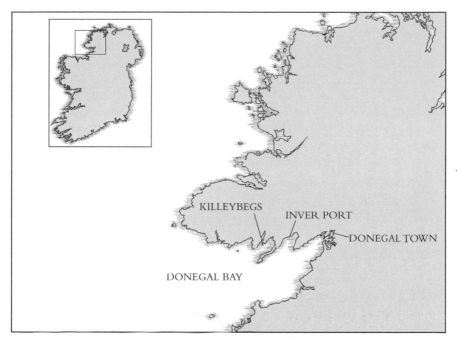

Donegal Bay, Ireland.

In 1750, Charles Smith described the fishery: 'the liver affords from twenty to one hundred gallons of oil. They are struck with harpoons and are well worth looking after'. It seems that by this time there were established uses and potential markets for considerable amounts of oil to be used in a variety of ways. The oil could be burned in the small 'cruisie' lamps that were the main source of illumination in the cottages of the coast, with the advantage that it was less smoky than other traditional fish oils, such as those obtained from herring or dogfish. Other uses included preserving timber and dressing wool, while the oil was often used as a salve or balm for injuries and aching limbs. But the most attractive prize must have been the development of a lucrative market for the high-grade oil for municipal lighting.

As Martin McGonigle reports in a paper on the establishment of a whaling and shark fishing station in Donegal Bay, it seems likely that this valuable commercial opportunity was the driving force for Andrew and Thomas Nesbitt when they established a large-scale whaling and shark-hunting station at Inver Port, near Killybegs, around 1760. In 1759 Thomas Nesbitt was already well established in both the herring fishery and hunting Basking Sharks. It seems logical to assume that the investment in plant and manpower was largely driven by his success in taking Basking Sharks, and the experience he had gained in harpoon fisheries and rendering down their livers for oil. In order to expand his operation, Nesbitt travelled to London to buy a whaling ship of 149 tons, which carried five whaleboats as catchers, and hired experienced men who had spent time whaling in Greenland. This substantial extra capacity in terms of men, gear and vessels to pursue whales also had significance for the hunting of sharks. As one reporter observed, 'the sun-fish are seldom caught, except by boats belonging to the whale fishery'. This was reflected in the capture of 42 Basking Sharks by the Nesbitt fleet in one week in 1761, an astonishing figure for that era, where other Basking Shark fisheries appear to have been limited to low-impact artisanal enterprises.

21

Such a professional attitude to the hunt was almost certainly the exception rather than the rule. Most of the men involved were farmers and fishermen, who, like their Hebridean counterparts, supplemented their income by seasonally taking part in the Basking Shark fishery. McNally reported that at the end of the eighteenth century Galway alone had between 40 and 50 boats involved. Small, open sailing vessels known as Galway hookers, the tough and seaworthy gaff-rigged craft that were the workhorses of the west coast and islands, were the preferred vessels for these men. Some even used the lightweight open curraghs, seemingly flimsy but amazingly seaworthy hide-covered boats, an act (as far as I am concerned) that must have required either extraordinary bravery or absolute desperation. McNally quotes The Galway Weekly Advertiser of May 1823:

> SUN FISHERY – We are happy to learn that the boats have been very successful in killing a number of sun-fish this season; the weather has been extremely favourable for the last three or four days; and we understand there are now upwards of one thousand boats upon the bank, seeking to take this most valuable fish, the consequence is, that no fresh fish of any kind is to be had in our market.

Even though there may be some doubt about the number of boats involved, it is evident that the fishery was viewed as being of such high importance to the local farmers/fishermen that they dropped everything (including fishing for other commercial species) to take part. The driving force must have been profit, the value of the liver oil driving men to take such chances that 'the poor fellows will risk even their lives to secure one of these fish'. The west coast of Ireland is a harsh and unforgiving place, which could support only subsistence agriculture, so the chance to secure such a valuable reward was apparently more than enough to draw those men away from tending to their farms, as was reported by the Commissioners of Fisheries: 'The pursuit of sun-fish or Basking Shark in the months of April and May employs a good many hands at a season particularly inconvenient … a considerable capital is applied to navigation, though very little to agriculture.'

The bank referred to in the Galway report is almost certainly the Sunfish Bank, and it was well-known to the fishermen of the coast that abundant shoals of Basking Sharks were to be found there in season, much larger than the scattered groups encountered closer to shore. In 1825 The National Fishing Company proposed the establishment of a fishing station at Inishbofin, an island off the coast of Connemara, precisely because of its proximity to the bank. The station was planned to support larger, more seaworthy vessels that could stay at sea in more challenging conditions and so pursue the seasonal opportunity afforded by the presence of the shoals of sun-fish. Construction of the station didn't go ahead, but the logic of using larger, more independent vessels to support small catcher craft was sound.

Although the west of Ireland did not have an abundance of larger vessels, they did have plenty of small craft and fishermen expert in handling them. Regional surveyor Hely Dutton reports that the small craft from Galway and Connemara took between 100 and 200 sharks there between 4 and 5 May 1815, which gives an indication of the size of the shoals. It is worth considering what a serious undertaking it would have been to fare 30 or 40 miles out to such an exposed site far from the rugged coastline of Achill Island. Almost out of sight of land on a clear day, in the ceaseless swell of the North Atlantic, exposed to rapidly changing weather and in a strong run of tide, this would not have been a place to linger except in the best of weather. As one local told the 1836 Commission of Enquiry:

Off the coast, about thirty miles or just within sight of the high land of Achill Island, is the ground called the Sunfish bank; we are on it with Sleavemore Achil, about E.N.E. per compass, seventy to ninety fathoms. The bank is remarkable for the break of the tide on it, with ebb and flood, and is supposed to be a ridge of land extending from the Blaskets to Erris Head, in about seventy fathoms. Half a mile further off we have fifteen fathoms more water, and the increase of depth is also considerable within it; the water outside deepens quickly to 100 fathoms and upwards; and the probability is, that the bank is nearer the edge of soundings.

And another confirmed:

There is a bank along the whole of the Galway coast, which extends from Inniskea, of the coast of Mayo, to the islands of Arran; about this bank, and still further to sea, are to be found the Basking Shark, or Sunfish.

'Remarkable for the break of the tide on it' and 'still further to sea'; these are the sort of remarks that would have been immediately picked up on by any mariner listening to such accounts. Given the distances and the slow speeds attainable by the small craft of the day, it would come as no surprise that boats going out to the bank might be expected to remain at sea for days on end, and the men in them to go without food for 24 hours or more while they played the sharks. The rewards must indeed have been great, and the potential kudos immense for the men who pursued the sun-fish for it to have been worthwhile. As local lore of the time pointed out 'A boatman was little esteemed long ago until he had spent a season or two at the Sunfish Bank'.

Given the dangers of hunting sharks at such a dangerous location, it should be noted with little surprise that when the fishery began to decline in the west of Ireland, crews stopped going out to the Sunfish bank. Instead, the hunters preferred to chase sharks that appeared in the bays and around the headlands along the coast. But as we shall see in a later chapter, that would not be the end of hunting the Basking Shark on the Sunfish Bank.

At around the same time, across the sea in the Highlands and Islands of the west coast of Scotland, the potential value of hunting the Basking Shark on a commercial basis hadn't gone unnoticed. Thomas Pennant clearly knew of their distribution in the North-east Atlantic when in 1772 he reported:

They inhabit most parts of the western coasts of the northern seas: Linnaeus says within the arctic circle: they are found lower, on the coast of Norway, about Orkney Isles, the Hebrides; and on the coast of Ireland in the Bay of Balishannon and on the west coast about Anglesea

As we saw earlier, the Norwegian fishery was already established by the time that Gunnerus described the Basking Shark, to which we can now add both Donegal Bay and Anglesey as existing hunting grounds. Pennant was equally aware of an experimental commercial fishery from the previous decade, established on the island of Canna in the Inner Hebrides. As in Ireland, and indeed elsewhere, it must surely be the case that Canna was selected on the basis of an existing resource of Basking Sharks nearby.

This experimental fishery was the brainchild of Donald MacLeod, tacksman of the island, who obtained a grant of £250 (over £30,000 in 2017) from the Commissioners for the

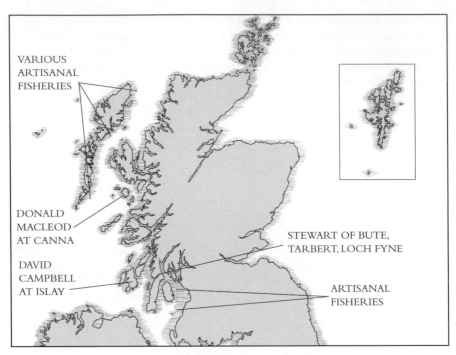

Early hunters and key hunting sites in Scotland.

Annexed Estates to buy a vessel and tackle to hunt the 'cairban' (the local Gaelic name for the Basking Shark) as well as to fish for cod and Ling (*Molva molva*). His grant application made the somewhat grandiose claim that he was 'the first that ever discovered the Cairban and proved successful in killing them; and that he has instructed several people on the west coast in the method of catching them'. If this is correct, then we can safely assume he had existing experience prior to 1766, his first season with his new vessel and equipment.

Due to the late arrival of his grant, he started the 1766 season very late in the season, but still managed to take seven sharks between Rum and Canna. Finding the vessel 'too expensive and unwieldy for the Kerban' he sold the large vessel and bought two smaller ones. In the 1767 season he took eight Basking Sharks in the same area, a disappointing figure for a whole season's effort. Seemingly depressed by the lack of success, not only in terms of shark hunting but also the cod and Ling fisheries, he reported to the Commissioners in 1768 that he was discontinuing his 'tryalls'. However, despite his pessimism, a group of Directors of the British Fisheries Society who visited the island in 1787 found that 'A great many Sun-fish or Basking Sharks are taken in these seas. The liver is oily, and the only part of the fish that is made use of'. They also found that the local people were 'a sober, quiet industrious race, and not contemptible fishers, especially of the sun-fish'. Perhaps Donald MacLeod was simply unlucky with his experimental fishery, and had a couple of bad seasons when few sharks were present?

Farther south in the Firth of Clyde, another dedicated fishery had developed at about the same time as the experimental fishery centred on Canna. The earliest mention of this fishery comes from Tarbert, Loch Fyne, in 1768 when Archibald Menzies, the Inspector-General of the Commissioners for the Annexed Estates, reported that 'One Stewart at Bute … has been pretty successful for some years past in this fishery with the Harpoon.' As confirmation of this report, Menzies personally witnessed a Basking Shark hunt in action near Tarbert where a nine-metre fish was finally brought in and yielded 192 gallons of liver oil. This report, combined with Pennant's report of the successful hunt in Loch Ranza, Arran in 1772, suggests that the local fishery was far from in its infancy, and was in fact an established activity practised over a wide area.

Archibald Menzies must have been impressed with the potential for this fishery, as in 1777 the commissioners agreed to provide a grant of £200 (£25,000 in 2017) to David Campbell of Shawfield, Islay. Once again, as in Canna, this was to fund an experimental fishery 'for herring, for cod, and for cairban, which come at different seasons'. No record exists concerning the success of this experiment, but given that the site was reasonably close to Basking Shark hotspots around Tiree and Colonsay, local abundance must once again have been the driver. However, if it had been a success, then undoubtedly there would have been associated reports to that effect. But equally, as at Canna, the (apparent) failure of these experiments could likely have been put down to the very exposed nature of both sites, open to the worst of the weather and the fierce local tides. As we shall later learn, these are wild places indeed, and not for the faint of heart. By comparison, the waters of the Firth of Clyde are very sheltered except from the south, and have much weaker tidal flows. This would have allowed for far easier conditions for small open boats and their crews to harpoon and capture the sharks.

But despite the challenges inherent in pursuing the Basking Shark in the exposed waters of the Sea of the Hebrides and the Minch, there is evidence that the courageous men of the islands were just as willing to face the risks as their counterparts in more sheltered waters. By the 1790s it would appear that the fishery had gained considerable traction, and both artisanal and more dedicated fisheries were in operation across much of the west coast of Scotland. Parish records from the Old Statistical Account of that era identify that of 175 parishes involved in sea fisheries, 10 discuss the presence of the Basking Shark, with a further three noting more oblique references that may also concern the species. Even out at remote St Kilda, the animal had not gone unnoticed, albeit in a more romantic sense in a song composed by a young couple around 1780, in which the admiring woman eulogises her beloved thus: 'Thou art my hero, thou art my basking sunfish,'. It would seem that in St Kilda, at least, the shark was viewed in a heroic, positive light at that time.

From Ayr in the south, through the Firth of Clyde and out through the Small Isles as far north as Harris in the Outer Hebrides, communities pursued the Basking Shark. Apparently the market had developed to the stage where fishing vessels made the journey from Peterhead on the east coast of Scotland to Barra in the southern part of the Outer Hebrides to pick up cargoes of dried cod and Ling and 'the oil from the sun-fish which they catch, which is sometimes considerable'. This must have been a lucrative trade, as it is a long, hard haul around the far north of Scotland from Peterhead to the Outer Hebrides in an engineless sailing vessel, even in summer. The same financial benefit must also have driven the crofters of the coast to take the considerable physical risk of pursuing the shark, as is well described in a report that appears in the Old Statistical Account from South Uist at that time that the Basking Shark:

> is a stupid and torpid kind of fish; he allows the harpooner often to feel him with his hand, before he darts at him. The inhabitants to the east side of the island, (such as are able to fit out boats, lines and harpoons,) have been for some years very

successful during the summer months, in this branch of business, owing entirely to the laudable exertions of the trustees for managing the fisheries of Scotland, in granting premiums to the owners of boats, that extract the greatest quantity of oil from the liver of the Basking Shark. The lucky adventurer in this fishing, should he chance to harpoon a large one, may have 9 or 10 barrels of liver, from which the return in clear oil is about 8 barrels.

Clearly this was no longer just a small-scale, opportunistic fishery to supply local needs. The use of subsidies from the earliest days confirms that there was a keen knowledge of the worth of the liver oil, and a willingness to use public money to develop the fishery wherever it could be prosecuted. The Basking Shark was now a prized commercial target from Ireland to Scotland and Norway, setting the scene for the slaughter to come. In an age when whaling, whalers and whale products were well known in all of those places, perhaps that's not surprising, and it seems logical to suggest that a parallel market for Basking Shark products was now well established.

A further indicator of this may be seen in the way that harpoon fisheries spread across all areas. Apart from the occasional animal bycaught in other fisheries or driven ashore by boats, it appears that by the time of the first mention of the commercial fisheries, the harpoon was the established weapon of choice. Fairfax suggests that the first harpoons to be employed in the Scottish fishery were simple two flue, forged iron head models of the type commonly used in whale fisheries, attached to a wooden shaft of around two metres in length inserted into the base of the head.

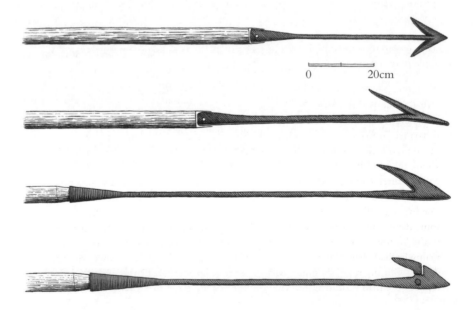

Types of hand-held commonly used by whalers in Scotland. The top, two-flue type, is the one most likely used in the early Scottish Basking Shark fishery.

Rather than being thrown at the shark from a distance, the usual method of harpooning would appear to have been to make a close approach to the shark, whereupon two men would drive the harpoon downward into the animal. This would be followed by a rapid retreat to avoid damage to the boat from the tail of the shark, lashing out in an effort to dive. Thomas Pennant provides a graphic description of a Basking Shark hunt from 1772, when he witnessed a shark being taken in Loch Ranza, Arran:

> They swim very deliberately with their two dorsal fins above water, and seem quiescent as if asleep. They are very tame or stupid; and permit the near approach of man; will suffer a boat to follow them without accelerating their motion, till it comes almost within contact, when a harpooner strikes his weapon into the fish as near the gills as possible: but they are often to (sic) insensible as not to move until the united strength of two men has forced in the harpoon deeper: as soon as they perceive themselves wounded, they fling up their tail and plunge headlong to the bottom, and frequently coil the rope around them in their agonies, attempting to disengage themselves from the weapon by rolling on the ground, for it is often found greatly bent. As soon as they discover their efforts are in vain, they swim away with amazing rapidity, and with such violence that a vessel of 70 tuns, has been towed by them against a fresh gale: they sometimes run off with 200 fathoms of line, and with two harpoons in them; and will find employ to the fishers for twelve and sometimes twenty-four hours before they are subdued. When killed they are either hauled on shore, or if at a distance, to the vessel's side. The liver (the only useful part) is taken out and melted into vessels provided for that purpose: a large fish will yield eight barrels of oil, and two of sediment, and prove a profitable capture.

As a later record recounted by Fairfax from Arran in 1875 confirms, this was in no way a safe or simple undertaking:

> A century ago there was more animated fishing at Arran than any I have mentioned – that of the Basking Shark or sun fish… The liver of a good sized fish yielded about eight barrels of oil, for the sake of which it was keenly pursued. All the fishermen in Arran were provided with harpoons and tackle for its capture; nor during the season of its visit did the Arran and Saltcoats packet ever cross without seeing them. A sharp look-out was kept, and no sooner was its well-known sail (two or three feet high) discerned than all was bustle and activity, and soon a boat was speeding toward it over the water. Four men manned a boat – two rowed, one managed the rope, and one handled the harpoon. Taking it was attended with some danger, for it is very powerful; and like the whale, on being struck it plunges to the bottom. On one occasion the rapidly uncoiling rope caught the foot of the man who had charge of it, and in a moment he was overboard, and down into the depths. 'Cut the rope,' shouted the oarsmen to John McDonald, the master. 'No,' he replied; 'the man is dead already, and the fish will do good to his widow.'

This vivid image shows a Basking Shark being hunted in Loch Ranza, Arran. Note that there are several open boats with hunting crews in the vicinity and the square-rigged vessel to the left appears to be some sort of mother ship. All of which suggests that the hunt was already well developed in 1772 when the hunt from which this image was drawn took place. From Pennant, A Tour in Scotland and Voyage to the Hebrides 1772, 1774.

The mechanics of the hunt

Judging from the illustration above, whaling-type harpoons were employed in this hunt, probably because they were readily available items at ship chandlers. However, in time, the two flue harpoon was gradually replaced across the range of hunting areas due to its propensity to pull out of the Basking Shark, as the flesh was softer than in whales. In Ireland, toggle-action harpoons, designed to penetrate further into the flesh and then spring open to provide massive resistance to removal, were being developed as described in 1824 by Hely Dutton:

> A very curious instrument in its construction; it is composed of a steel spear barbed: for almost half of its length it is grooved at one side; through the side of the groove there is a strong rivet, on which an iron handle turns; at one end of this handle there is a socket for the reception of a wooden pole, to which a strong rope line is made fast: the spear is launched with the handle closed up in the groove. When the fish finds himself wounded he flies off, which disengages the handle in the groove, and as it opens and turns across in the wound, it forms a barb of the whole length of the handle, from which no exertion of the fish can extricate it.

McNally reports that 'at the Irish Industrial Exhibition of 1853 at least two manufacturers of fishing equipment exhibited harpoons, and the Commissioners of fisheries had on display a sun-fish harpoon from County Galway'. According to reports that he gleaned, the harpoons used in the Aran Islands had a soft iron shaft that would bend readily when the sharks dived and rolled on the bottom. This helped to reduce the leverage on the shaft and made it less likely to pull out of the shark. It is clear that much had been learned from the hunt over the decades, which had allowed the equipment to be optimised for hunting Basking Sharks, as opposed to simply using cast-off whaling gear.

The final approach to a big shark must have been a nerve-shattering experience for the crew of a small open boat. It should come as no great surprise that lives were lost on occasion, or that the crews were highly respected for their bravery. From Brabazon, The Deep Sea and Coast Fisheries of Ireland, 1848.

The above illustration, taken from Wallop Brabazon's account of the Irish fishery in 1848, shows the harpooner standing in the bows of the catcher boat with the harpoon held aloft as if ready to be thrust into the shark like a javelin. Given the critical importance of accurate placement of the harpoon to ensure that it would not pull out, this must surely only have been attempted in ideal, calm conditions, whilst at other times the more traditional downward thrust must have been used. Brabazon notes this:

> These fish are most powerful in the water, and if harpooned in the shoulder they are very hard to kill, often carrying off the whole harpoon line, but experienced harpooners strike them in the body, near the dorsal fin, rather low down, where it will go through into the intestines, or near the vertebrae towards the tail.

Due to the considerable skill and risk involved, it should therefore come as no surprise that the harpooneer was the most highly paid of the crew. As was traditional with fisheries, crew were paid on a 'share' basis, and the harpooneer received a share 50% greater than the other crew members. But all played a vital role in the hunt, including the oarsmen who were responsible for driving the boat away from the shark at the moment the harpoon struck home; 'They must be struck with great caution, as they will stave in the boat with a blow of their tail, if it is at all within reach'. The illustration from the Pennant account shows the catcher boat stern-to the shark, with the harpooneer in the stern facing the shark, perhaps to avoid the dangers of the tail and to facilitate an escape from the scene of the strike, but this may owe more to artistic licence than the reality of the hunt.

An interesting aside note from the very first days of both the Irish and Scottish fisheries was the early development of gun-fired harpoons. Arthur Young reported that Thomas Nesbitt of Donegal Bay had spent three years and a huge amount of expenditure developing this weapon, albeit for the main purpose of whaling, but he cannot have failed to see the potential for his device to be profitably employed against Basking Sharks. Whatever his intent, The Royal Society for the Encouragement of Arts, Manufactures and Commerce credited him with the invention of the first swivel-gun harpoon around 1760. Over in Scotland, the much referred to (but seldom seen) 'Stewart at Bute' certainly had the Basking Shark in his sights by 1768 with his own version of the harpoon gun, as Archibald Menzies explained:

> One Stewart at Bute is fitting out a boat with a small swivel at the Bow, he proposes fixing a harpoon to a piece of wood which is to serve as a Wadd to the Gun. This he is to fire into the fish, whether or not this will answer Experience will only show.

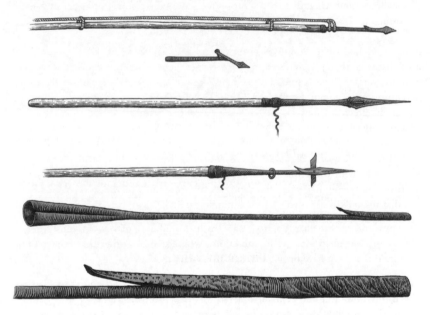

A variety of hand-held harpoons were used in the early Irish Basking Shark fishery.

As no further reference is made to the use of the harpoon gun to kill Basking Sharks until the twentieth century, perhaps these early experiments proved fruitless or were simply too costly to justify their purchase and adoption for such an unpredictably available prey.

Once brought to the surface, the fishermen would finish the shark off with strikes from a lance, and then the shark would be drawn alongside the catcher boat to be livered, or towed to an accompanying larger vessel for the liver to be extracted. In the illustrations from both Pennant and Brabazon's accounts, there are larger sailing vessels in attendance nearby, which suggests the latter option may well have occurred, at least in the more organised fisheries. In the more artisanal fisheries it would appear that the liver was often extracted alongside the catcher vessel. This must have been pretty precarious in anything other than the calmest conditions, as in some cases the shark could have exceeded the length of the vessel.

The shark would be lashed fore and aft alongside the vessel belly up and then cut open to expose the liver for removal and loading aboard. Without the buoyancy of the liver the carcass would have been just dead weight, and as McNally reported, 'the ropes must be slipped with great dexterity, or it would bring down the boat'. On other occasions, and most likely when close to shore, the shark could be towed to a beach to be livered, a far safer option for a small boat, and one that was less likely to result in loss of life or liver.

Given the high value of the liver, the methods used at the time to render it down to oil were rudimentary to say the least, especially in the more artisanal fisheries. The Commission of Inquiry in Ireland in 1836 heard that:

> Whenever the boats succeeded in taking one or two fish, they returned with the liver, the sun extracting the oil, which was baled out of the boat with the sea water; and when they arrived on shore. The most rude means were used in separating the remainder of the oil from the liver. Casks for containing the oil were not previously in readiness; and when there was a successful season, the price of casks became exorbitant.

The basic method of rendering the liver once ashore consisted of cutting it into smaller chunks that were then placed in boiling water, when the oil would be released. It would then float to the surface to be skimmed off and placed into large storage containers. According to Hely Dutton it was important that the liver be processed as soon as possible, or quality would be adversely affected. Prompt processing improved 'the goodness of the oil for burning, as producing a brighter flame, and more free from any offensive smell'. Additional refinement through allowing adequate time for the sediment to settle, or even filtering through cloth, would enhance the quality of the oil further. It is clear, though, that in Ireland a casual attitude was the norm (with the possible exception of the more professional and better equipped concerns), and it seems likely that the same lack of care applied on the west coast of Scotland. Given that the bulk of the trade went for municipal lighting, perhaps that is unsurprising, as there was no apparent premium for better grades of oil.

Financial values

That the basking Shark fishery could be lucrative both in Scotland and Ireland is beyond doubt. In his 1776 account, Pennant reported that the profit from a large liver could be as much as £20 (roughly £2,500 in 2017). Reporting from the early part of the nineteenth century, James Hardiman gave an average liver value of £20–£30 (£1,360–£2,040 in 2017) for Ireland. In terms of value in the marketplace, Hely Dutton reported at around the same time that sun-fish oil retailed at between £4–£6 per 30-gallon barrel (£272–£408 in 2017), with a liver yielding between four and 12 barrels, although he pointed out that prices fluctuated according to supply and demand. The prices he quoted may have seen a recent hike following a government subsidy to improve the quality of sun-fish oil to encourage the development of an export market. Hely Dutton also stated that the two main production centres at Westport and Galway had a combined output of around 25,000 gallons, with a value of £5,000 (£340,000 in 2017), to which should be added the additional production from the artisanal fishermen of the coast.

Costs to the fishermen were relatively low, especially when set against the financial rewards. Around 1780 in Scotland, Fairfax quoted costs of £20 for a small wherry of 15–20 tons, and crew wages for three men and a boy for a summer season of around £7. Gear costs were similarly low, Brabazon reporting that in the 1840s in Ireland harpoons cost 10s. each, while 200 fathoms (1,200 feet) of harpoon line cost £3. Obviously these costs still represented significant sums, but as in many cases they were covered by subsidies, and the successful capture of one shark alone would have been enough to cover a season's wages and gear costs. Against this would have to be allowed losses of harpoons and lines, and, as we have seen, even men, but in such an age of poverty the potential financial benefits must have represented a bonanza to the people of the coasts.

The cessation of the fishery

Yet by around 1830, the fishery in both Scotland and Ireland was in steep decline. That such an apparently thriving and potentially lucrative fishery should have declined so dramatically in such a short period seems remarkable. Undoubtedly there was no one single factor that led to this situation, rather various circumstances intertwined to create a situation that spared the Basking Shark.

The first of these was hinted at from the earliest stage, when Donald MacLeod of Canna reported back to the Commissioners for the Annexed Estates on the outcome of his subsidised 'tryalls': 'On the whole my opinion is that from the tryalls I made the Kerban and Cod and Lyng fishing will become no considerable length on this coast, without a considerable encouragement from the publick.'

By which, he clearly meant further subsidies, bounties and grants from the public purse. MacLeod, like the Nesbitts in Ireland, saw the future of the fishery as a well-organised, mechanised affair with substantial vessels and catcher boats, dedicated staff and (perhaps) shore facilities. Each of those fisheries was also dependent on other seasonal prey (cod, whales) to provide an alternative income to develop a sustainable business. Yet the fact that none of the three major subsidised fisheries – MacLeod of Canna, Campbell of Islay and the Nesbitts of Donegal Bay – made any kind of long-term success of their activities, suggests that even with further subsidies, their fisheries faced substantial difficulties, not least the seasonal nature of the fishery and the vagaries of weather and shark abundance. As such, it is hard to imagine the fishery as being a reliable financial proposition to attract outside investors or major subsidies.

Even the small-scale, artisanal fishermen needed grants to buy the basic equipment – harpoons, lines – that would otherwise have been unaffordable to peasant farmers and fishermen. But compared with the substantial sums required to buy fleets of boats needed to build and man a commercial fishery, they were small beer. The small-scale local fisheries could operate with whatever boats they had to hand, plus locally available manpower, and could continue with the assistance of small grants for gear, as long as sharks were plentiful. For those farmer/fishermen operators, efforts to exploit the short season (generally agreed to be April–June) had to fit in around their existing land-based activities such as tending to their livestock and crops, but that was not an impossibility. Hunting the Basking Shark offered a considerable bonus for them, and the availability of official bounties for the best catches were the icing on the cake. Ultimately, these low-cost, artisanal operators had the best chance of achieving a sustainable business.

The bounties varied between Ireland and Scotland, but the aim was the same – to support and propagate the fishery. As we have seen earlier, in Ireland bounties were introduced (£3 per ton) to help develop the export trade in sun-fish oil through improving the quality of the product. In Scotland the bounties were based on the volume of oil produced per vessel, as reported by Fairfax originating from Barra in 1818:

Competition for carban premium.
'Mary' of Sandray in Barra 112 gallons
'Mary' of Barra 186 gallons.

But these attractive bounties were in all cases short-lived, and their withdrawal soon had a dramatic effect on even the small-scale fisheries, as reported in the New Statistical Account from South Uist in the early 1800s: 'About thirty years ago, there was a considerable fishing carried on in dog-fish, sun-fish or basking-shark. But since the bounty on the oil of these fishes has been withdrawn, this branch of industry has been altogether discontinued.' Or from Barra around 1840: 'The people of Barray were in former years very successful in harpooning *cearbans* or sail fish, from which they extracted a good deal of oil, and received a premium from the Board of Trustees for Fisheries; but this productive source of wealth has been discontinued.'

Why were the bounties discontinued? Fairfax suggests that the end of the Napoleonic wars and the opening up of the Sperm Whale fishery in the Pacific meant that quantities of whale oil (the main competitor to Basking Shark oil) were now available, and that coupled with Free Trade policies, the bounties no longer made economic sense. As a result, by around 1830 the shark fishery was in steep decline. McNally suggests that alternative oils such as rapeseed oil were becoming more abundant and thus the price paid for sun-fish oil had fallen at around the same time. In any case, paraffin would soon be available (c. 1850), which would further undermine the use of other oils for lighting.

Around the same time, a substantial reduction in the number of sharks appearing seasonally both in Ireland and in Scotland became a further contribution to that decline. In reports given to the Commissioners in 1836 it was confirmed that the Claddagh fishermen had taken only one shark in 1834, and none in 1835. The Mayor of Galway complained that the fishery had been poor for several years, a view that was supported by reports from other districts.

One additional factor may have been a greater appreciation of the risks involved. Lives were being lost in this risky fishery, and despite the renown attached to the brave fishermen, who could blame them for seeking less dangerous prey? One disaster in particular had a dramatic effect on the morale of a local population, and largely brought the fishery in that area to a close.

In 1873, two boats from the lonely islands of the west Connacht coast set off to pursue a shoal of sharks sighted between the islands of Inishbofin and Inishark. The harpooneer, Michael O'Holloran, was an experienced shark hunter and struck the first harpoon into a large Basking Shark. Eventually the shark was brought to the surface and as a precaution against it breaking free, a second harpoon was set in to the shark. Suddenly the first line broke, the shark sounded (dived deep) and the second line flew out over the gunwale until a kink developed in the line and seized in the groove in the gunwale. Before the line could be cut, the boat was capsized and the crew thrown into the water; five of the nine men were drowned, the other four being rescued by the crew of the second boat.

A few days later the Inspector of Irish Fisheries, Thomas F. Brady, visited the island and reported on the drastic effect of the accident on local morale:

No words of mine could describe the scene of woe and desolation. I saw the four able, hardy fishermen who had been rescued that day from the jaws of death, turn pale at the sight of a sun-fish. This is no exaggeration. There was a number of them about – three had already been harpooned. The best harpooneer on the island, who had been on board the ill-fated boat on the day of the sad occurrence, quailed at the idea of going to spear a fish. I saw him afterwards, when I was several miles out to sea north-west of Shark Head, turn pale and hang over the gunwale of the boat when we came up to the boat that had one of these monsters secured and lashed. There was no pretence about this, he could not feign what I saw. The recollection of what had passed a few days before struck terror to his very heart.

Despite the raising of a relief fund for the bereaved families, clearly the spirit had gone out of the community, at least, to pursue the Basking Shark. During the rest of the season only three sharks were harpooned, and two of them got away when faulty gear broke. The one fish captured (of nine metres) towed the catcher vessel 12 miles out to sea and was only brought to Inishbofin by the combined efforts of four boats and 26 men. Apart from an occasional capture, the era of the great hunt for the sun-fish was drawing to an end in Ireland – for the time being, anyway.

Across the water in the Inner Hebrides, the minister of Tiree and Coll, a site where sharks had once been plentiful, reported that by 1840 the Basking Shark had rarely been seen in the previous four decades. Further south, the Clyde fishery seems to have all but ceased by the early 1800s, with the herring fishery being the only major fishery reported as being active. No mention of the Basking Shark fishery is made except in dismissive terms. Once again this seems largely to have been driven by a collapse in numbers.

But what caused such a sudden and widespread decline? Surely this is a question that must have perplexed the fishermen of those days just as much as it does any latter day scientist or natural history enthusiast.

The Canadian experience

In an era when shipping had opened up world trade, word soon travelled amongst seafarers about the distribution and abundance of potential prey. As ships developed trade routes along the Pacific coast of Canada, observers aboard them recorded the native fauna, with an especial interest in marine life, so it is no surprise that a creature the size of the Basking Shark would appear in the records kept by those early explorers. Scott Wallace and Brian Gisborne, in *The Basking Shark, the slaughter of BC's gentle giants*, suggest that the first of these records probably comes from the purser of the fur trader 'Columbia', one John Hoskins, in 1791, who made an entry in the ship's log of 'whales and large sharks' off Estevan Point on the west coast of Vancouver Island. Given that these 'very large' sharks were sighted at the surface, it seems unlikely that they could be anything other than Basking Sharks. Hoskins also mentions in a later entry that the local First Nations people 'also get sharks which are here very large', perhaps the earliest record of a possible Basking Shark hunt from the area.

Wallace and Gisborne point to a number of early records that seem to indicate the presence of Basking Sharks throughout the area. These include a report from 1795 that seems likely to have been of Basking Sharks, from the Yankee sloop *Union* under 19-year-old captain John Boit. Sailing farther north, off the north-western point of the Queen Charlotte Islands, he noted in the ship's logbook that there were 'many large whales and sharks around'. Further evidence of probable Basking Shark sightings throughout this area came from Scottish fur trader and surgeon William Fraser Tolmie, who was posted by the Hudson Bay Company to the west coast mainland town of Fort McLoughlin (near Bella Bella). In June 1835 he recorded 'large sharks seen daily in the McLoughlin Bay'. Given the summer period, the shallow water of the bay, and the surface sighting of the animals, these must surely have been Basking Sharks.

Wallace and Gisborne report that William Eddy Banfield traded with the First Nation people on the west coast of Vancouver Island during the 1850s and recorded that in Clayoquot Sound, 'whales, dogfish and sharks are caught in large quantities here, for their oil'. That these sharks might be Basking Sharks is borne out by a reference they identify from the same era from anthropologist Philip Drucker, who, in his book *The Northern and Central Nootkan Tribes* outlined the way that the local tribes hunted sharks:

> In the 1850's or thereabouts the dogfish oil industry grew up. From then until about the end of the century many men devoted a large part of their time in late spring and summer to fishing for dogfish…or harpooning big 'mud sharks'. For the big sharks, the sealing harpoon was used with one or two sealskin floats (of the type used for whaling) on the line. There seems to have been no fear of sharks nor any feeling that hunting them was dangerous or difficult. The creatures of course, were not wary and did not require such cautious stalking as seal, sea otter, or sea lion.

Wallace and Gisborne point out that 'mud shark' was the local name for the Bluntnose Sixgill Shark (*Hexanchus griseus*), yet those sharks are bottom-dwellers, and therefore the harpoon would be an unsuitable weapon to hunt them with. But anyone who has been around Basking Sharks will attest to how easy it is (they are 'not wary') to approach and harpoon them. They also point out that later records identified Clayoquot and Nootka Sounds as hotspots for the Basking Shark. In a recent book, First Hereditary Chief Earl Maquinna George of Ahousaht confirmed that: 'There is a stream in Clayoquot Sound called "shark creek" which is named after the Nuu-Chah-Nulth (formerly Nootka) place name, uumach-aqtlnlt, which is the translation for Basking Shark.'

As a result, it would be hard to imagine that these reports can refer to any other shark species than the Basking Shark, and that they were indeed hunted along the west coast of Vancouver Island.

Yet at the same time, Wallace and Gisborne note that naturalist Dr George Suckley suggested that Basking Sharks were rare, at least in Puget Sound to the south. In 1860 he reported that:

> A very large shark was captured at Port Discovery in December 1856.
> My informants told me that from its liver four barrels of oil were extracted!
> Large sharks are very rare in Puget Sound; so rare that it is not improbable that they are stragglers which have followed the warm 'Pacific gulfstream' from more southern regions.

The only shark that could have yielded that amount of oil from its liver would be a Basking Shark. Dr Suckley's claim that they are 'very rare' in the region was directly challenged in 1862 by prominent local historian and polymath James G. Swan who issued a vigorous rebuttal in the local newspaper, the *Washington Standard*, reported by Wallace and Gibson:

> They (large sharks) must have swarmed here since the Doctor (Suckley) made that report, for at this time there are regular shark fisheries on different parts of the waters of the Sound; one in particular at the mouth of the Snohomish River

in Possession Sound, and another on Hood's canal, both of which produced an amount of oil sufficient to induce several companies to prosecute the business. During the present summer while cruising in the waters of the Straits (of Juan de Fuca) and along the coast both north and south say fifty miles each way, I noticed a great number of sharks, in fact so plentiful were they, that instead of being as Suckley says 'very rare', I have seen the water apparently alive with them particularly off San Juan or Patchina Harbour (south west Vancouver Island), nearly opposite Neeah Bay (Washington). Either Dr. Suckley was most grossly misinformed, or else as I before remarked, sharks have swarmed in these waters since his observations were made. The Indians seldom capture them, for although their livers make a large amount of oil, they are difficult and dangerous customers to attack in canoes.

The vivid nature of this account will be recognised by anyone with experience of Basking Sharks as a true eyewitness report and not a second-hand impression from afar. Swan's report must be viewed as the more credible, as is reinforced by this further account from 1868:

A very large species of shark, known among whalemen as 'bone shark'[1], is occasionally killed by Makahs, and its liver yields great quantities of oil. I saw one in October 1862, killed in Neeah Bay, twenty-six feet long, and its liver yielded nearly seven barrels of oil, or over two hundred gallons. These sharks are very abundant during the summer and fall, but the Indians rarely attack them except when they come into shore to feed which they do at certain times. They are easily seen by the long dorsal fin projecting above the water, and, as they appear to be quite sluggish in their movements, are readily killed with harpoons or lances. The flesh is never eaten.

Swan also reported a Makah whaler undergoing an intensive ritual of starving, not sleeping and bathing in the ocean for six days in order to prove his fitness to be a whaler. Doubting that he could endure such a regime for more than two days, the Makah man admitted that 'if that would not suffice to make him a whaleman he could kill sharks.'

Farther north, later records from the Queen Charlotte Islands from geologist and ethnographer George Mercer Dawson (and others) confirmed that the species was still present in those waters. In 1878 Dawson left Virago Sound at the north end of the Islands aboard a vessel and encountered 'a very large shark, which followed the boat for some distance, occasionally showing its back fin above the water. Length estimated at twenty feet.' And in his published report of 1880 he expanded on that short account:

Large sharks abound on the northern and western coasts, and are much feared by the Haidas, who allege that they frequently break their canoes and eat the unfortunate occupants. No instance of this kind is known to me, but they fear to attack these creatures. When, however, one of them is stranded, or found from any cause in a moribund state, they are not slow to take advantage of its condition, and from the liver extract a large quantity of oil.

---

[1] Interestingly, a name believed to have been used on the US east coast in the early nineteenth century.

In 1879, Alexander Mackenzie of the Hudson Bay Company based in Masset on the Queen Charlotte Islands reported an abundance of 'sharks of large size' during the dogfish season between July and September. And in the 1897 *Year Book of British Columbia*, the Basking Shark finally gets a named mention, as being 'plentiful in Queen Charlotte Sound during the summer months. It attains a great size, is perfectly harmless, and so tame that while basking it may be touched by hand.'

Scott Wallace and Brian Gisborne use these records well to make a credible and coherent argument that the Basking Shark was widespread and numerous for at least a century along the north-west coast of the United States and the waters off the west coast of Canada. Between Puget Sound and the northern extremity of the Queen Charlotte Islands there appear to have been Basking Sharks everywhere. As in Scotland and Ireland, there were particular 'hotspots' where the creatures were seasonally abundant, and around these sites some communities adapted whaling techniques to capture the sharks and exploit their rich livers. Equally as in Scotland and Ireland, some communities such as the Nootka pursued the hunt more aggressively than others (such as the Makah), but that may simply reflect greater experience from pursuing whales than any inherent lack of courage. It is also clear from James Swan's account that there was not only an indigenous, opportunistic hunt for the Basking Shark, but that there were several competing companies pursuing it on a commercial basis.

The overall picture of Basking Shark distribution and numbers from the west coast of Canada was positive at that time, with no sign of any serious decline. But as we shall see, that wasn't to last. Within a century the Basking Shark would seldom be seen in those waters.

Chapter 3
Changing Views

During the early years of the twentieth century the Basking Shark seems to have disappeared almost entirely from the social and cultural radar of the countries and communities that had previously hunted it. Perhaps this silence might be explained by a periodic lack of abundance as well as the risks involved in hunting the creature, but there might be another reason for the lack of reports. Perhaps the diminished value of their liver oil in an era of abundant whale- and hydrocarbon-based oils simply rendered the hunting of Basking Sharks uneconomical, at least on a commercial basis. That lack of financial incentive alone might have led to the fading from collective memory of the knowledge, custom and practice of hunting the shark amongst the coastal people, leaving only a distant, unwelcome memory of a thoroughly hazardous enterprise.

Plans to resurrect the hunt on a more organised commercial basis occurred from time to time, such as in British Columbia in 1910, in a press report discovered by Wallace and Gisborne outlining that:

> sharks in abundance are found in the waters off Vancouver and Queen Charlotte
> Islands, numbers of the fish averaging twenty feet in length. The fishing will
> be pursued from sailing boats, the catches being brought in and treated at the
> whaling stations.

No further reports appeared to indicate whether the proposed operation went ahead, but the lack of information suggests that it wasn't necessarily successful.

However, the potential for an industrial type of shark hunting operation didn't go away, as was evidenced by the proposal forwarded by Sidney Ruck of the Consolidated Whaling Company for a 'Shark Industry to be Developed on a Large Scale' reported in the Victoria B.C. *Times* of 1921. Referring to a new reduction plant in Barkley Sound where 'the huge Basking Sharks abound … in large schools', it suggests the sharks were to be hunted using harpoons shot from guns in the manner of whaling. Evidently blessed with a vivid imagination, Mr Ruck went on to describe how 'sun sharks…. race up and down (the Alberni Canal) in schools of thousands' and that their numbers were so dense that 'a month ago, one of the coastal steamers ran into such a school of these big fellows that, packed tightly against the sides of the boat and around her bow, they stopped her completely.'

Wallace and Gisborne lament the fact that no evaluation of the extent of these fisheries is possible, largely because there was no differentiation in the market data in that era relating to Basking Shark-specific products, whether as oil or fishmeal. Many different types of fish would have been rendered down to contribute to the overall figures held in the fisheries statistics, making it impossible to estimate the number of Basking Sharks killed by these operations. As a result, no analysis can be made of what effect those fisheries might have had on the population, or even allow an estimation of what the pre-exploitation population might have been. But they point out that fish oils were attaining high values in the period 1910–1940, indicating that Basking Shark oil would still have been a valuable commodity, and deduce that even 'a small ongoing fishery over a thirty-year period would have long-term consequences to the population.'

Over in Ireland, active hunting for the Basking Shark had ceased for the time being. The animals were still sighted in many of their old haunts, although Kenneth McNally reports that the 'species appears to have been less abundant than formerly'. However, a growing threat to the species was becoming apparent: that of accidental net entanglement ('bycatch'), particularly in the offshore salmon nets set around the Irish coastline. Not only did the nets cause a steady, attritional rate of shark mortality, but of course caused a major headache for the fishermen involved, through damage to – and loss of – their nets. Between 1912 and 1915, 10 sharks of between four and seven metres were bycaught in Ardmore Bay, County Waterford and many more were sighted in this area not previously noted for Basking Sharks. Such occurrences naturally led to suggestions that if sharks were becoming increasingly abundant then the moribund Irish fishery might be resurrected, but to no avail – it would be 30-odd years before a 'real' fishery for the shark would once more exist in Ireland.

The ensuing decades up until World War II shifted the perception of the Basking Shark in the public eye in a number of ways, some prosaic, one tragic, none positive. The niggling concern over bycatch became more of an issue as fishermen from the island of Barra in the Outer Hebrides (1922) to the west coast of Canada (1937) reported increasing economic damage from interactions with Basking Sharks, and began to publicly clamour for action to reduce their losses.

In Ireland the traditional fishery saw a momentary artificial revival through Robert Flaherty's seminal film 'Man of Aran'. Flaherty had the extraordinary good fortune to find an islander called Pat Mullen to assist him with the production of the film. Mullen had personal knowledge of the ancient hunt from his grandfather, who had been an active participant in the previous century. He even retrieved two rusty harpoons from the fireplace of his grandfather's cottage, that were to act as patterns for a Galway blacksmith to make copies for use in the film.

The resulting piece of film, although showing a re-enactment of the original hunt, is still extraordinarily evocative, as the tiny curragh closes in on the shark and the hand-held harpoon is thrust home. It isn't hard to understand why fishermen from St Kilda to Achill Island who were engaged in the hunt were revered in their communities, and honoured in folklore and song. The message was unmistakable – the heroes were the hunters, not the shark.

It is interesting to note that the ensuing years and absence from the fray had not diminished the ability or courage of the hunting men of Aran one iota. And, in fact, it actually encouraged them to continue. From killing one shark solely for the purpose of the film, they went on to kill nearly 20 more (and benefited from the saved oil during the lean years of World War II).

During the 1930s there was apparently a steady increase in Basking Shark abundance in British waters, and once again this caused much grumbling amongst the local herring fishermen owing to the risk the sharks posed to their nets. However, this increase in numbers also resulted in a number of bizarre proposals that aimed to profit from the increased presence. In 1934 an article appeared in the *Glasgow Herald* promoting a 'new sport' in the Firth of Clyde, in which the public were invited to join a ketch-rigged Grimsby pilot boat operating out of Greenock and spend a week harpooning sharks.

The increased abundance of Basking Sharks in the Firth of Clyde didn't just attract the attention of would-be big game hunters though. Occurring as they did in a popular seasonal holiday area, Basking Sharks were a great draw for tourists, and small boats would often put to sea to watch the sharks. This understandable curiosity led to an accident in September 1937 that was to enter the history of the species as the Carradale Incident, and which dramatically changed the image of the Basking Shark for many years to come.

Lying on the eastern shore of the Kintyre peninsula, Carradale looks out onto Kilbrannan Sound and the Isle of Arran. A popular summer destination for tourists from Glasgow, the township consists of three elements: the fishing harbour at Port Crannaich to the north, the sheltered bay at Port Righ in the middle and the broad sands of Carradale Bay to the south.

Captain Angus Brown was born and raised in Carradale, but lived in Swansea where he was master of a steamship called the *Duchess*. Up on holiday with his wife and two young children Jessica (10) and Neil (six), on 1 September 1937, Angus went out in a small (4.5m) clinker-built wooden sailing dinghy called the *Eagle* belonging to his brother Robert, who hired the boat out when it wasn't being used for lobster fishing. The weather appears to have been par for that time of year, a mixture of overcast skies and a breeze from the south.

There had been a considerable number of Basking Sharks in the sound off Port Righ, and observers reported that the sharks had been 'leaping out of the water non-stop' all day. As such activity appears obviously hazardous to small craft, it seems strange that anyone would have thought to take a small boat out anywhere in the vicinity of breaching sharks. Maybe their curiosity simply got the better of them, but if so, it was to prove a fatal mistake.

Angus and Robert Brown were accompanied by a local youth Donald MacDonald, aged 14, together with Jessica and Neil. One minute they were sailing along 'going like a clipper', the next they drew to halt when the main halyard that held the sail up parted and the sail fell down. Turning around to row back to Port Righ, the next thing they knew, they were all in the water. Angus and Robert managed to right the boat, but it capsized again, leaving them floundering, trying desperately to hold on to the boat. Angus Brown was dragged up on to the boat, but died almost immediately, probably due to a heart attack. Boats set off from the shore immediately to assist the survivors, rescuing Donald and Jessica, and recovering the bodies of Angus and Neil. Robert Brown was nowhere to be seen, and it was some days before his body was recovered.

Eyewitnesses ashore confirmed that a Basking Shark was the most likely cause of the accident. Young Colin Oman confirmed that he had seen a big splash and then the boat was on its side. Another holidaymaker saw a shark breach: 'the sight of which was such an amazing one it was only when we noticed through the haze that the mast of the *Eagle* had disappeared that we suspected that something was wrong and gave the alarm.'

When brought ashore later, the *Eagle* was found to have a broken tabernacle (a large bracket that the mast stands in), but contrary to some newspaper reports, there was no hole in the side. Donald MacDonald, who surely would have known, suggested that the hull was 'bruised and cracked', but otherwise intact.

This does not suggest a direct attack, more a glancing blow of some kind, as any Basking Shark taking it into its mind to directly ram a small wooden boat would undoubtedly have reduced the boat to matchwood. On the other hand, a breaching Basking Shark could strike the gunwale of the boat on its way up, and simply flip it over. A shark descending from a breach and landing on a small boat would have destroyed the boat and probably killed several of the occupants, therefore it would seem that the impact was neither severe nor direct. It also seems clear that what happened was totally without warning, Donald MacDonald recalling the following day that 'the boat seemed to stand right up on her end', a description that coincides with the idea of a sudden upward strike as opposed to a blow from the tail that would have knocked the boat sideways.

Basking Sharks tend to breach most regularly when there is a shoal present, which suggests that such behaviour has some social purpose. Having observed hundreds of breaches over the years of our studies I can confirm that the animals have no spatial sense or apparent awareness of the presence of boats – the act is totally random. I have had several near misses with breaching sharks, one erupting like a missile out of nowhere, so close to our yacht that the splash on re-entry sprayed water all over our boat, and I have interviewed two fishermen who had sharks damage their boats when breaching, both of which sounded accidental. Having spent considerable time in the presence of thousands of sharks, I have never seen any sign of overt aggression on the part of a Basking Shark. I have often had a shark approach our yacht, and come right up close to our transom, but never had one touch the boat.

And yet, Colin Oman's father, one of the first of the rescuers to reach the stricken boat around 15 minutes after the incident, confirmed that the shark was still going around the boat and lashing its tail. A week later, a fishing boat on its way back to Campbeltown (to the south of Carradale) was subjected to what definitely sounds like an attack by a Basking Shark, when a crew member: 'Saw a large shark charge at the boat. The shark struck the propeller a glancing blow. The stern of the boat was lifted three feet out of the water by the impact and came down again with a crash, fortunately on an even keel ...'

A couple of days afterwards, the passenger steamer *Dalriada* was attempting to enter Carradale through a sea, according to the master, 'alive with sharks'. Most of them moved aside for the ship, except for one massive shark that 'made straight for the *Dalriada*. It circled round and round the steamer leaping high out of the water and lashing its tail furiously.' Fortunately, it stopped short of attacking the steamer outright.

Denis Fairfax quite reasonably poses the rather worrying question as to whether there might be 'rogue' Basking Sharks that might on occasion charge a boat. As I have already stated, personally I have never experienced anything resembling such an action. But I have heard one credible eyewitness report from a trawler of a Basking Shark that rammed their boat, causing considerable damage, although it didn't compromise the boat's safety. And I have observed shoals of sharks engaged in forms of what is believed to be courtship behaviour that certainly appeared to have an underlying element of aggression about it.

Such shoals of sharks sometimes form a 'wheel' beneath the surface of the water, where they swim in tight formation in a circle. It is occasionally possible to see dozens of sharks engaged in these wheels, and what is strange about them is that they appear to have ceased feeding altogether during this activity, swimming rapidly with their mouths closed.

Divers I know who have been close to a wheel report that the sharks seem to be far more aware of their presence than is usually the case and have even behaved 'aggressively', suddenly swimming directly towards them. A highly experienced friend of mine told me that he was only too glad to get out of the water after experiencing one such occasion.

So is it possible that could there be some form of latent aggression that manifests itself only when shoals of sharks are present and engaged in courtship-like activities? If that were to be the case, then the Carradale Incident loses its claim to accident status and becomes something altogether more sinister – a shark attack. My best guess remains that this was simply a tragic accident, caused by a random breach capsizing the *Eagle*, but I'd have to accept that I might be wrong. As we shall see, there are still many mysteries about the Basking Shark that remain unanswered, and this may well be one of them.

The changing perspective, oceans apart

In a sense, the Carradale Incident compounded a change in the public perception of the Basking Shark across its habitual range. That the creature could now be said to pose a direct threat to man added to the growing outcry from fishermen concerning the damage caused by the shark to their fishing nets and gear. And that latter complaint could be heard wherever the sharks appeared.

As we've already seen, in 1937 fishermen in Rivers Inlet on the mainland coast of British Columbia began complaining about the problem of Basking Sharks damaging their salmon gill nets. It seems likely that in parallel with the waters around Britain, a periodic upswing in shark abundance was underway, and in the following years the mood in Canada changed dramatically. As Scott Wallace and Brian Gisborne remark: 'Regardless of the cause, one thing became clear: the Basking Shark was no longer described as a "gentle giant" but had now become a "great brute" and a "deep-sea pirate" that needed to be destroyed.'

Simultaneously, the fishermen in the Firth of Clyde in Scotland were complaining that an extraordinary number of sharks were entering their waters to the extent that they were 'infesting Clyde waters'. This deluge of complaints escalated until it was raised in the House of Commons in May 1939 by local Members of Parliament, who portrayed a pitiful scene of frightened fishermen and holidaymakers, afraid to use the waters of the Clyde due to the presence of abnormal numbers of sharks. A particular demand of the MPs was that the Secretary of State for Scotland should take action to destroy Basking Sharks and stop 'their depredations on the herring shoals', those worthy politicians apparently being unaware that herring were not, in fact, plankton and that Basking Sharks fed exclusively on the latter.

As might be imagined, in May 1939 the Government had other things to occupy its time, and fobbed off the complaints with only the promise that the Fishery Board's cruisers (a type of fisheries protection vessel) would continue to destroy Basking Sharks by the 'methods that have proved successful in the past', a reference to 1938 when one of the cruisers had killed 50 sharks by the simple expedient of ramming them when sighted at the surface.

Further demands followed from the MP for Greenock, for targeted action over a wider area that included the west coast of Scotland and the coast of Northern Ireland. The Secretary of State treated this demand with a degree of levity, asking the member to 'let him know the whereabouts of the sharks'. As there were compensation programmes in place for damage or loss of fishing gear that could be directly attributed to Basking Sharks, the Government obviously took the view that this was the most appropriate method to deal with the problem, rather than direct the fisheries cruisers to embark upon a wild goose chase to find and destroy sharks.

But perhaps reports of the taking of such drastic action found their way across the ocean to British Columbia, where by 1942 the problem of Basking Sharks damaging nets around Rivers Inlet appeared to have escalated dramatically. Wallace and Gisborne report a litany of complaints from around the region, all clamouring for action. The Rivers Inlet Sockeye Salmon fishery has been the third most valuable salmon fishery in British Columbia for over 125 years, with thousands of fishermen working seasonally to supply 10 canneries at its peak. As the fishery has a very short season, damage to nets or gear could have a major impact on the profitability of the companies that prosecuted the fishery, and it was this that dictated the call for action, prompting what a headline in the *Province* newspaper described as 'War declared on sharks'. That this was no small matter appears to be confirmed by the net boss for the Canadian Fishing Company, who reported that 'Hundreds of huge Basking Sharks' had caused 'thousands of dollars worth of damage to gillnets in the Rivers Inlet district', with the loss of '70 nets in a period of seven days' by his company's boats alone. Obviously in the long term this was untenable, and the call for drastic action met little opposition.

Wallace and Gisborne quote an article in the *Province* (also featured in *The Fisherman*) that outlined a plan to destroy Basking Sharks that had been put into action. This included a boat operated by BC Packers (nicknamed the 'razor-billed shark slasher') that had been modified with the attachment of a 'sharp steel ram' attached to the bow; this was designed to 'cut the sleeping monsters down as they lay on the surface'. Six sharks had already been 'removed from life' by this device, the report concluded. Oddly, there are no further records of the success or failure of 'the slasher', so we cannot establish whether further sharks fell victim to its blade.

It is clear, however, that events were reaching a desperate level. In a parallel development, fishermen and fishing companies in the Rivers Inlet area established a 'cooperative compensation scheme'. Fishermen paid annual fees into a fund, with their payments matched by the fisheries companies to pay out in the event of any fisherman suffering net damage or loss. That such a fund was necessary eloquently underlines the severity of the situation for the salmon industry.

Nor were the sharks going away. In June 1944 it was once more reported that 'giant sharks are again annoying sockeye salmon fishermen at Namu', and that those sharks were 'much bigger than in other years'. In July 1948 *The Fisherman* reported that sharks were 'a major problem in Smiths Inlet again'. But that seems to have heralded a drastic tipping point in shark presence, as from that moment onward there were no further reports of sightings of Basking Sharks in the region.

It is hard to pinpoint an exact cause for such a complete disappearance. A change in plankton distribution, perhaps? Or maybe the 'slasher' was put to far greater use than the historical record implies. And just how many sharks lost their lives due to the admittedly high levels of accidental entanglement? Or maybe all three impacts combined at one level or another to remove the Basking Shark from the equation altogether? The almost total absence of sharks from the region to this day ensures that a serious question remains on the table: who or what was responsible? As we shall see later from elsewhere on the west coast of North America, this was not the only site to suffer such an apparent extirpation of the Basking Shark. Maybe in time a pattern would emerge that would help to identify the cause.

Chapter 4
Anthony Watkins,
the Pioneer in Scotland

The Carradale Incident had one unintended consequence: it attracted the interest of a man imbued with a need for action, who would become the model for a totally different breed of shark hunter.

Anthony Watkins had adventure running through his veins. Born into a moderately wealthy family who took climbing holidays in the Alps, Tyrol and Lake District, he had been brought up to enjoy the wild outdoors from an early age. His older brother, Henry 'Gino' Watkins, had become the darling of the polar exploration world in 1931, having made a 600-mile open boat voyage around the south coast of Greenland for which he was awarded a Founder's Medal by the august Royal Geographical Society of London. Tragically, Gino Watkins was to lose his life the following year in Greenland, drowned when hunting for seals from a kayak at just 25 years of age.

His younger brother must have shared at least some of that spirit of restlessness, as by the time he reached the age of 25 he had already worked in five jobs in as many countries, including one in Bolivia and another in Jamaica. But by the late 1930s he was making an effort to settle down and found himself a dull but well-paid job as a clerk in the City of London. Bored and missing a little excitement in his life, one day his eye was caught by an entry in a ledger for the sale of two hundredweights of shark liver oil to a firm in Scotland. Interest piqued, he thought back to a recent newspaper report of the loss of several people in a yacht in Scotland – the Carradale Incident – that had been blamed on the Basking Shark.

The young man already had some knowledge of sharks, as in Jamaica his favourite pastime at weekends had been shark fishing. He knew from having visited an American shark fishing station that there were other markets beyond the liver oil, including tanning the hide to make shagreen, and rendering the flesh down to make fishmeal for fertiliser or livestock feed. And, he reasoned, as there were these huge Basking Sharks swimming around Scotland in their droves, why had nobody thought to hunt them?

Even at this early stage it is apparent that Watkins was a man with a practical as well as a wild bent (and a nicely dry sense of humour, as is apparent in his book *The Sea My Hunting Ground*), and his approach to this germ of a project was both methodical and thorough. One of his first actions was to contact Dr Seccombe Hett, a retired surgeon and keen amateur zoologist who had been studying Basking Sharks in the Firth of Clyde, and had played a major role in unravelling the cause of the Carradale Incident. Dr Hett turned out to be an ideal sounding board, with personal experience of having tried to catch a Basking Shark, using a hand-held harpoon from a small motor launch called the *Myrtle* in the company of his son and one Dan Davies, the owner and skipper of the boat. Despite having lost dozens of hand harpoons and hundreds of metres of expensive yacht manila rope, he and his son finally managed to get a harpoon to hold and, after 15 hours being towed around the Clyde, had at last caught a shark.

Dr Hett was able to explain to Watkins a considerable amount of detail about the anatomy of the shark (having dissected the female he had harpooned) but little more that might help the younger man establish whether a fishery was viable or not. He was also able to arrange the charter of the *Myrtle*, and was offered valuable advice on the need for better harpoons, having had considerable trouble with his original models made by a local blacksmith.

There's a term for people who find themselves taken over by an overwhelming fascination with sharks – Shark Fever. One might be forgiven for thinking that Dr Hett himself was more than a little touched by this affliction, but if indeed he was, it didn't stop him giving his young protégé some excellent advice, 'Don't underestimate Basking Sharks. They're more like whales than fish. Watch yourself, and don't let Davies get carried away by his enthusiasm. You'll find out he's inclined to be reckless.'

Judging by what was to unfold, Davies wasn't the only one for whom the description of 'reckless' might apply. But Watkins took Dr Hett's sound advice, and managed to track down a small engineering firm in London that made hand harpoons to the same design as those used by the old American whalers in pursuit of Sperm Whales. Watkins recognised that the ideal harpoons would have been the exact models employed in the final years of the Irish fishery, but lacking time, had to settle for what he could obtain to hand. To carry the metal harpoon heads he had wooden shafts made from two-inch diameter ash, with lead shot in the hollowed-out rear end of the shaft to achieve perfect balance. The whole made for a three-metre-long harpoon, attached to 200m of 50mm diameter yacht manila rope. On the face of it, this made for pretty much unbreakable tackle for securing a Basking Shark, and Watkins considered himself ready for the fray.

At the same time, Watkins went about discovering as much as possible about potential markets for shark products obtainable from the Basking Shark. Already aware of the market for liver oil, he explored the possibilities of both medicinal and industrial oil, and found companies willing to sample and assay the oil to determine its suitability. J.M. Davidson of Glasgow, who would go on to become the major buyer of shark products throughout the Scottish fishery, agreed to try out the flesh for fishmeal, although they considered it unlikely that the flesh could be sold for human consumption, pointing out that 'at the moment it was hard enough to sell even the best quality herring and cod'.

Watkins pursued his enquiries far and wide, and found a tannery in Scotland who agreed to try a sample of Basking Shark skin, and even a Norwegian company who had previously dealt with the skin of the Greenland Shark (*Somniosus microcephalus*). He tried to find a market for the fins via a Chinese restaurant in Piccadilly, although this came to nothing due to language difficulties, leading him to conclude that 'the Chinese seem to share with the

English an ability to live in a country indefinitely without ever mastering the language'.

The opportunity to be his own boss appealed greatly, and he felt that the response from potential buyers had been positive enough to merit further exploration of his idea, in spite of the vagaries that he had already identified; he clearly considered that there was money to be made hunting the shark. So, together with his good friend Tommy Wewage-Smith, he boarded the night train to Glasgow with all of the harpoons and other paraphernalia. He was ready to make his first acquaintance with the Basking Shark, and embark on a wild ride he would never forget.

A reality check

His first impressions of Dan Davies and the *Myrtle* were mixed, Davies being a powerful and capable young man but the *Myrtle* being in less than perfect order, in obvious need of a re-fit and with a recurrent engine problem that would later come to haunt them. Davies's crewman was a dependable Aberdonian by the name of Harry Webster, and the two men were to prove to be staunch shipmates in the adventures that followed, even if the *Myrtle* had seen better days. There were Basking Sharks reported to be around the seabird island of Ailsa Craig in the southern Firth of Clyde, and the weather was perfect: flat calm with the sea like a mirror. Travelling steadily through schools of porpoises and diving gannets – positive signs of life beneath the still surface – the anticipation was at fever pitch, until: 'Shark!' Watkins sighted his first Basking Shark.

After stalking the shark for some time, the *Myrtle* closed in and Davies hurled the harpoon. After a few seconds the shark expressed its displeasure by whacking the bow of the boat with its tail before heading straight for the bottom. The boat's engine was stopped, the rope whipped out and the crew waited until the shark had reached the bottom before making the rope fast. And then 'the *Myrtle*'s bows whipped round, and she began to surge ahead, slowly at first, then with increasing speed.' The Clyde sleigh-ride was on.

It wasn't to last. After a few hours, the motion of the boat changed and it was clear that the shark had got away. Hauling in the rope, Watkins noted that the last 30 feet or so were covered in 'a thick, black and evil-smelling slime', and it was frayed and worn from abrasion against the shark's skin. Finally, the remains of the harpoon were brought on deck. The shaft had snapped off and the harpoon itself was bent into an L shape – the new harpoons were obviously not up to the job. The harpoon was straightened and another shaft fitted, and once more the *Myrtle* set off in pursuit of a Basking Shark. A further shark was soon sighted, the harpoon thrust home, but after a few hours the result was the same. The harpoon pulled out, just as it did with the following three sharks. Sometime after midnight it was a thoroughly disappointed crew that tied up alongside in Campbeltown at the south end of the Kintyre peninsula.

The problem seemed simple enough. Either they needed better harpoons, or a lighter boat. Designing and fabricating improved harpoons was out of the question in the time allowed. At just over 12 metres in length, the *Myrtle* must have weighed many tonnes, which, combined with a considerable amount of water resistance from the hull, would contribute to the drag on the harpoon in the shark's back, and the combined effect had obviously proved to be too much. A smaller boat would weigh far less and the drag would also be greatly reduced. But where would they find a suitable boat at such short notice? Then the thought struck Anthony Watkins: 'We ourselves had a smaller boat, the *Myrtle*'s dinghy, only eight feet in length. If we could do all the harpooning and playing from this, the strain on the line would negligible.'

For a few minutes the argument swung back and forth. What if the shark lashed its tail against the dinghy as it had done to the *Myrtle*? Watkins reckoned that with smart boat handling they should have just enough time to get clear of the tail. What if the shark actually intended to hit the *Myrtle* with its tail? Watkins couldn't believe that a shark could have the intelligence to use its tail as a weapon. The final call had to be from the skipper, Dan Davies, who not only had the most experience of harpooning sharks, but of course was the owner of the dinghy. Pausing only to fill his pipe, and ponder for a moment, he declared: 'Yes. It's a good idea. Should be interesting too.'

May you live in interesting times

Finding no sharks down at Ailsa Craig the following day, Davies altered course for Kilbrannan Sound, the wide stretch of water between Kintyre and the Isle of Arran, right to the site of the tragic Carradale Incident. This proved to be the right move, and the wild plan of action was soon in place, with the *Myrtle*'s tiny punt soon in pursuit of a Basking Shark with Anthony Watkins at the oars and Dan Davies ready with the harpoon. After a number of false starts that clearly gave Watkins food for thought, eyeing up the massive creatures at such close proximity, the harpoon was finally struck home: 'Something like a small depth-charge went off just under the stern. A tail rose clear above the surface and towered high over us, before crashing down on the water with a noise like a clap of thunder'

A few moments later, both men were in the water. As any sailor knows, rope has an uncanny ability to develop tangles and stiff, heavy rope is the worst of all, especially when flying out through a rowlock designed to guide the rope. In no time at all a knot formed at the rowlock, jammed, and the bow of the dinghy was pulled straight underwater. Watkins and Davies found themselves battling to get clear of the flailing rope, still attached to a very disgruntled shark heading at full speed for the bottom of the sea. It was then that Davies admitted that he couldn't swim. Dr Hett's warning about his recklessness had been grounded, in fact, it was clear. But clinging to an oar, Davies pronounced himself quite alright, and the two men, holding on to the oar, set off for the *Myrtle*, which was sitting dead in the water several hundred metres away, with the crew unable to start the engine.

Whereupon the shoal of sharks started to take an apparent interest in the two struggling men – or so it seemed. Half a dozen sharks swam right up to them:

> For a few seconds the sharks sniffed around us, like a pack of dogs around a tree. They were so close that we could have kicked them in the face, and I seriously considered doing this. I decided against, as it seemed that we probably had more to fear from a lot of frightened sharks than a lot of curious ones. Dan had evidently reached the same conclusion, for he too had frozen into immobility. Suddenly, as if disappointed with what they had found, the whole shoal sank out of sight. We breathed again.

Finally making it to the still inert *Myrtle*, the two men were soon safely aboard, to be greeted by two very shaken crewmen. To Watkins's satisfaction even Dan Davies admitted, 'We had scared the daylights out of ourselves'. They had been very lucky, and the laughter that followed must have been a release of tension as much as seeing the funny side of the situation. But now they were in the middle of several shoals of sharks, the dinghy had disappeared, presumably still attached to the shark and now somewhere in the depths. There was nothing for it but to return to Campbeltown, regroup and decide what – if anything – further could be done.

Half an hour later, a thump on the deck brought Anthony Watkins back topsides to witness their upturned dinghy doing a steady three knots, still very evidently attached to the shark. Coming alongside the capsized dinghy, they managed to right it and bail some of the water out before one of the crew jumped back in and completed that task. It then took four cold, wet hours to sort out the tangled rope and clear the knot at the rowlock, all the time with the shark pulling the dinghy along like a child's sleigh. Just when it looked like the shark might be theirs, the harpoon, perhaps weakened in its hold in the flesh by the work on the rope, pulled free. Retrieving the bent harpoon, the four men disconsolately set off for Carradale for the night, to sleep and contemplate their failure.

If at first you don't succeed

A lesser man (some might also say a wiser one) would have called it a day right there and then. Given the problems with the *Myrtle*'s engine, the difficulties with the rope, the alarming approach of the sharks and the great good fortune they had in being recovered, it would have been the easiest thing in the world to simply pack up and go home. Except, perhaps for that atavistic call – Shark Fever. And by now Anthony Watkins and his crew had all been well and truly gripped by it.

No sooner were they ashore than Watkins was listening to reports, asking questions and finding out as much as he could about the life and habits of the creature that had so recently nearly ended his life. There was obviously no question of giving up, the only decision now being how best to continue. Despite a few misgivings, Watkins was won over by Davies and Wewage-Smith to the idea of continuing the hunt from the dinghy. Davies said: 'I know yesterday was a bit of a flop, but we proved two useful things. One is that sharks won't go for a small boat; the other is that you can avoid the tail after harpooning.' To which Tommy Wewage-Smith added: 'We'll never catch a shark without using the dinghy, and we'll look pretty good fools if we don't.' And so, despite a few lingering concerns on Watkins's part, the die was cast. Back into the dinghy they would go.

With the *Myrtle*'s engine running once more, the crew set out again to scour Kilbrannan Sound, seeing only a couple of breaches in the distance to give any sign that the sharks were still in the vicinity. Finally, off Skipness Point, a fin appeared, and Watkins and Davies clambered into the dinghy once more. The fin proved elusive, though, until:

> Suddenly there was a mighty rushing sound and a huge glistening shape rose out of the water at a distance of less than forty yards, stood poised on its tail, high as a three storeyed house, for a fraction of a second, then fell forward in a thunderous belly-flop to disappear in a great mountain of foam.

> A series of giant ripples began to strike us, and the boat rocked violently, shipping water on both sides

Both men were well and truly shaken this time, and the thought of the Carradale Incident and its tragic outcome was hard to put out of their minds. But the shark didn't show again, and the search continued. Later that day off Lochranza on the Isle of Arran they found themselves in the middle of a group of Basking Sharks, one of as many as 20 shoals in sight spread out over two or three square miles. Picking the nearest shoal, they made their way to a position behind the rearmost shark, a big animal showing a metre of fin above the water, and rowing steadily, Watkins manoeuvred the little boat to the shark so that its bow was almost touching the fin. Davies raised the harpoon and flung it home, the stock standing up 'like a mast' from the shark's back. A quick pull on the oars and they were clear of the danger of the tail. The shark dived, then surfaced and turning directly towards the boat, charged straight at them. At the last moment, the shark ducked under the dinghy and dived for the bottom, the rope running out smoothly and easily this time, before they checked the line and settled down to what was to be the longest ride of their lives.

At first all went well. The *Myrtle* stayed in close contact and they were able to 'play' the shark by sitting on the rope and allowing it to ease out when the shark found a deeper hole, keeping tension on the rope, but not enough to work the harpoon loose. After a short while, though, the *Myrtle* closed with the dinghy and Tommy asked Dan to take over aboard her as the infernal engine was once again threatening to cut out, and the two men swapped places. Whereupon the *Myrtle*'s engine stopped, and the dinghy drew steadily away from her, causing Tommy to remark presciently, 'Blast that engine, I don't trust it one little bit. It's going to conk out for keeps one day soon.'

All the time the shark was steadily moving south, and by evening they were well down Kilbrannan Sound, and Carradale came into view. Watkins took a break aboard the comparative luxury of the *Myrtle*, to have some supper and stretch out his limbs, and took over from Davies at the wheel. The *Myrtle*'s engine, an early petrol-paraffin device, disliked running at the slow speeds that were necessary to stay close to the dinghy and coughed and spluttered endlessly in protest, sometimes ominously cutting out altogether. Dan Davies being the only one with the magic touch to get it started again, there was now no question of him taking a stint in the dinghy, and with night falling, all concerned must have had deep misgivings about the unreliable and poorly maintained engine in the *Myrtle*. If it cut out and took a long time to start, the dinghy might be out of sight in the dark before they got under way again – and what then? And then, just when they least needed it, the weather took a turn for the worse, and the wind began to get up.

Swapping places once again with Wewage-Smith, Watkins settled into the cramped dinghy. Sensibly, he took a lighted hurricane lamp with him, so that the men aboard the *Myrtle* could keep an eye on their whereabouts. Before long they were in open water south of Arran, passing the lighthouse on Island Davaar at the entrance to Campbeltown Loch and shelter, and still the shark pulled resolutely onward.

And then the *Myrtle*'s engine did what it had threatened to do all day. It stopped, finally, and the two men watched its navigation lights slowly recede into the distance until they finally disappeared altogether. The light on the island of Sanda off the southern tip of the Kintyre peninsula blinked on their right at less than a quarter of a mile distance, and then Watkins became aware of something that he most certainly would have preferred not to see. The shark had altered course.

> Suddenly I noticed that something was wrong. Harry's sturdy form in the bows was
> swinging round towards the light. The harpoon line, instead of pointing straight
> ahead, was angled out to the right. We were turning, we were rounding Sanda.
> Instead of continuing down south, we were turning west around the Mull of
> Kintyre and into the open Atlantic.

Instantly realising the awful implications of this development, Watkins hoisted the hurricane lamp on one of the oars, to make it visible to the men in the *Myrtle* at a far greater distance. By now Watkins and Webster could see the bleak outline of the Mull, and hear the waves breaking on the rocks. And even in this most awful of predicaments, Watkins paused to marvel at the incredible internal navigation system that the shark had employed to navigate its way past blind alleys like Campbeltown Loch down to escape from the Clyde and the savage wound in its back, turning for deep water and home in its hour of need.

What an awful place

It's worth digressing here for a moment to consider the desperate piece of water they were entering. The Mull of Kintyre has a fearsome reputation amongst sailors, and rightly so. Tides can run at up to five knots close to the infamous Mull, as the contents of the Irish Sea fight their way in and out of the narrows (12 nautical miles across) between the Mull and the Antrim coast at the northern end of what is called the North Channel. The race off the Mull is particularly nasty, and can generate wicked over-falls and even square waves when a fresh wind meets an opposing tide. I have sailed through those waters many times in yachts of between 12 and 24 metres, and have always been very glad to see them behind me. Having been caught once in difficult conditions in the notorious MacDonnell race off Rathlin Island, I can promise you that if you are in the wrong place at the wrong time you won't forget it. Suffice to say that to be towed by a shark into those waters in the middle of the night in a dinghy less than three metres long would not be high on my 'to do' list.

Meanwhile, back in the dinghy, the two men began to feel the change in the sea as they lost the shelter of the land. To their immense good fortune, the wind was very light, 'a faint puff from the west', just enough to raise a few white horses, but there was the never-sleeping Atlantic's swell breaking on the rocks inshore of them. By letting out extra rope and sitting in the stern, they were able to let the bows raise a little, and 'found that our little boat then rode the rollers beautifully, running up to the crests and diving down into the troughs as if we were on an unending switchback.' Undoubtedly they were extraordinarily lucky to be where they were in such benign conditions.

And the shark kept driving on, not turning north as Watkins hoped, but continuing on, towards America:

> The outline of the Mull blurred and merged into the darkness and the sound of the sea breaking onto the rocks died away in the distance. The only sound now was the lapping of the water under the hull and the hiss of the waves as they swished by us. There was no land and no lights in sight, no stars and no moon. Outside the little circle of our own light there was nothing. It wasn't even possible to distinguish sky from sea.

And still no sign of the *Myrtle*. They were still heading west out into the mighty Atlantic. Watkins was becoming slightly nervous: 'Have you ever been out here before, Harry?' 'Och, no, Mr Davies only hires out his boat for the Clyde. The *Myrtle*'s no' a boat for the Atlantic.' At which point I can only imagine an intelligent man like Watkins asking himself: if the *Myrtle* was not a boat for the Atlantic, then what did that imply about the dinghy?

Finally, dawn came, and the arriving light gave them a better idea of the direction in which they were headed, slightly north of west. Towards Greenland, not America, as if that was any consolation. But the daylight revealed no comfort in the form of the *Myrtle*, shipping or land. The sky was dark and lowering, although there was still little wind. In this lonely place, with only the shark for company, even the bravest of men could have been forgiven for expressing their fear, but Watkins and Webster maintained the proverbial stiff upper lip.

> I looked at Harry. There was no emotion in his face. It might have been carved from the granite of his home town.
> 'This is a bore,' I said, conscious that the remark was hardly an apt one in the circumstances.
> 'Aye, it's a damn nuisance,' said Harry, doing a little better but still not making a really adequate comment.

The dark sky foretold bad weather, and Watkins recognised the urgent need to do something positive to alleviate their plight. The shark seemed to be tiring at last, and the speed of the tow was now dead slow. He had an idea:

> 'Well, Harry,' I said, 'what do you think we should do?'
> 'Och, I dinna ken. I'm not used to situations like this.' He spoke as if I had spent most of my life being towed about the oceans of the world by sharks.
> 'It seems to me,' I said, 'that our best chance is to see if we can get the shark to the surface and then try and turn him back to land. He seems to be tiring. And obviously we can't go on like this.'
> 'Aye, that's a good idea,' agreed Harry.

So they set about pulling the shark toward the surface. On the face of it that might seem an impossibility, but Watkins realised that when the shark was swimming along, as long as it was not swimming downwards it should be quite easy to lift. Although it was backbreaking work, after an hour or so, with the line jerking wildly and the black slime on the rope stinging their raw hands, they knew they had the shark close to the surface. Finally, a giant fin rose out of the sea just ahead of the dinghy's bow. Now, to try and turn the shark around.

Seizing the oar, Watkins stood up, leant forward and jabbed the shark in the right eye.

> His head sheared off violently to the left and his tail gave a flurry of indignation astern. The boat was jerked sideways through the water, and I sat down abruptly. It was all very frightening, but when it was over we had turned through a quarter of a circle.
> Without giving him time to get back on course, I stood up and jabbed him again in the eye. Again he sheared away, and we were whisked round a little farther. I was prepared for it this time, and did not even have to sit down to regain my balance. A third and a fourth poke, and he was completely turned round and heading due east, as near as we could estimate.

They were headed back towards land and safety. An occasional prod from the oar, and the shark maintained the correct course. Watkins even had time to compare the shark as a motive power favourably with the wretched engine of the *Myrtle*! But the dark sky finally revealed its intent, and the wind began to steadily increase. The little dinghy began to ship water over the stern as well as the bows, and the two men began to bail more rapidly.

The seaworthiness of even tiny boats never ceases to amaze, as Watkins was to remark to Harry:

'She may take more than we think.'
'Aye, most boats do,' said Harry.
It was a profound truth. Very few boats, no matter what their size, are ever tested to
their limits. People are apt to assume that the limit is just beyond the roughest they
have ever been in, or the worst that they know similar craft have endured. When
it gets wilder than this, they become anxious, even though they may be riding the
seas well and their boat could, in fact, take far worse.

Amen to that. And so hanging on to the shark line and bailing furiously, occasionally prodding the poor shark to maintain the landward course, the little dinghy staggered along in the worsening weather and rising seas. By mid-afternoon the weather seemed to have stabilised and their spirits began to rise with the thought that – perhaps – no, maybe – they might survive. '"Harry," I said, "I believe our troubles are nearly over. I'm sure we shall see land soon." "Aye, it canna be far off," said Harry, and his voice was a cheerful as mine.'

Then an aircraft appeared, seemingly doing exercises, making a zigzag course towards them. After a few minutes it became clear that it would pass over them, and when it did it dived down and began to circle the small boat and its lucky occupants. Not that Watkins felt that way about it: '"Don't wave," I said in an agitated voice, "or they may think we're in some sort of trouble and call out a lifeboat. And stop bailing. You make it look as if we're sinking."'

As if to underline their nonchalance, Harry produced a damp pack of cigarettes, and both men lit up. Watkins produced a book from his oilskin pocket and pointedly started to read it. Not that any of this fooled the aircrew.

Shortly afterwards, now much buoyed by this visitation, but still determined to make the shore under their one-shark-power engine, Harry pointed to an approaching ship. '"I think she's a lifeboat." Said Harry after a pause, "Aye. She's a lifeboat right enough."'

Spooked by the rumble of the twin engines, the shark dived for the bottom once more. A crewman threw them a rope, which Watkins promptly threw back 'with an apologetic air, shouting above the wind. "Awfully sorry you've been called out. It's all a mistake. We're in no sort of trouble. Just fishing, that's all."'

This met with a predictable and entirely justified response, and the coxswain told them in no uncertain Anglo Saxon terms what they should do:

I will not give his exact words. Lifeboat Coxswains are honest and God-fearing
folk, but at times they tend to express themselves with a picturesque vigour which
it would be unfair, and indeed against the law, to repeat verbatim in print. However,
the gist of his remarks was quite clear. Would we please cut the line, and come
aboard.

Refusing indignantly, Watkins demanded that they be left alone, and the lifeboat cease to disturb their fish. A conference ensued aboard the lifeboat, and eventually the coxswain returned and speaking more reasonably, Watkins assumed that he had seen reason and was about to leave them alone. But no, as he was to later learn, the coxswain had 'simply come to the conclusion that we were lunatics and that it was best to humour us.'

The coxswain then played his ace: 'We were called out early this morning by the *Myrtle*, and we've been looking for you ever since. If you refuse to come aboard we shall have to stand by. And it's Sunday, our free day.' Game, set and match to the coxswain.

But before boarding, Watkins borrowed a buoy from the lifeboat and attached it to the line with the hope that it could be found later and the shark retrieved. Once aboard, he obtained the exact position from the coxswain with every intention of returning for his shark.

And so the lifeboat made its way back to Campbeltown, where: 'A large crowd had assembled on the jetty, and we pushed through them with some difficulty. They were more simply anxious to know whether we were all right, though one patriarchal fisherman asked us what else was to be expected if we fished on the Sabbath.'

Maybe it was time to give up?

Having read this far you already know the answer to that, and no sooner had the two big, bold fishermen found the *Myrtle* than Watkins was demanding to know how soon they could put to sea. "Five minutes," said Dan, with what seemed to the rest of us a touching faith in the *Myrtle*'s engine. We retired below and half an hour later we heard that uneven and hesitant chug which told us that we could cast off.'

Rounding the Mull it was clear that the weather had once again worsened, and the *Myrtle*, no sea boat, designed only for operating in sheltered waters, began to show her shortcomings. Long and narrow, with high topsides, as soon as they turned north and took the seas on the beam she began rolling like a pig, and was soon shipping some good solid dollops of water over the decks where her lightly built hatches and skylights welcomed the water below. Then she began to leak seriously, and the crew got out the buckets and started bailing in earnest.

By now even Watkins was concerned and demanded to know what was up: '"Hey," I bellowed at the top of my voice, "what's going on below?" Dan poked his head up. "There's no special leak, it's simply that this battering is opening up the seams."'

Watkins reasoned that if Dan hadn't told him to head for shore, they couldn't be sinking, but:

> On the other hand, I was beginning to know Dan. He blamed himself (quite wrongly) that the shark was not already safe on a beach somewhere. His blood was up, he was determined to find the shark at all costs. I had a feeling that he would continue looking until the *Myrtle* sank beneath him.
>
> I looked at the sea. It seemed to be boiling with fury and malice, while the wind shrieked and screamed at us menacingly. I pictured what it would be like if we all four had to take to the small boat. It had been bad enough with only two of us in it, and a far milder sea than this. I spun the wheel over and headed for the Mull and Campbeltown.
>
> Immediately the aggrieved face of Tommy appeared from below. 'What the hell are you doing?' he shouted.
> 'Going back home as fast as I can,' I answered shortly.
> 'But what about the shark?'
> 'Damn the shark.'
> 'But good God, Tony, what's the matter with you? Can't you take a few inches of water below without losing your nerve?'

Up until this stage it would be hard not to view the poor old *Myrtle* as a latter day *Pequod*, with not one but three Captain Ahab's in command. Now they were down to only two. But the right decision had been made, and the unhappy crew once more made their way back to the comforts of Campbeltown.

Unsurprisingly, after nearly 48 hours without sleep, battered by the weather, shoulders aching from the strain of pulling up the shark, hands raw from the slime and the harsh rope, Watkins felt tired. But as any mariner knows, after such a hiding there is only one possible antidote – a good pub. And so on arrival at Campbeltown the worthy crew set off up the pier with the intention of adjourning to the Argyll Arms and whisky, warm or cold, only to find that the iniquitous Scottish licensing laws had removed even that pleasure from the agenda. It had been a long, hard couple of days, and this discovery must have felt like the final insult. Almost enough to make you give up on the idea of chasing Basking Sharks, whether for excitement or perhaps in the hope of making a living.

But not quite, apparently. Although the weather kept them in port the following day, they used the time wisely to sort out the worst of the leaking seams of the wooden hull of the *Myrtle,* while a mechanic fiddled around with the engine, making the boat at least serviceable if not perfect. They later held a conference on ways they could improve their shark hunting technique – there was obviously no thought of giving up. If it had been me I'd have walked away from the whole idea on the spot and taken up something less risky, like bowling or billiards. But Shark Fever is a strange thing, and once infected with one of the more virulent strains, the patient may find that there are few cures beyond drowning or bankruptcy. Which may in some way explain why two days later, Watkins once more found himself in the sub-three-metre-long dinghy being towed by a Basking Shark. Fortunately for him and his companions, on this occasion the shark stayed in sheltered waters around Arran and didn't head for the open Atlantic, which might have proved fatal.

This time they were trying out a new refinement of attaching buoys to the harpoon line in an effort to tire the shark out, the buoys creating considerable resistance to forward motion without creating sufficient drag to pull the harpoon out of the animal. At last, during the second night of their travels, the shark tired and slowly Watkins and Wewage-Smith hauled it to the surface. With a rope ready to lash the dangerous tail, and so render the shark powerless to escape, the two men prepared for the final assault:

> I began to pull in, and inch by inch worked the shark up. The line vibrated with his movements, but he seemed incapable of making any really violent effort. After a few more hauls I could dimly make out below us something like a white pulsating cloud. It didn't look like a shark at all. It looked more like a great jellyfish or an octopus, and was as big as a medium-sized house.
> 'What the hell are we hooked on to?' gasped Tommy in my ear.
> 'Stand by with the tail rope,' I said, 'and for Pete's sake get it on right away.' We were dealing with a stricken dying giant, but if he could summon up even a fraction of his former strength he could still hammer us down into the sea as easily as man could drive a nail into butter.
>
> The first attempt was a total failure, and the shark went for the bottom once more.
> 'Not as easy as I thought,' I gasped, 'perhaps we should wait until it gets light.'
> 'Nonsense,' said Tommy cheerfully. 'I'm beginning to get the hang of this cowboy business. I'll do better next time.'

After two more failed attempts to lasso the tail, Tommy finally got it right, and with the tail under control, the battle was all but over. Passing the rope to the *Myrtle*, the triumphant team set off for Tarbert, a sleepy little port in Loch Fyne, north of Arran. Struggling to get the *Myrtle* alongside the quay with a still active shark wrenching at the ropes made them look amateurish:

> A fisherman, unable to see the shark on the far side, watched our efforts with undisguised contempt. Suddenly he saw the shark, and his mouth opened.
> 'Begod, have you caught one of *them*?' He said with awe.
> 'Aye,' said Dan. 'Do you think anyone would mind if we beached it here?'
> 'No, no. This is a fishing port. You can land any fish you like here.'

So with the help of some of the fishermen, they pulled the shark to the water's edge and waited for the tide to fall to begin the dissection of the shark. At low water the vast fish was completely exposed to view, and a large crowd gathered to watch the proceedings, led by a worried harbour-master:

> Who inquired anxiously how long we proposed to leave him there. Basking Sharks, even when fresh, make their presence felt – all the trippers had handkerchiefs to their noses and so also did some of the less robust fishermen – and it was obvious that in a day or two Tarbert would become uninhabitable. We promised that we would tow out the remains within twenty-four hours.

Taking samples of the skin, Watkins noticed how rough it was, and also how the colour and markings of the shark, so vivid and obvious in life, faded to an almost uniform slate grey as the shark dried out. The abrasive skin was so rough that, if touched with a bare hand, the skin was punctured and the black slime 'set up an instantaneous and violent irritation.'

Measuring the shark out of the water proved interesting, too, as pacing it out on deck when the shark was alongside had suggested that the shark was well over nine metres in length. Now the tape measure confirmed the animal to be just under eight metres overall; still a big shark but no monster.

Cutting the shark open proved to be a big hit with the crowd:

> The next step was to open up the shark. This entailed making a long slit down from the gills almost to the tail, and then cutting downwards from each end of the slit, so that the wall of the stomach opened like a door and the insides slid out and spread themselves over an area big enough to build a house on. I had inadvertently pierced the big intestine and hundreds of gallons of partially digested feeding, a revolting thick red liquid, poured out in cascades and flooded the whole of the harbour. The crowd cheered. The harbour-master winced.

The two lobes of the liver, each six metres in length, stood out 'like ivory tusks', and potentially as valuable. Watkins estimated the combined weight of the two lobes at 14 hundredweights (711kg). The total amount of flesh was estimated at around two tons (2,032kg), and Watkins couldn't resist running an immediate 'taste test', giving some flesh to:

> An Airedale who had been watching our activities with a hungry eye. As an afterthought I gave the owner my address, asking if he would let me know if the

dog suffered any ill effects. He replied in some indignation that if it did, I should hear, not from him, but from his lawyer. I said that I intended to try some of the meat myself, but he was unconvinced. (We did try it. It was pretty nasty).

Dan Davies had set up a simple oil extraction plant to enable them to produce the required sample of a drum of oil, and a sample barrel of liver was prepared for dispatch, as was a barrel of raw liver, a barrel of flesh and two skin specimens. Watkins had achieved his goal, and would soon learn whether there was a market for the shark products, and most important of all, whether there was money in it – Shark Fever or not, their busman's holiday was over.

Turning the concept into reality

Watkins returned to work full of ideas and half-formed plans:

> 'Ah, back again, refreshed I hope,' my boss welcomed me benignly a few days later. 'Let's see, what were you doing? Fishing in Scotland, eh? Lucky fellow. Best sport in the world. Catch any big ones?'
> 'One fair-sized fish, sir,' I answered.
> 'Splendid, splendid, you must tell me about it later.' He was a busy man, and he had only a limited time to spend on being nice to the boys.

Slowly the reports came in from the potential buyers. The first, from the liver, confirmed the lack of Vitamins A and D, and so ruled out the use of the oil for medicinal purposes, and therefore also potentially high prices. The second, regarding the oil, confirmed an offer to purchase the oil at around £25 per ton (£1,150 at 2017 values). The third, regarding the flesh, confirmed an expected value of £7 per ton (£322 per ton in 2017) as fishmeal, but ruled out its sale for human consumption: 'Between the lines I thought I detected a certain heartfelt conviction on this point that suggested he had, like me, tried it out on himself.'

The reports from the two tanneries were negative, indicating that the cost of curing and treating the skin was prohibitive, and there was no market for the finished product in any case.

So there was a market for the oil and flesh, and the returns looked attractive, even given that there were so far only two viable products. But there still remained a thousand and one practical details to examine before any further progress could be made. In this department, Watkins yet again showed considerable insight, focusing on the weather, and its impact on the ability to catch sharks at the surface. Studying the weather records for the west coast of Scotland from the archives of *The Times*, Watkins made a record of the wind speed on every day between May and August for the previous five years:

> A day with a five m.p.h. wind I put into the category of a 'good' shark-fishing day, 10 m.p.h. was a 'possible' and 15 m.p.h. was 'bad. I found that out of the 120 days of the season, on average one could expect 34 good days, 56 possible and 30 bad days.

This seemingly off-hand calculation parallels almost exactly the parameters used for modern boat based surveys, where 15mph (Beaufort Scale 4) is the cut-off point at which a survey has to be abandoned due to the difficulty of accurately sighting and recording Basking Sharks at the surface. Watkins possessed an inventive and analytical mind.

 56

The cost of setting up to deal with the two marketable products, oil and fishmeal, was his next concern. His first attempt at costing a suitable factory was a shock to the system, as the company he contacted who specialised in plant of this nature came back with a price of £5,500 (c. £350,000 in 2017) plus transport costs for the machinery to be moved to the site. As the site was likely to have to be about as remote as it possibly could be, the latter cost would not be a negligible factor. A probable total price of £7,000 (c. £450,000 in 2017) was agreed after negotiation. The idea of scale and shape of the operation was beginning to take shape in his mind:

> There would be this shore factory somewhere in the Clyde, serviced by a fleet of up to half a dozen catchers. These would be ordinary fishing boats mounted with harpoon guns. In the winter the whole outfit would go over to herring, for the fishmeal plant could process herring as well as sharks. The cost would be something like this:

Shore factory	£ 7,000
6 catchers, complete with guns and equipment	£ 6,000
Working capital	£ 5,000
Total	£18,000 (c. £1,150,000 in 2017)

> Now came the question, could I raise this sum? My own resources, with some outside help, were £1,500, which left a balance of £16,500 to be raised from other sources.
>
> I made a tentative enquiry into the world of finance, and found the answer was a short and snappy 'no'.

It seemed that most people considered his commercial ideas to be as mad as his initial attempts to catch a Basking Shark, and it was obvious that he would have to drastically re-evaluate his plans if he were to get them off the ground at all. Maybe the best way to start would be small and build the business from there. His revised budget reflected this new reality, and was far more modest:

Factory	£ 600
1 catcher	£ 600
Working capital	£ 300
Total	£1,500 (c. £95,000 in 2017)

This was still a considerable sum of money for a young man to risk, and the amount of working capital was cutting it fine, to say the least. Watkins reflected that if it all went badly wrong he should very rapidly be out of money and a job, but on the other hand, he had a chance to be his own boss in a 'profoundly interesting little business' and that 'there was even a sporting chance that it might be a huge success and I might grow rich.' Furthermore:

> On the whole I thought it was a good bet. I was not backing a certainty, nor even a favourite. It was more like raising a jackpot before the draw with a four-card flush, and betting afterwards no matter what you draw. As a gambit it loses, I dare say, as often as it wins, but it adds excitement to the game, and when it comes off it's very satisfying.

Anthony Watkins certainly liked to gamble, and he knew that fortune favours the brave. You have to admire his nerve. And his sense of proportion: 'I thought the matter over for some time, and then told my boss I had found a promising opening in the oils and fats business in Scotland. He took the blow well.'

Putting it all into practice

Watkins's first purchase was a Norwegian made Kongsberg Farbenfabrik 37mm harpoon gun that fired a harpoon weighing 17lb. These guns were very similar to those used by the Norwegian inshore whalers in pursuit of the Minke Whale (*Balaenoptera acutorostrata*) and were suitable for boats over 11m in length. The gun, together with a dozen harpoons, ammunition, duty, insurance and freight cost £180 (£8,225 in 2017) and could be delivered within three months to Glasgow, leaving time to mount the gun on a soon-to-be-purchased catcher.

A visit to J.M. Davidson at his offices proved fortuitous. Not only did Davidson take the young man seriously, but felt that there was no reason why the business should not succeed. He also suggested an idea that would save Watkins much money: that Watkins should build his own factory with local labour and second-hand machinery, and even came up with the name of the man who had built his own fishmeal plant, James Agnew of Carluke.

Watkins wasted no time in getting to Carluke, arriving at Agnew's chicken farm that evening. Initially reluctant, as soon as Watkins mentioned his proposed prey, Agnew changed his tune: 'Sharks! You mean those great big fellows, Basking Sharks, I think you call them? The kind that towed those two young lunatics out into the Atlantic last summer?'

It seemed that Watkins's fame had preceded him. But he knew instinctively that he had found the right man to help him build his factory. Things were progressing well. Now to find a suitable boat.

This looked easy. There were at that time hundreds of boats that might suit his requirements in the Clyde fishing ports. The local boats were all similar in design. Technically known as Loch Fyne Skiffs, they were simple, sturdy wooden boats that were very seaworthy. Motor driven, most of them with Kelvin petrol-paraffin engines, they were equipped with cramped living quarters up forward and very few concessions to luxury. All were fitted with power-driven winches for hauling nets that could (in future) be put to good use in hauling up sharks, so they were ideal boats for his purpose.

Having whittled down an extensive list of possible vessels, Watkins eventually plumped for the *Dusky Maid,* a 12-metre model with a 36h.p. Kelvin engine. She seemed in good condition, and the engine ran well, so he settled down to the delicate art of haggling, and closed the deal at £400 (£18,300 in 2017). Moving down to Girvan, Watkins set about fitting the new gun and additional modifications to turn the boat into a practical shark catcher.

Watkins had originally planned to place his factory on Arran, but the owner of the island, the Duke of Montrose, was having none of it: 'The answer was an emphatic negative. Reading between the lines, one could almost see His Grace swelling with indignation at the suggestion that such a trade might be conducted on his island.'

The stretch of the Kintyre peninsula opposite Arran was owned by the Forestry Commission, who were far less squeamish about the idea, and were happy to rent any available space to him. Carradale seemed ideal, as it had the best communications and infrastructure, including a pier with a daily steamer visit. Eventually, a small inlet called Blackport 400 metres north of the pier was selected, being deemed far enough away from the popular beaches not to wreck the holiday trade with the stench of shark and the extraction of their oil.

But the lawyers he employed advised that even that concession to decency might not be enough, and that one sole complainant could block his plans. Watkins set about securing local support for his enterprise, moving in to the guesthouse belonging to the senior member of the MacDougall clan — a smart psychological move, to say the least. Watkins told Neil MacDougall his plans and talked of employment and opportunity. MacDougall affirmed that he could find a dozen crews in the village as the summer herring fishery was scarcely paying. Watkins then broached the delicate subject of the site of the factory, and asked if anyone might object:

> 'Aye, old MacBride will. He owns the last house along there. He'll object right
> enough. Aye, he'll object very strongly. He's a very difficult man. He's a foreigner,
> you know.' A note of disapproval came into Mr MacDougall's voice.
> 'Really?' I said with surprise, for MacBride sounded Scottish enough.
> 'Aye, the MacBrides come from Pirnmill. They've only been in Carradale for two
> generations. Still, don't you worry. I'll talk him round.'
>
> Mr MacDougall was as good as his word, and within a few days I became the
> lessee of Blackport, with official authorization to build and operate a shark factory
> there. Pirnmill, I discovered later, was six miles from Carradale.

Watkins and Agnew settled down to draw up the blueprint for the factory. As is the nature of these things, the project was already over budget due to extra costs incurred during the conversion of the *Dusky Maid* to a shark catcher. This in turn demanded that the factory be kept to the absolute bare essentials. Scouring shipbreakers and scrap merchants for the necessary machinery, the factory slowly came together.

Through Neil MacDougall, an experienced herring skipper, Bob Ritchie, was taken on, and Neil's youngest son Alec joined as engineer, both to eventually crew the *Dusky Maid*. But first they had to help build the factory, which was no simple matter in such a remote location, most of the plant having to be floated in on rafts. As one of the main items was a one-ton boiler, every man was needed to help manhandle and lever it up into position. Finally, the factory was ready, built at the extraordinarily low price of £570 (c. £36,000 in 2017). James Agnew had proved to be a real find and had done a magnificent job. Now all Watkins needed was a foreman to run the plant.

Neil MacDougall introduced him to a Carradale man who had gone to sea as an engineer and wanted to return home to his wife and family. His name was John Paterson, and Watkins knew he was on to a good thing: 'He was a few years older than me, a tall, spare man with a saturnine face and a sardonic sense of humour. I have been on the whole very lucky in business, but never more so than when I took on John Paterson.'

1939 – the first season

With the factory now complete, Watkins's small enterprise was ready to open for business, but ominously no sharks had yet been sighted in the Clyde:

> Rather than hang around waiting for them to turn up, I decided to go north and
> explore the Outer Hebrides' fishing grounds. I had no precise information of the
> movement of the sharks up there, but there was evidence that at times they were
> extremely numerous, far more numerous than in the Clyde.

Having just built a factory at considerable expense, one might well be tempted to ask why a man as smart as Watkins might have done so in the Clyde with the knowledge that sharks were more plentiful elsewhere. In time he would certainly come to reflect on this himself.

But for the meantime, he and the crew – Bob Ritchie as skipper, Alec MacDougall as engineer, John Paterson and Anthony Watkins – had plenty to do to prepare for the journey north. They set about filling the fish hold of the *Dusky Maid* with barrels to take the raw liver for the return journey to the factory, four 220m coils of 50mm diameter yacht manila for harpoon lines, 12 harpoons and the required ammunition. By the time they had finished there was no room to spare.

Perhaps now with a greater awareness of the damage that bad weather off the Mull of Kintyre could inflict on a small boat, they decided to take the safe route north via the pretty Crinan Canal. The Canal passes through some glorious scenery, and saves a lot of time and trouble for smaller craft, but it is hard work 'bunting' open the many lock-gates, and Watkins found it all rather boring.

Emerging from the final lock at Crinan Basin (opposite the north of Jura), the wind was raining and blowing hard, justification enough for the detour through Crinan avoiding rounding the Mull. Even in the sheltered waters of the Sound of Jura conditions were wild, which makes their next move even more inexplicable to me. As Watkins explained:

> At the entrance to the Sound of Jura is the notorious Corryvreckan, the meeting-point of three powerful tidal races which rush together to form a giant whirlpool. With the big Atlantic rollers crashing in through the gap between Jura and Scarba, and the wind whipping the surface up into a frenzy of rage, Corryvreckan appeared like a huge cauldron of boiling water stretching as far as the eye could see.
>
> I was at the wheel for the ignoble reason that I was afraid I might be seasick unless I had something to occupy my mind. I had not been to sea since the previous summer, and I always take a little time to get over an initial squeamishness. 'Good God,' I exclaimed, shocked at the scene. 'Can we go through that?'
> 'Och, aye,' said Bob in his soft, imperturbable voice. 'Keep her straight ahead. I've been through on worse days than this.'
>
> We plunged over the rim of the cauldron, and the deck almost disappeared in swirling foam. The *Dusky Maid* began to twist and kick and buck, now standing on her head and now on her stern, in a series of movements which I had never before felt in a boat. Generally, there is at least some regularity, some rhythm. When you go up you know that you will soon begin to go down. But here the boat was tossed about violently and unpredictably by the conflicting forces. The screw seemed to be thrashing the air fully half the time; yet we made good progress and no serious quantity of water was coming aboard.

Given that once through the Corryvreckan they altered course for Oban, and there's a far easier and safer route there via the Sound of Luing, it's tempting to think that Bob Ritchie was testing the mettle of the *Dusky Maid* and her crew, perhaps to see what he had let himself in for. Or maybe he, too, was bored by the passage through the Crinan Canal, and needed a bit of excitement.

Leaving Oban the next day, they made a smooth passage down the Sound of Mull:

> But as soon as we emerged into the open sea to round Ardnamurchan Point we had a battering almost as bad as the previous day in Corryvreckan. This lasted all

the way to Mallaig, and I arrived in a thoroughly depressed frame of mind. As a potential shark fishing ground this wild and tempestuous coast compared very unfavourably with that well-protected and friendly stretch of water, the Firth of Clyde. Here the distances were so vast that I scarcely knew where to begin the search. Sharks might be anywhere within an area nearly as large as the English Channel. And until the wind abated, wherever they were, they would skulk below the surface.

To which I can only concur, and say welcome to the Hebrides.

The *Dusky Maid* and her demoralised crew wisely sat out the storm in Mallaig, and then set off for Skye and the Outer Isles, and it was here that Watkins began to question the existing fund of lore on the migratory habits of the Basking Shark. He had previously accepted the hypothesis presented in most of the books of the day that suggested that the Basking Sharks arrived in British waters from the south, and steadily migrated north. If this information was correct, Watkins believed that if he carried on up the west coast of Skye before crossing the Minch to the Outer Hebrides and then continued north along their shores he should at some stage encounter the migrating sharks.

Setting off from Mallaig, the *Dusky Maid* and her crew would have been treated to magnificent views of Eigg and Rum as they rounded Point of Sleat at the southern tip of Skye, then the stark beauty of the Cuillins as they altered course to head up the west coast of the island. Passing the entrance to Loch Scavaig, the most dramatic of all of the sea lochs, they passed the pretty little island of Soay, where they encountered an old man and a boy in a small boat tending a stake net for salmon. Watkins decided to hail them to see whether they had any information on sharks.

'Seen any sharks around here.' I called out as soon as we were in hailing distance. He rowed up to us and hung on to the gunwale with one hand, a gnarled and grizzled old man, the grandfather, or even great-grandfather, of the boy, I guessed. 'Aye, there were plenty of them here last week, just before yon storm. They were busting through my net most every day.' He looked inquiringly at the hooded gun in the bows.

'That's a harpoon gun.' I answered his unspoken query. 'We're after sharks. Do you think it worth our while waiting here to see if they return?'

'Well, I shouldn'a think so. We're seldom troubled by them for more than a week or two at a time.'

We went on, cheered by this definite information. We were only a week or so behind the sharks, and we should surely catch up with them in a few days if we continued north.

Heading on up the Skye shore, Watkins went ashore to speak to the lighthouse keeper at Neist Point, the promontory to the north of Moonen Bay. Hoping to glean more information about sharks, to his surprise the lighthouse keeper showed him an entry in his log that recorded the presence of large numbers of sharks in the Bay in the last week of April, which would mean that the sharks had been there *before* they had appeared off Soay, which would in turn indicate that perhaps they had been moving south, not north. Watkins put this down to perhaps being merely a local movement, and the *Dusky Maid* set off for the Outer Isles.

In the lee of the islands the sea was calmer and there was more life about, but still no sharks were sighted. Stopping in the fishing port of Stornoway to refuel, Watkins went round the crews of the herring drifters asking about shark sightings. To his amazement he learned that in the middle of April the sharks had been very common off Tolsta Head, well to the north of Stornoway:

> This meant that all the time I had been travelling north, the sharks had been travelling south. It didn't tie up in any way with the textbooks. So, far from migrating north, the sharks had first appeared in the extreme north and had then moved south, reaching Moonen in late April and the southern shore of Skye in mid-May.

The fishermen also told him that the sharks had been plentiful for the last five or six years, that they never stayed in the same locality for more than a week or two, and were seldom seen after June. This wasn't necessarily what he wanted to hear:

> Somewhat disheartened, we left Stornoway and turned south again. The problem of locating sharks up here was very different from the Clyde, and so far I had not discovered the key to their movements. Indeed, it was beginning to look as if there was no key. It seemed that they simply arrived somewhere off the west coast in Spring, moved up and down the coast unpredictably and departed again soon after midsummer.

Heading south down the chain of the Outer Isles in fine conditions in a sharkless sea must have been frustrating. Scouring the sea for elusive fins can try the patience of even the most determined observer, but their patience was finally rewarded just north of Lochboisdale on the island of South Uist. '"Sharks ahead!" Bob Ritchie suddenly said in an unemotional voice.'

Surely they must have felt that their luck was about to take a turn for the better. The group of sharks Bob had sighted were well up, and presented an easy target. Watkins readied the gun. The approach was too slow, then too fast, and then the gun was discharged with 'a thunderous crash.' But instead of the line snaking out wildly, it just hung down vertically into the depths. Recovering the harpoon, there were no tell-tale signs of flesh on the barbs, although there was slime on the rope. It was clear that he had missed the shark completely, at point blank range.

The gun was re-loaded and the pursuit began again, with Bob Ritchie this time promising to take more care at the helm. Carefully stalking a shark, Ritchie put the boat right behind the animal, and Watkins fired again – with the same result. The harpoon just dangled in the water: '"It must be the sights," I said. "I know damn well they were near enough dead on the target when I fired. I'll have to test them. They seemed all right in Tarbert."'

Testing the sights by the simple method of firing the gun at a ball of newspaper showed that whatever the problem was, it was not the sights. When the harpoon was retrieved the paper was neatly skewered on the line.

By now there were several shoals up at the surface, so another approach was made:

> I got in a nice easy shot and to my utter astonishment, the harpoon shot up in the air, spinning like a boomerang, until it was jerked up short by the forerunner and plummeted into the sea.

'What the hell's going on?' I asked in amazement.

I saw what happened,' said Alec. 'The harpoon hit the shark's back, and bounced off. These bloody fish must be made of concrete.'

Rightly surmising that the sharks were no more solid than the whales the gun and harpoons had been designed to kill, Watkins turned his eye to the rope. Perhaps the heavy line they were using was deflecting the harpoon as it hit the water? Testing this theory by substituting a light line for the heavy one showed that this was indeed the case. The harpoon stuck fast in the shark, and the line ripped out until it snapped. They had lost a harpoon but found the problem. They would have to find a lighter line that would not deflect the harpoon, but would still be strong enough to hold and retrieve the shark. More experimentation was required.

The disconsolate crew headed slowly into Lochboisdale for the night, and the next morning Watkins was on the phone arranging for several 220m ropes of varying thicknesses to be sent from Glasgow. It was here that he discovered one of the truisms of operating in such a remote place – everything takes at least a week to get to you. But determination and forward thinking were both fundamental elements of Watkins's personality, and he put the time spent waiting for the rope arrival to good use:

> During this time, the weather was magnificent, and every day I tortured myself by going out in the *Dusky Maid* and counting sharks. This was not just masochism. I wanted to get a line on the shark position up north and an idea of how a commercial shark fishery would work out, and observing sharks systematically, and keeping a careful record of my observations, was really just as important as catching them.
>
> To give myself an occupation in the evening – for Lochboisdale, though a magnificent natural harbour, is a tiny village with only a handful of inhabitants – I wrote letters to harbour-masters, lighthouse-keepers and fishery officers in the Scillies, Ireland and Norway, asking for information about sharks. With these answers, and my own observations and inquiries in Scotland, I hoped to be able to work out accurately the timing and course of their seasonal migrations.

When the rope finally reached them, they wasted no time in getting out to sea to test their theory. The first shot, with 80mm diameter rope, had the usual effect, and the harpoon just bounced off the shark. Changing down to a 75mm diameter rope was no better. Finally, they achieved success with the next size down, the 60mm diameter rope and, well struck, the shark dived for the bottom. Putting the rope on to the winch, they began to haul the shark in.

The 60mm diameter rope had an estimated breaking strain of around three tonnes, but Watkins was concerned that it might not be strong enough to take too much punishment, and was just on the point of easing the rope to let the shark tire itself out on the bottom, when suddenly the rope parted, and the shark escaped. Watkins mused quietly:

> I could see my companions were bitterly disappointed, but I myself was quite cheerful. We were making some progress, even if we had actually not caught anything. These teething troubles were to be expected, I comforted myself. My previous experience of hand harpooning had taught me that it is never as easy to catch a shark as one imagines.

Back into the fray the next day, and for the first time in 10 days there was not a shark in sight. Finally, off the little island of Eriskay, a solitary fin broke the surface. But this shark was in no mood to be harpooned, and repeatedly dived as soon as the *Dusky Maid* tried to approach. The crew tried every trick in the book in approaching the shark: slow and quiet, fast and furious, but always with the same end result – the shark dived before they were within 20 metres.

Watkins reasoned that this was no accident, especially as they were always approaching from dead astern. He had checked the sphere of vision of the shark he had killed in Tarbert the previous year, and knew that there was no way the shark could see them. So some other form of 'early warning' system had to be at play, and he believed it to be the lateral lines along the shark's body:

> The exact way they function is not known, but it supposed that they act as a sort of radar in spotting low frequency vibrations. To a fish, which spends much of its life in a world of absolute darkness and absolute silence, sight and hearing are comparatively unimportant. It must rely mainly on some other sense to detect objects in the water.

Finally, the shark made a fatal mistake, and stayed up at the surface long enough for the *Dusky Maid* to close in. The gun boomed, the harpoon struck home and the shark tore off into the depths. Slowly but surely they played the shark, all the time dragging it inexorably towards the surface and death. Watkins was ready in the bows with the tail sling, as he explained: 'Once this was securely round his tail and made fast to the Samson post just behind the gun, the battle would be over'. But it was still far from over as far as the shark was concerned:

> Soon the line, instead of running down steadily into the depths, began to vibrate and tremble with the frantic movements of the shark, and I knew he was very close. I peered down into the green and murky water, and now I could dimly discern the dark, writhing shadow. It rose steadily, assuming a more definite shape, and my jaw dropped. It was very different from the sharks I had seen come up after twenty-four hours or more on a hand harpoon. Then their movements had been slow and ponderous, no more than the dying struggles of exhausted monsters. But here was a shark in full possession of all his strength and vitality twisting and turning and fighting with every ounce of his weight.

And still capable of trying to evade its fate. Twice Watkins tried to get the tail sling over the tail and both times failed, but not before the mighty Basking Shark had dealt its tormentors a few blows. The tail 'came whistling through the air, and I jumped back just before it slammed down on the gunners platform with the force of a pile driver and then struck the harpoon gun a glancing blow and sending it spinning round like a Catherine wheel.' Finally, the shark 'turned a giant cart-wheel under the bows, the line slackened and the body was clear of the water, then jerked tight as the full weight of the fish came on it, and snapped like a thread of cotton.'

> We looked glumly at each other, soaked to the skin and distinctly chastened by this lively display of aerobatics. The gun was still spinning around merrily, cascades of water were pouring out from the scuppers, and I saw that several of the planks of the gunners platform had been shattered.

With three of his 12 harpoons lost and not a single shark to show for it, the stakes were getting higher, as a keen cards player like Anthony Watkins would have known. Searching the sea south of Barra, they finally spotted a fin and closed in once more, this time getting an easy shot. And finally the lessons that had been learned paid off. Playing the shark, tiring it out and after several attempts, the tail sling was on and the 'battle was over'.

Towing the shark into Lochboisdale to beach it for later removal of the liver gave the crew more pause for thought, as even though the *Dusky Maid* made a far better towing vessel than the yacht-like *Myrtle*, progress with the shark alongside was desperately slow. Out again the following day, and with another shark successfully lashed alongside, they found that the *Dusky Maid* was only slightly faster than 'the average shark cruising on the surface' and it took them an age to get up to another fin for a further assault. With a shark lashed on either side, the *Dusky Maid* could make very little progress at all, and so returned to Lochboisdale to join the first shark on the beach. Clearly the difficulties presented by the need to tow the sharks to a fixed base were going to prove a major problem in the future.

Further difficulties arose during the cutting up process. Despite the long-handled knives Watkins had had made, 'modelled on the whalemen's flensing knives' cutting out the harpoons with their double 150mm barbs extended into the flesh 'was still quite a job.' The harpoons, basically whaling models, although efficient enough, could be better designed to catch sharks.

All in all, Watkins had every reason to feel modestly confident. They were learning the ropes well, and could catch sharks. But out on the water once more, the sharks still proved elusive, before they finally caught a big shark in very shallow water after a protracted stalking session. This shark put up a terrific battle, and once they got the tail sling on, showed that this was not always the end of the fight:

> The *Dusky Maid* was buffeted around like a tin can on a dog's tail. The Samson post bent with the strain, something I would have said was an absolute impossibility. Something had to give, and, perhaps fortunately, it was the tail-rope. It was a four-inch manilla rope with a breaking strain of seven tons.

But the crew's new found skills proved to be too much for even such 'a bonny fighter' as one of the crew called the shark, and with a second and a third tail rope in place, they put into Castlebay, Barra.

Using the shark as a fender against the rough pilings of the pier must have seemed a good idea at the time, but it didn't make for sound sleep, at least for Anthony Watkins:

> He was still very much alive, and heaved and rolled incessantly throughout the night. The boat bobbed up and down as though she was anchored in the open sea, and there was a loud, continuous scraping sound as the rough skin brushed up against the outer planks. It did not seem to affect my companions who slept profoundly, even hoggishly. One is, I think, entitled to use this expression to describe the slumber of robuster souls under conditions which keep our refined and sensitive selves awake. Both for our convenience and efficiency, as well as for ordinary humanitarian reasons, I felt, we must carry some implement to kill a shark as soon as we had him safely alongside.
>
> We beached him early next morning. He was still alive and struggling violently, though as soon as he was aground and partly out of the water he appeared immediately to give up the ghost.

Although Watkins's humanitarian instincts were in the right place, he must have instantly realised just how difficult it would prove to be to find a way to humanely kill the shark given what he saw next:

> When we came to gut him I found that the harpoon had gone smack through his spine, completely shattering it. This terrible injury, which would kill any ordinary fish outright, had apparently failed to have any effect on him whatsoever.

Confident Watkins may have been, but he was about to run into another immutable law of the Basking Shark. Just as quickly as they can arrive, they can disappear. They only caught one more shark in the Hebrides and, despite combing the waters all the way north to Stornoway and down south via the Skye shore, not a fin was to be found. 'Within the space of a week the sharks had vanished without trace.'

Returning to Mallaig, Watkins received a telegram:

> 'Return Clyde immediately plenty of sharks here.'
> 'Aha', I thought, 'so that's where they've gone to.'
> Then I looked at the telegram again and saw it was dated over three weeks ago.
> That meant it was sent when sharks were also plentiful up north. Their movements were not so simple as I had thought.

The Clyde factory

Back in the Clyde, Watkins learned that although Basking Sharks had indeed been plentiful in the Kilbrannan Sound a few weeks before, they had disappeared for the moment. As Watkins remarked, 'This was bad news. What if they had done the same as the Hebrides sharks and vanished altogether?'

More worryingly, he realised that although the journey to the Hebrides had been thoroughly useful, the bills were mounting up, and the sale of the oil from the livers would not meet them. A quick calculation in his cashbook showed he had just £20 in the bank, just enough for two weeks' wages, and it became clear that he had started up with a far from sufficient amount of working capital. Unless he started to make money immediately he would have to close down altogether.

With the little factory now operational and producing liver oil, Watkins boarded the *Dusky Maid* and headed for his old hunting grounds around Ailsa Craig. But after a week covering the whole of the Clyde estuary not a shark had been sighted. Some skippers he talked to suggested: 'Sharks? Och, they're away up north the now.' Watkins knew this not to be the case, as he had seen plenty in August of the previous year in the Clyde, but nonetheless the current dearth of sharks was very worrying.

The following week started poorly, too, despite perfect weather. For three days they sailed through perfectly calm seas, from dawn till dusk.

> Suddenly I heard a stamp on the deck. I stuck my head up without undue excitement, for I had quite given up hope. A quick look round confirmed that there were no sharks. We were out of sight of land.
> 'What's up now?' I growled.
> 'I can smell shark,' said Alec.

'Aye, there's shark around, right enough,' added Bob. 'And no very far off.'
I sniffed unbelievingly, and immediately got it, a distinct and quite unmistakable
whiff of shark. They must be very close, I thought. Water acts as a complete barrier
to the human nostril. What we were smelling was slime from their bodies which
must have drifted to the surface and come into direct contact with the air.

Then about fifty yards ahead I saw a dark mass moving slowly below the
surface. No fin was visible, but it could only be a shark. Bob had seen it, too, and
was steering towards it.

As we approached, other shapes came dimly into view, until we could see
through the bright clear water the whole shoal in formation below us like a flotilla
of submarines.

But this was a flotilla of submarines that never surfaced, and as dusk fell, the *Dusky Maid*
turned towards the Ayrshire shore to spend the night in Dunure. The following morning,
they retraced their steps via a reciprocal bearing and this time found the shoal at the surface.
Within a couple of hours two sharks were alongside. By the end of the week they had caught
five sharks.

Watkins had by now a fair idea of his running costs, and knew that his break-even figure
was four sharks per week. Five sharks and he was into profit. The difficulty he currently faced
was that the position of the sharks, right out in the middle of the channel between Arran
and Ayrshire, was too far from the factory to make the long, slow trip back worthwhile. This
necessitated beaching the carcasses, which meant waiting for the tides to do so, as well as the
need to return to the beach to liver the sharks. And there was always the likelihood that when
the *Dusky Maid* returned to the previous position of the shoal, the shoal had gone down or
moved on. And then there was always the weather, ready to go bad the minute things were
looking up.

In a bid to short-circuit the process, they began livering the sharks at sea, which proved
far easier than they had expected. Once opened up, the liver could be cut out in pieces and
hoisted aboard. But the problem of removing the harpoon from the flesh remained. If the
harpoon had gone clean through the body there was no problem, but if it was stuck in the
body, then 'a mining operation was needed to extract it'. And as Watkins remarked:

> To add to the difficulties, and to harrow my own feelings if not my companions',
> the shark struggled fiercely throughout the operation, even when it had been
> opened up like a kipper and lost all semblance to a living fish.
>
> I tried hard to find some way of killing them as soon as they were captured.
> Firstly, I tried a shotgun with heavy buckshot. As soon as the shark was secured
> alongside I put a shot into its brain. I had previously dissected a shark's head, so
> I knew exactly where to aim. The shark rolled its eyes, but went on heaving and
> contorting its vast body as energetically as ever. I put six more shots into the brain
> cavity and must have completely obliterated all trace of the brain, which is scarcely
> larger than a golf-ball. It had no effect. The shark struggled as fiercely as ever. I then
> severed the spine just behind the head with an axe. I was not particularly surprised
> that this was a failure, as the spine had already been shattered behind the dorsal
> fin by the harpoon. I then tried hitting the shark on the head with a seven-pound
> sledgehammer, and went on hitting until the head had turned into a bloody pulp
> and I was completely exhausted. The shark went on struggling. When the dissection
> began it died no more quickly than any of the others.

Watkins put this down to the Basking Shark being 'spinal controlled', where not all motor functions originate in the brain, but in some cases are handled in the spinal column.

> A shark, in short, is motivated mainly by reflex actions, so that even without a brain, when it is 'dead' by any ordinary standard, its body will continue to receive stimuli from the outside world and will react to them normally and even with apparent intelligence. If we had released our brainless shark it would probably have swum off and resumed its normal life, rejoining a shoal, perhaps even performing the sexual act. It might 'live' for some days in this state.

Watkins' fishermen companions were far less concerned about the death of a shark than he was, citing a comparison with the 'millions of herring they had seen slowly suffocating to death'. In any case, hard to stomach though the final acts were, as we've already seen, herring fishermen suffered great losses in terms of nets and gear from Basking Sharks, and so might well have seen the death of one of their number as a thoroughly good thing. And shark conservation (or even conservation in general) was not even a matter for consideration in those days, in an era when most seafarers actively disliked sharks. Watkins may have found the final death throes of the shark distasteful and wished to hasten the end, but he was emphatically not about to stop killing the sharks now that his enterprise was finally getting off the ground:

> For there is no reason to suppose that sharks suffer more than other fish just because they are bigger. If anything they probably suffer less, for they have a more primitive nervous system than the bony fishes.

So the refinements of the process of slaughter continued to be those that made life easier to process the sharks, such as a move towards gutting the creatures in shallow water. Watkins had found that once aground, the sharks couldn't move and the liver and (more importantly) the harpoon could be cut out far more easily by a man in thigh-high waders. No more time wasted waiting for tides, meant more time out hunting.

Matters continued to improve financially when the shoals finally began to approach Arran, and the Kilbrannan Sound in particular, where the *Dusky Maid* could simply throw the tail ropes ashore at the factory for the carcasses to be winched onto the beach for gutting. The scene was a gruesome one:

> By the evening the factory was a delightful shambles. Eight elephantine carcasses lay on the beach. Six of them had been split open and gutted, and rivers of blood and guts had cascaded down into the sea and turned it into a thick tomato soup, over which hundreds of gulls fought and shrieked. From the factory chimney, doors and windows poured dense clouds of noisome vapour from the bubbling livers.

It sounds like a scene from Brueghel, and only men who had become inured to the stench could have withstood it for long. That certainly was the case with Watkins, who in any case knew that the stench had a monetary value. But exposure to the malodorous reek didn't please everyone. It was now high season in that summer of 1939, and Watkins was far from being alone in the boarding house as it filled up with holidaymakers. After a while, Watkins noticed that his entry to the dining room precipitated a rush to the door:

One evening I went into the high tea after a hard days work at the factory. As usual, there was a general exodus from the dining-room, some even leaving a part of their meal untouched. The only exception was a stout, red-faced man at the next table, who went on munching unconcernedly. I decided to bring matters to a head.

'Look here, what on earth is the matter?' I said in an aggrieved voice.

'Don't you know?' said the red-faced man in genuine surprise.

'No, I don't,' I snapped.

He paused thoughtfully for a few seconds and then said:

'Well, to tell the truth, old chap, you stink like the devil. Though I don't see cause myself to make all this fuss. Mind you,' he added, for he was evidently a fair minded man, 'I work in a slaughter-house. I can stand bad smells better than most folk.'

After that I took precautions and went through the most elaborate cleansing process when the working day was over, and gradually I was received back into polite society.

As the business began to become more secure, and 'less of a hand-to-mouth affair', Watkins allowed himself to think of future plans. Up until this point he had only been exploiting the livers, and the carcasses were simply towed out to sea and sunk, 'a complete waste of good material for fish meal.' Shipping the flesh in barrels to Glasgow proved to be unprofitable, so if the flesh was to be utilised it would require the installation of suitable plant at the factory to do the processing. And the requirements of the plant would be based on its ability to deal with the largest sharks.

Watkins had already made exact measurements of 50 sharks and had arrived at a potential maximum total length of a fraction over nine metres, the largest he had caught so far in fact being just under nine metres. Consulting with Bob Ritchie and Alec MacDougall as to whether there were bigger sharks (but they had simply been unable to catch them, or they had broken free), all were agreed that this was inconceivable. The argument over the maximum size continues to this day, but in fairness to Watkins, he did his homework, and as we shall later see, proved his point forcefully with one of his future competitors.

Watkins considered that by installing a fishmeal plant at the factory, and buying and fitting out more catchers, the business could be developed, at least in the Clyde. When the shoals were further afield in the larger Firth of Clyde, one catcher could be devoted to gutting the sharks and bringing the liver back to the factory in barrels, but other catchers could stay with the shoals and keep working. This was all well and good, but it does seem extraordinary that such an obviously intelligent and diligent man didn't wonder about the potential impact of his activities on a localised population in the longer term, or whether the recent abundance in those waters was just a flash in the pan.

Instead, he concentrated his thoughts on the mysteries of the shark's migration, and the possibility of exploiting the northern fishing grounds in the Minch. Watkins had by now given up on the idea expressed in the text books of a northerly migration. Responses to the letters he had sent from Lochboisdale tended to bear this out, and instead suggested:

A westerly or inshore movement in the spring on a wide front stretching from the Scillies to the Lofoten Islands, followed by a withdrawal into deep water beginning in midsummer. The Clyde, which had never seemed to fit in with northerly migration theory, fitted neatly into this picture. Sharks arrived there later than in the Outer Hebrides because it was farther west, and they stayed later because the Clyde is exceptionally rich in summer plankton.

This all pointed to the reality that there was no place in the north where you could reliably expect to find sharks for any length of time, although as we shall later learn, that might well be an over-simplification. But his analysis that the only practical way to open up that area would be via the use of a factory ship, to follow the catchers and deal with the catch on the move, was correct. Once again, Watkins showed the considerable acuity that would make his the most successful of the larger operations, and make him perhaps the only one of all of the shark hunters that made more than just a living out of the Scottish Basking Shark fishery. Even more remarkably, he made money at the far lower prices being offered for the liver oil in the pre-war era. It seems likely that none of the others could have achieved that, as we shall see.

So, for the moment, it would only be fair to allow him a moment of quiet satisfaction as he looked out from the bustling factory at Carradale on the sight and sound of the *Dusky Maid* harpooning a shark just offshore:

> I looked at her through field-glasses and saw a small black triangle just in front of her bows. There was a puff of smoke, and the fin erupted into a fountain of foam and spray, in the middle of which a wicked black tail lashed about angrily. Seconds later came the boom of the gun over the water.
> 'Looks like we're going to have a busy afternoon,' remarked John with satisfaction, and I grunted happily in assent. I could not honestly conceive of a better way of spending an afternoon than gutting sharks and boiling livers.
> It seemed almost too good to be true.

It was. At that moment, John Paterson's small daughter arrived, bearing a telegram addressed to one Second-Lieutenant A. Watkins. A couple of years before Watkins had joined the Supplementary Reserve of Officers, as 'a patriotic thing to do'. Now with the world on the brink of turmoil in the shape of World War II, he was being called up. For the next six years he would see action across the world in command of a tank, giving the Clyde and Hebridean sharks a temporary reprieve. The summer was over – for now.

Chapter 5
The Post-War Hunt

In the spring of 1945, Anthony Watkins was in command of a Sherman tank, leading No. 3 Squadron of the 1st Armoured Battalion, the Coldstream Guards, battling their way into the heart of Germany against a defeated enemy still capable of putting up a fight. During a lull in the fighting the squadron mail arrived, delivering him a vivid reminder of another world far beyond the conflict. A letter from Mr Hugh Highgate, 'a very venerable figure in the world of oils and fats', outlined the possibility of a vastly changed business environment for his shark hunting concern:

> Thank goodness it looks as if this war will soon be over. I wonder how long it will be before you return home and can get your shark-fishing business re-started. It will, I think, be some time before the whaling companies and the copra and groundnut plantations get into their stride, and the demand for edible oils should greatly exceed the supply for some years. It is possible that you will be able to get as much as £100 per ton for your shark oil.

That was four times the price he had received in his pre-war operation, and sent him into a momentary reverie:

> A hundred pounds a ton, I thought; that was a fantastic figure. At that price shark-fishing had a dazzling future. It would be worthwhile going for it in a big way. More catchers, certainly, and possibly a factory ship instead of my old shore station, so that the whole Scottish west coast could be opened up, and possibly the Irish coast as well. Such an outfit would cost, say, £200 a week to run. And if it produced twenty tons of oil a week, quite a possible figure, that would be worth £2,000. A profit of £1,800 (around £70,000 at 2015 values), in a week....

Watkins couldn't wait to put his plans into action, but in the meantime His Majesty still had need of his services, and it wasn't until November that he finally emerged onto the streets of London as a civilian once more. Time to review the options for the expanded venture before the potential start to a new shark- hunting season in 1946.

The first question was what to do with the Carradale factory. It had lain dormant throughout the war, and much of the machinery would now be in poor condition, and, as a result of various new Factory Acts passed during the war, would require much expenditure and updating in order to be able to operate once more. Even then, the major problem of a static factory remained in place. It would always be too far from all but one major hunting ground, and Watkins very much had his eye on the 'magnificent northern fishing grounds of the Outer Hebrides.'

Then there was the question of the necessary capital to buy the two additional catcher vessels and the factory ship that his hastily conceived plan called for. Fortunately, that was less of a problem, as throughout the war the *Dusky Maid* had diverted from hunting sharks to sustaining the food supplies of the beleaguered nation by fishing for herring in the Firth of Clyde. Under the command of his good friend, old Mr MacDougall, she had made more money than ever before in her working life, and the profits had been steadily accruing in the Company's bank account.

Before leaving London for Carradale, and determined to be up and running for the 1946 season, Watkins contacted the Kongsberg factory in Norway and ordered two new harpoon guns, 36 harpoons and 500 rounds of ammunition. Here he received the first inkling that the post-war world was a changed one, in negative as well as positive ways. The Kongsberg factory was in complete disarray, no guns or harpoons were in stock, and they had no idea when production would recommence. All he could do was plead that the order should be fulfilled by February, and hope for the best.

After a warm reception back in Carradale, Watkins settled back in at the MacDougall's boarding house. As he wrote, 'it was almost as if the war had not happened.'

But during his discussions with John Paterson he discovered that there were many further changes that had occurred as a result of the war. For example, it appeared that while securing crews would be no problem, suitable boats were in short supply. Herring fishing was currently making enormous profits, and no new boats had been built during the war. Most of the better boats had been commandeered by the navy during the conflict, and were only slowly being returned to their owners. Vessels like the *Dusky Maid* were being sold at around £1,500, more than three times what he had paid for her. A new vessel of around 16.5m equipped with an 88 horsepower diesel engine would cost £5,000 (*c.* £200,000 at 2017 values). And crew wages had more than doubled from their pre-war values to £5 per week. This news naturally took some of the gloss off Watkins's rosy view of the potential profitability, but even so, with oil prices still expected to be around four times their previous value, there was still plenty to be optimistic about.

Leaving John Paterson to look around for a couple of catcher vessels, Watkins set off for Peterhead in the hope of finding a suitable steam drifter to convert into his factory ship. Steam drifters were also herring fishing boats, but were around twice the size of the Loch Fyne skiff type (like the *Dusky Maid*), and therefore had much more available deck space to install the lifting and handling equipment required for hoisting the shark on deck for livering, as well as sufficient room for installing the plant to render the livers down to oil. Most of the boats on offer proved to be on their last legs, and it took some 'diligent weeding out' before he found a good boat, the *Gloamin'*, which as he remarked was a 'terrible name, but a satisfactory ship.'

Taking charge of the design and specification for the conversion himself, Watkins created a complete inventory of the materials required for the various tradesmen to work with – as and when they could both be found. Many men were still being held back in the services, the various shipyards were overwhelmed with work, and iron and wood were both in short supply. In order to keep the various men at work on the conversion and not poached by other concerns, Watkins was obliged to stay on site. Back on the west coast, John Paterson had identified a couple of possible catcher boats that might be worth inspecting. On his first trip to the Clyde, Watkins purchased the *Paragon,* newer and slightly larger than the *Dusky Maid,* and on the second visit the *Perseverance,* an older but carefully maintained model. Watkins now had a solid little fleet of catcher craft, each just over 12m in length and all equipped with relatively modern Kelvin diesel engines of 44h.p. Watkins left them in action Herring fishing under John Paterson until the time came to convert them into shark catchers.

Then came a serious blow to his plans – Kongsberg would be unable to supply any guns, ammunition or harpoons before August at the earliest. This was a major setback, as even though he still had one gun from the pre-war hunt, somehow the ammunition and harpoons had gone missing during the war. Six weeks before the start of the 1946 season, his fleet was lined up ready to go but was unable to set out due to lack of the most vital element of equipment they would need, the harpoon guns. It looked like the whole enterprise might have to be put on hold for a further year, a bitter disappointment given the hard work and substantial amount of capital that had been invested so far.

Once again though, Watkins showed enterprise and an unwillingness to admit defeat. Having seen his tank squadron on the receiving end of the lightweight German anti-tank weapon, the *Panzerfaust,* he toyed with the idea of developing a harpoon gun along those lines, but was soon advised that this was a dead end due to the weapon's inability to develop sufficient velocity to propel a heavy harpoon. However, a London firm's ballistic expert recommended that they try cutting down old two-pounder anti-tank guns that were being sold for scrap by the War Office. These guns would need considerable modification, but Watkins reckoned they were about the right calibre and he knew that ammunition was available for them in unlimited quantities. As there was no other option available at such short notice, he ordered three guns for conversion.

Harpoons were available, though, as the original mould he had developed pre-war was still in existence. What could not be had, however, were steel shafts that would fit the barrels of the new guns, so wooden sticks were ordered in the hope that they would be adequately strong. The guns, once they had arrived, looked and felt clumsy after the purpose-made Kongsberg gun, but would have to do. As soon as one of the guns was mounted on the *Dusky Maid,* it was tested with acceptable results, albeit not on a shark. Last minute glitches with the conversion of the *Gloamin'* tried Watkins's patience to the utmost, and necessitated his presence in Peterhead constantly to see through the final touches. Finally, he and the *Gloamin'* were on their way through the Caledonian Canal to meet up with the catcher fleet in Mallaig.

It's worth considering what a momentous development this constituted. For the first time in well over 150 years a dedicated shark-hunting fleet was about to go into action in Scotland. Also for the first time, this was a fleet equipped to deal with sharks on an industrial basis, capable of pursuing a shoal of them over long distances and able to harpoon serious numbers of prey. On the basis of Watkins's sketchy plan, he reckoned to be able to produce 20 tons of liver oil a week without too much difficulty (granted he could find the shoals). Working on an approximate basis of three average-sized sharks to yield one ton of liver oil, to achieve this target value would require them killing around 60 Basking Sharks a week – a quantum leap compared with their previous efforts. If this slaughter could be sustained over a 20-week season split between the Outer Hebrides and the Firth of Clyde it would represent a take of 1,200 sharks. And still no mention of any kind of sustainable catch levels – perhaps, as is so often tragically the case, it was simply taken for granted that 'there were still more fish in the sea.'

But Watkins had every reason to feel confident, at least on the financial front. As he remarked: 'Before the war I had had a job sometimes trying to sell the oil. Now the dealers were clamouring for it, bidding against each other, even offering to buy my whole season's output in advance.'

If the guns performed as was hoped, and the weather was better than average, then there was everything to play for. But the crews were green, the guns hadn't been tested in anger and with all the rest of the equipment on the *Gloamin'* still untested, there was plenty that could still go wrong.

And now, approaching Mallaig, he was about to discover that there was one more variable that he had not foreseen – competition, in the form of Gavin Maxwell and his Isle of Soay Shark Fisheries venture. And their worlds were about to collide.

The *Paragon* had developed a problem with her fuel pump that had to be fixed before the fleet could operate effectively, and as Mallaig was the last port with repair facilities there was no time to waste. Watkins went looking for a mechanic he had known well before the war, and was told that he was at work on the *Sea Leopard*, an ex-Admiralty Harbour Defence Motor Launch (HDML) he had noticed in the harbour. On catching up with the mechanic, Watkins guilelessly made conversation:

> 'It's nice to see the yachts back again,' I observed chattily, as we walked away towards the *Paragon*. 'Makes one realise that the war is over.'
> 'Aye, it is that. This one is going in for sharks, you know.'
> 'Really,' I said with deep interest. 'Who's the owner?'
> 'Major Gavin Maxwell. He was stationed up here most of the war, and he got interested in sharks. He's just built a factory on Soay.'
> 'On Soay,' I echoed in surprise. This was a small island south of Skye. It was some way from the best shark-fishing ground, and, one would have thought, expensive and inconvenient from the administration point of view. 'That's an odd place to choose. Is he really going in for it seriously?'
> 'Well, I think the Major is doing it more for sport than anything else. He's the brother of a baronet, you know. He doesna need money.'
> I was rightly not convinced by this argument. Nowadays, as I suspected, younger brothers of baronets needed money as much as the next man.

An accurate assessment indeed, and just how prescient those words proved to be would become all too apparent in the very near future.

Watkins decided to make a courtesy call on Maxwell, with a view to discussing the possibility of doing business together. The meeting didn't go well. Watkins was led below to meet:

> A fair-haired, sharp-featured man of about my own age. He greeted me, I thought, with more than a touch of hostility.
> 'What I came for is this,' I explained, disconcerted by the atmosphere. 'As you know, I only use the livers. Perhaps we could do a deal over my carcasses. That is, of course, if you produce fish-meal at your factory.'
> 'My factory,' said Maxwell, 'hopes to produce nine different products among them fish-meal. However, I'm naturally not anxious to encourage you to stay up here. I

should have thought it far better for both of us if you stuck to your own area in the Clyde. Here am I, building up an important new industry, and you----'

'But good Lord,' I interrupted, 'if there aren't enough sharks up here for both of us, we're both wasting our time in this business. We shan't interfere with each other. I shall be fishing mostly on the other side of the Minch, nowhere near your factory. There are more sharks there, as a rule', I added, perhaps not very tactfully.

'I have plotted the main stream of the northerly migration as going up the Skye coast, right past my factory,' corrected Maxwell.

'Do you mean to say you haven't discovered yet that the old northerly migration theory is a myth?' I looked at him in genuine amazement. Even though all of his preliminary investigations must have been carried out in wartime conditions, it seemed surprising that he had not discovered this key fact.

'On the contrary, my researches have confirmed the truth of the theory, which is, you will find, supported in every textbook.'

'That doesn't mean much. Every textbook talks of sharks growing to forty feet in length. Have you ever caught one of that size?'

'As it happens, no. But all I've caught so far is a few sharks for research purposes.'

'I'll have a sporting bet of ten shillings you don't ever catch a thirty-footer.'

'Ten shillings.' said Maxwell. 'Why not make it ten pounds?'

'Twenty if you wish,' I said shortly, and on this we parted.

This whole exchange was freighted with portent for the future. Watkins methodical, constantly evaluating and challenging existing knowledge with an eye for the commercial advantage, Maxwell, lazy in his thinking, yet at the same time over-ambitious. These were two very different men, so it might come as no surprise that they failed to get along at this initial meeting, although (perhaps surprisingly) they would later become friends of a sort.

It was also true (as Watkins had rightly sensed) that Maxwell was well aware of his pre-war shark-fishing efforts, and indeed that Watkins intended to recommence that fishery after the war. Maxwell acknowledged as much in his own book *Harpoon at a Venture*, outlining his conversations with his friend John Lorne-Campbell, laird of the island of Canna, and a man with a major interest in all types of fisheries in the Hebrides. Lorne-Campbell had himself been interested enough in the potential of shark-fishing to make enquiries about oil prices, likely prompted by having himself heard of Watkins's efforts in the Hebrides pre-war, as much as his fishery in the Firth of Clyde. Yet Maxwell in his book simply states that Lorne-Campbell told him that Watkins; 'had caught sharks in Loch Fyne during the three years before the war.'

What might be termed the 'fish telegraph' works wonderfully well throughout the fishing world, and gossip about new fisheries, gear modifications, successes and failures changes hands with ease and alacrity. Watkins quite rightly would have wanted to keep his hard-won learning to himself, but he would have been simply unable to conceal where and how successful he had been in hunting sharks from his fellow fishermen – the fish telegraph would have seen to that. Yet Maxwell, at least from this meeting, claimed to have had no idea of Watkins's successes or failures, and even claimed to be 'building up an important new industry', as if Watkins and his fleet were simply a figment of his imagination.

In fairness to Maxwell, his own account of the meeting acknowledged some doubts of his own, especially concerning the wisdom of relying on a shore factory. Maxwell asked Watkins how long it took to lift a shark aboard the *Gloamin'*:

'Oh we reckon to have a shark aboard the *Gloamin'* half an hour after harpooning, and the liver into the steam barrels in another half-hour. I could never waste time towing back to a shore factory – one's got to be able to deal with everything at sea.' This was uncomfortably near my original conviction. 'Then what will you do with the carcass after the liver's out?'

'Dump it straight overboard.'

I said that I thought that the number of rotting carcasses would clear the rest of the sharks out of the area.

'It's possible, but not likely. If they do move on I can follow them.' The missing words 'and you can't' seemed as clear as if they had been spoken.

This was an honest appraisal that went straight to the heart of the matter for Maxwell's fledgling enterprise. Why build a factory on Soay, of all places? How did he get to this juncture, despite his concerns over the need for a fixed base? In itself, this was problem enough, but there were further weaknesses in his thinking that would very soon come back to haunt him. But perhaps we should step back for a moment and encounter the highly unusual man who had just entered the fray.

Gavin Maxwell

Gavin Maxwell was born in July 1914, third son of Colonel Aymer Maxwell of the Grenadier Guards and Lady Mary Percy, fifth daughter of the seventh Duke of Northumberland, at their country house at Elrig, Wigtownshire in the lowlands of Scotland. Within three months of his birth, his father was killed in the first German artillery barrage of the First World War, only a couple of hours after his arrival in Antwerp. Thereafter, Gavin Maxwell would be raised by his doting and dominating mother, sharing a bed with her until the age of eight.

Maxwell's biographer Douglas Botting described his subject thus: 'Gavin was persistently and cataclysmically accident-prone throughout his life, so much a prey to misfortune, that his life has something of the quality of a Greek tragedy.'

From the very first, Maxwell was a sickly individual, suffering a difficult birth and a 'delicate and ailing constitution'. He probably suffered from 'what is now called "bi-polar illness" a form of clinical manic-depression, that was to exert a considerable influence on his patterns of behaviour in later adolescent and adult life.'

None of which was helped by his isolated childhood at remote Elrig, idyllic though it was, which simply didn't allow him to develop the kind of self-defence mechanisms that would have stood him in good stead in adulthood. As Botting recalls, by the age of 10 when he was incarcerated in an English boarding school, he had met fewer than 10 other children, and three of those were his siblings. As Maxwell himself remarked: 'This was the beginning of the breakdown of my image of what life was, going to school in England seemed nothing but a violent disruption, something terrible.'

And more than that, Gavin Maxwell missed the freedom of the wild lands he had grown up in, and his fascination with their wild inhabitants. In the company of Hannam, the estate gamekeeper at Elrig, his interests soon turned 'from bird-nesting and butterfly-catching to guns and shooting, which before long became as obsessive a passion for him as it had been for his father. He saw no contradiction in loving the creatures he killed, and from now on he was to be accepted by those few friends he found as a sort of 'Mowgli with a gun'. Not only did he love shooting, but he also excelled at it, finding an outlet for his passions that might well have led him to his future appointment with the Basking Shark.

Illness was to disrupt his school years and to dog him throughout his life. Only weeks into his basic officer training with the Scots Guards, he fell seriously ill with the first of many duodenal ulcer attacks and had to return home to convalesce. By the time he completed his officer training the war was already well underway, and apart from skills that had their roots in his shooting background, there was little to distinguish him from his fellow officers.

During the London Blitz, he and his men were stationed at Blackwall, opposite East India Dock, in a key area for the German bombers to target. One night during an air raid, Maxwell was doing the rounds of his perimeter when he heard a bomb falling close by. The bomb didn't explode, so he set off to try and find the crater, eventually finding it in a churchyard, realising with horror that the crypt beneath was in use as an air raid shelter:

> I ran down the long winding steps and struggled with the door. As it burst open
> under my weight I was hit by a stifling wave of air so noisome that I retched
> even at its first impact. The temperature was that of a Kew hot-house, the stench
> indescribable. As I became accustomed to the dim light I saw that the stone floor
> was swimming in urine, and between the packed forms were piles of excrement
> and vomit. One hundred and twelve people had been in that airless crypt for seven
> hours. They were not anxious to be disturbed; abusive voices, thick with sleep,
> told me to close the doors. I had just time to open both wide before I was myself
> hopelessly sick, helplessly and endlessly.

On his return to his base, an encounter with a Guardsman from the Hebrides spoke to him of another world, somewhere in the distant future, and Maxwell told a fellow officer, 'I've made a resolution. If I'm alive when the war's over I'm going to buy an island in the Hebrides and return there for life; no aeroplanes, no bombs, no commanding officer, no rusty dannert wire.'

Spreading a map of Scotland on the floor, the two men lay down to peruse it as closely as possible, with a childlike view to picking the ideal island.

> We spoke of Hyskeir, Rona, Canna, Staffa; in my mind were high-pluming seas
> bursting upon Atlantic cliffs and booming thunderously into tunnelled caverns;
> eider ducks among the surf; gannets fishing in deep blue water; and, landward, the
> scent of turf smoke.
>
> After an hour there were rings drawn around several islands. I had drawn an
> extra red ring around the island of Soay, an island unknown to either of us, below
> the Cuillin of Skye. We were still playing at make-believe; Soay was my island valley
> of Avalon, and Avalon was all the world away. Presently the sirens sounded, and
> down the river the guns began again.

Before Avalon could be found though, the war had to be fought; but no sooner had Maxwell been posted to a fighting battalion than his old ulcer problem reared its ugly head once more, necessitating six weeks of hospital treatment and a spell of rest at Monreith, the family seat near his home at Elrig. This latest bout of illness led to his being downgraded to Category C by the Army Medical Board, meaning he was fit for home service only, putting an end to his dreams of combat action with the Guards. But although this ruled him out of active service, the Army recognised his skills as a sniper, expertise in fieldcraft and camouflage, and that he was an excellent instructor. As a result, his application to become an instructor with the Special Operations Executive (SOE), set up to train agents to operate behind enemy lines in occupied Europe, was accepted and he entered training to equip him for his new role.

Part of his training sent him to the arduous paramilitary course at Special Training School (STS) 21 at Arisaig on the rugged West Highland coast, almost within sight of Soay. In due course, he would complete the training successfully, despite his poor health, and return to Arisaig as an instructor on the same course. Eventually he would rise to command STS 24 at Inverie House in the wilds of Knoydart (north of Mallaig), with a special interest in teaching fieldcraft and small arms skills to the agents in training, and later STS 22 (the Wireless Telegraphy Training School) at Rubhana Lodge in Morar, and become the chief instructor at STS22a (the Foreign Weapons School) at Glasnacardoch. Evidently he could excel in the right environment – and he'd found his wild-playground paradise.

Days off were very rare, so it was to be two years from the date of his original plan in London's docklands before he had a chance to visit the island of Soay. Maxwell had befriended the officer in command of a small yacht called the *Risor* that was used to deliver amphibious training to the SOE agents. It was to prove a momentous occasion, as he was to later recall:

> I remember that it was a blue day, hot and still, and that it was lit for me with something of the vivid anticipation that belongs to childhood. My companions, whose home the yacht had been in peacetime, were wholly delightful, and the yacht itself had the orderly comfort of a neat cottage. We sailed from Mallaig in the morning. The islands swam in a pale blue sea, Eigg and Canna and Rhum with white puffballs of cloud balanced above its peaks. There was not the faintest breath of wind, and the whole length of Sleat was mirrored in a still sea dotted with resting birds.
>
> In a little over an hour we rounded the point of Sleat and headed due for Soay, on the same course that I was to follow times without number in all winds and weathers for four years.
>
> We crept cautiously into Camus na Gall, Soay's east bay, the leadsman calling soundings from the bows. The yacht's captain, a stranger to northern waters, had the navigational guide in his hand, a long bleat of warning that makes one wonder at any stranger sailing the Hebrides without a pilot. At 'By the mark, five,' he gave the order to let go, and the anchor rattled out noisily into the stillness. We were perhaps a quarter of a mile out from the shore of the bay, a gravel-and-boulder shore with a dozen or more houses lined just above the tidemark, some slate-roofed and some of the older turf-roofed dwellings with rounded walls. Smoke came from a few of the chimneys, but there was no sign of a human being, nor did any appear as we lowered the dinghy and rowed shorewards. We pulled it up in a run that had been cleared of larger boulders, and still no-one appeared from the houses.

With only two hours to explore the island, Maxwell set off to cover as much of the terrain as possible, taking in the narrow inlet that formed the west harbour that would later become the centre of operations of his shark-fishing enterprise. All was peace, and he clearly fell head over heels for the tiny island, dominated by the black rock massif of the Cuillins on one side, where, 'Even then, in the heat of a still July afternoon, white tendrils of mist moved sluggishly among the heights and the glacial nakedness of the corries.' Down by the shore:

Where the island ran out to a promontory, the water reflected the dense scrub of birch and oak; breaking the reflection, two black guillemots sailed in the water, small black-and-white birds with sealing wax-red bills; through the ripples their splayed legs showed the same colour. Gannets were fishing in the sound, snow-white against the immense dark of the mountains: shaggy cattle were cropping the rushes nearby. The sound of their jaws, the low slap of the tide, and the hum of the bees in the heather seemed only part of an immense and permanent stillness.

Returning to the yacht, he met one of the crew waiting to embark, carrying a fish box containing three large lobsters and a box of eggs. Maxwell asked him if he'd bought them:

He shook his head. 'They wouldn't sell anything, it being Sunday. That's why nobody came down from the houses – they're Seceders.'
    As he rowed back to the yacht he went on to tell me that he had had tea in one of the houses and had heard something of the island's troubles: its inadequate communications and transport, its decreasing population, and the absence of state sympathy. He added, 'There seems to be a good deal of ill feeling between the families, and they mostly seem to be related. The people I was with seemed to think a resident landlord would do the place a lot of good.'

Maxwell might have given pause at this stage to consider the implications of what he had just heard, and reflect upon what he might be taking on. Not just an island, but the responsibility for its occupants, too. Given the apparent ill feeling that existed between the populace, Soay might have been better left to its own devices, but Maxwell had made his mind up (and lost his heart), and soon entered into protracted negotiations with the current landlord, Flora Macleod of Macleod, becoming the owner of the island of Soay one year later.
    By now he reckoned that he would survive the war, and recognised that he no longer simply wished to retire to the island, needing far greater stimulus than simple island life: 'The years of hard work and organization had become habit, breeding, as with so many others, a restlessness, an impatience with former interests and ambitions, and a desire for application and achievement. There was no clear way to the satisfaction of these cravings on Soay.'
    He was aware that he would need to come up with some new means to make the island pay for itself. Maxwell had paid £900 (£36,000 in 2017 terms) with money borrowed from his mother for the island. This sum was to include the salmon fishing rights to the coast, that until then had for many years been leased to Robert Powrie, who owned or leased many similar fisheries on both coasts of Scotland. Maxwell was to discover that his usual appalling luck had struck again, as somehow the lease for the salmon fishery had been renewed by Powrie for a further eight years during the negotiations to purchase the island, and, as a result, 'the door to the only obvious work on Soay had been slammed in my face. Without the introduction of a new industry, it was difficult to see how the island could be developed or improved.' But what possible 'industry' could be imagined for such a remote and impractical place? Or for such an unskilled, difficult, disaffected workforce?
    As he was to ruefully recall later, of that fateful day of his first visit to Soay:

This was the last time that I would ever see Soay as I saw it then, as an untroubled island with a single and beautiful face. When next I came to it I came as its owner, and the owner of all its troubles, internecine feuds, frustrations and problems; when

> I go there now it is with a fierce and bitter nostalgia, and when I walk across that
> narrow neck of the island I can hardly bear to look at the azaleas that I planted by
> the path side.

The die was cast, and for good or ill the island was now his responsibility. He had to come
up with something.

By the spring of 1944 Maxwell had bought a nine-metre-long open lobster fishing boat
called the *Gannet*, on which he spent all of his free time exploring the labyrinthine channels
around Arisaig, and occasionally farther afield. After D-Day on 6 June 1944, the demands
on his time were greatly reduced, and although he had to be back at HQ each night, the
days were his to explore as he wished, and he wanted 'nothing but to be where I was,' in the
Western Isles of Scotland.

> Those brazen days I spent in my boat, exploring the coast and the islands from
> Mull to the narrows of Skye, slipping imperceptibly back into a world I had almost
> forgotten, dreamlike and shining. I used to visit the seal rocks and spend hours
> watching the seals; sit among the burrows of a puffin colony and see the birds come
> and go, unafraid, from their nests; fish for conger eels by moonlight; catch mackerel
> and lobsters; and for the first time I saw a Basking Shark at close quarters.

On one such outing, Maxwell had with him a fat but enormously strong local man called
'Foxy' Gillies, and they were returning home from Glenelg:

> It was late afternoon; the sky was paling, and the hills turning to deep plum, their
> edges sharp and hard, as though cut from cardboard. We were about a mile off
> Isle Ornsay Lighthouse, heading southward over a still, pale sea, when I noticed
> something breaking the surface thirty yards from the boat. At first it was no more
> than a ripple with a dark centre. The centre became a small triangle, black and shiny,
> with a slight forward movement, leaving a light wake in the still water. The triangle
> grew until I was looking at a huge fin, a yard high and as long at the base. It seemed
> monstrous, this great black sail, the only visible thing upon limitless miles of pallid
> water. A few seconds later the notched fin appeared some twenty feet astern of the
> first, moving in a leisurely way from side to side.

It took a moment before Maxwell's brain could grasp that the two fins belonged to the
same creature, and engendered a mixture of excitement and fear. He also felt that there
was a purpose to this apparition, 'as though this were a moment for which I had been
unconsciously waiting for a long time.'

Having only seen a Basking Shark once before, and then from a distance, Maxwell was
simply unable to guess what lay below the surface of the water. Foxy was able to supply some
basic information on the local names for the shark – 'muldoan', 'sailfish', 'sunfish' and the
Gaelic name *cearbhan*. He also related the problems they caused for the herring fishermen,
damaging nets and small boats, and that their liver oil had once been pursued by the crofters
and fishermen along the shores for lamp oil. All this while they closed with the giant fish.

> The first clear view of a Basking Shark is terrifying. One may speak glibly of fish
> twenty, thirty, forty feet long, but until one looks down upon a living adult Basking
> Shark in clear water, the figures are meaningless and without implication. The bulk

appears simply unbelievable. It is not possible to think of what one is looking at as a fish. It is longer than a London bus; it does not have scales like an ordinary fish; its movements are gigantic, ponderous and unfamiliar; it seems a creature from a prehistoric world, of which the first sight is as unexpected, and in some ways shocking, as that of a dinosaur or iguanodon would be.

Maxwell had a Breda light machine gun mounted in the bows of the *Gannet*, ostensibly to shoot up mines that he found floating during his travels, but also to enable him to tackle any U-boat he might encounter, fanciful though that may seem. Foxy said, 'Try him with the gun, Major.'

In an action that may well confound those who later anointed him as a staunch conservationist, Maxwell did just that, closing with the shark and firing a single burst of 30 rounds into the animal at close range, seeing the mass of small white marks left by the bullets take their toll:

A great undulating movement seemed to surge through him, and near the stern of the boat the tail shot clear of the water. Its width was a man's height; it lashed outward away from the boat and returned, missing Foxy's head by inches, to land with a tremendous slam upon the gunwale of the stern cockpit. It swung backwards and hit the sea, flinging up a fountain of water that drenched us to the skin.

The shark was back on the surface in less than a minute. Six times we closed in; I had fired three hundred rounds into what was now a broad white target on his side. At the last burst he sank in a great turmoil of water, and it was ten minutes before the fin surfaced again. Now it seemed to me that he was wallowing and out of control, the fin lying at an acute angle. I thought he was mortally wounded, if not actually dead.

Foxy then had the bright notion of trying to secure the shark via the fin with the ship's boat hook. He stood up on the foredeck and Maxwell steered the *Gannet* up close to the fish:

I felt the bows bump the against the shark's body; the Foxy took a tremendous swipe with the full force of his eighteen stone. I could see the hook bite deep into the base of the apparently helplessly rolling fin. There was just time for Foxy's triumphant shout of "Got the b------," then the boat hook was torn from his hands and those gorilla-like arms were waving wildly in a frantic effort to keep balance, as shark and boat hook disappeared in a boil of white water.

Ruefully retrieving the boat hook that shot from the sea's surface 10 minutes later, the two men must have marvelled at the power and endurance of the creature they had just tried to dispatch. And for Gavin Maxwell, the realisation that he, too, had Shark Fever, and his fate would forever be linked to this encounter:

The fly spotted room at Blackwall, one golden day on Soay, the mystery and excitement of that chance encounter at Isle Ornsay – these were the first stepping-stones across the ford; from them my feet went on inevitably to the next stone and the next, and when I turned to look back the stream had risen and covered them, and it was too late to return.

The fateful dream was gathering momentum. Maxwell set about discovering as much as he could about the Basking Shark, mainly from the best source available, the herring fishermen in Mallaig. These men generally agreed that the sharks had not been common in local waters before the 1930s, but had steadily become more numerous and regular in their occurrence over the last 15 years. Maxwell was so intrigued, he acquired two harpoons with the intention of using them on the next shark he encountered. These were traditional whaling pattern harpoons, intended to be fired from the same type of muzzle-loading whaling guns that he himself later employed as his standard equipment, albeit with very different harpoons. However, these harpoons they modified for hand-held use, by lashing them to boat hooks to enable them to be driven into the shark from the deck of the *Gannet*.

As it was late in the season, it was a while before he saw another shark, but one finally turned up off Point of Sleat lighthouse on the southern tip of Skye. Foxy Gillies suddenly shouted, 'Major, Major, sharks!' and pointed to a small bay inshore. Maxwell soon spotted the tell-tale flash of light off a shark's smooth, wet fin, then another, until he could see half a dozen sharks in the bay.

All was chaos for a moment as the crew prepared for action. Foxy, as the strongest member of the crew, had been selected to drive the harpoon home, and stood in the bow while Maxwell homed the boat in on the nearest shark. Finally,

> Foxy drove the boat hook down with all his force into the water a few inches to the near side of it. I could see the boat hook shudder with the impact upon the solid mass below the surface, and saw Foxy pushing on the shaft for a final thrust. Then came the fountain of water shooting up from the sea, and the shark's tail, obscured by spray, lashed down upon the water with several tremendous slaps.
>
> Through all this Foxy was shouting, 'Got him this time. Right in the b------ this time!' and he was dancing to keep his feet clear of the rope as it whipped out at tremendous speed from the coil in the hold. It was a full coil, eighty fathoms, and we were only in eight or nine fathoms of water. I expected to see the rope change direction for the open sea and began to turn the boat to lessen the wrench when the rope's end was reached. But the rope did not change direction, and suddenly it stopped running out and went slack. Tentatively I began to haul in; a fathom came in without resistance—
> 'Is he still there?'
> 'I'm afraid not. I think---'
> As I spoke the rope was whisked from my hands as though attached to an express train, and the palms of my hands were skinned. This time about four more fathoms went out, and again it went slack. I began to haul in more cautiously this time and ready to drop the rope at the first signs of life at the other end of it. Three fathoms, four, six – then I felt the slight drag of the harpoons weight and knew that the shark was lost.

Bits of boat hook littered the surface around them, and the harpoon itself was mangled beyond recognition. Small fragments of flesh on the shaft of the harpoon showed that it had penetrated about a foot into the shark. Foxy was disgusted: 'Ach to hell! The harpoon was no good. It would have been better putting salt on his tail. And I was in the bastard, fair in him. I tell you, Major, there's not a f------g one in Scotland could have put it in further.'

Maxwell had just discovered the sheer power of his prey, as Anthony Watkins and others had done before him, but he was far from dismayed, and spent the rest of the journey sketching out ideas for a better harpoon. Shark Fever had certainly affected him: 'A firm determination to catch a shark was growing in me; it seemed a challenge. And then, quite suddenly – without, I think, any conscious build-up – I thought that here was the industry for Soay, the occupation I required, new and utterly absorbing.'

So on the basis of a first attempt to harpoon a shark that had ended in abject failure, a monumental decision was made that would in time end in financial ruin for Gavin Maxwell. It is almost impossible to identify the first of the many catastrophic mistakes that would leave him virtually penniless in a remarkably short period of time, but the Rubicon had now been crossed, at least in his mind, and there was no turning back. His fate was sealed.

Chapter 6
The Flag Drops

Gavin Maxwell devoted a significant amount of thought towards the design and development of an improved harpoon suitable for use on Basking Sharks, and within a week the first version had been fabricated in Mallaig. Although this was still a hand-held device, it displayed much of the thinking that was further developed in his gun-propelled harpoons, notably the concept of movable barbs that would open once the harpoon was home in the shark, making it almost impossible to pull out. But as it was now late September the season was over and there was no chance to try out the new design.

By November 1944 Maxwell was a civilian once more, and had finally made up his mind to go ahead with his plan of an experimental commercial shark fishery, based on Soay. To fund the new project, Maxwell returned to his mother, and secured an advance against his future inheritance on her death of £11,000 (c. £450,000 in 2017) as working capital, a far from insignificant amount.

With his first purchase for the new venture, he immediately made a massive mistake: 'I had taken the first false step and bought a worthless and entirely unsuitable boat.'

Maxwell recognised that he needed at least one sizeable craft as a catcher vessel, with the range and durability to cope with catching and handling large sharks. The same boat could also do duty for carrying materials to Soay for use in the construction of the factory. In his defence, he claimed, quite fairly, that: 'I was in a desperate hurry to get the whole venture started the following year; I was trying to do a great many things at once, and as a result I made some serious blunders and put the project under a handicap from the outset.'

Which would have been forgivable if he'd learnt from that first, catastrophic error. But as we shall see, Maxwell never seemed to learn from his mistakes, and was therefore doomed to make them again and again. This was not to be last time he would buy a vessel sight unseen with disastrous results. Not his fault, though, as he reasoned:

> That I bought a largish boat without seeing it sounds imbecile, but I had no expert knowledge and did not feel that my ignorant personal inspection could serve any good purpose. I sent an expert to survey an advertised boat and accepted his assurance that she was a bargain at a thousand pounds [around £40,000 in 2017].

She would have been expensive at as many shillings.

The *Dove* was an ex-sailing drift-net boat of the Stornoway fishing fleet; at forty-five years old she was still younger than many of her sister ships. She was a seventy foot 'zulu' lugsail rigged, with two Kelvin paraffin engines, a sixty and a thirty horsepower. For a year it was as if these two vied with each other as makers of trouble and delay, and there seemed a tacit understanding between them that in no circumstances would they work simultaneously.

Although he mentions sending an 'expert', Maxwell doesn't make it clear whether by that he means that he employed a qualified marine surveyor to inspect the *Dove*. If that was the case, then the question might reasonably be asked whether legal redress might not have been sought after such a disastrously inadequate survey. And it was not as if Maxwell didn't have good connections amongst his friends in Mallaig, who could have been despatched to inspect her and done a good job of assessing the condition and suitability of the *Dove* at low or no cost – and saved him an expensive and demoralising error.

But having bought the wreck, Maxwell 'naïvely' left her to carry on the winter herring fishing until he needed her. In a short time, she had run up a substantial loss, and then she was involved in a bizarre collision outside Stornoway harbour with a similar vessel called the *Lews Castle*. A year of litigation ensued, but the cause of the accident remained obscure:

> The only indisputable facts were that the *Lews Castle* was a total loss, sinking with all gear in a little under a quarter of an hour, while the *Dove* had only slight damage about her bows. It was a fortunate accident for the *Lews Castle*, older and more heavily insured than the *Dove*.
>
> The details of marine law are intricate, and to the usual complication of legal phraseology is added a babel of spiky nautical terms, many of them archaic and otherwise in disuse; the eye rattles and bumps over whole pages that contain no recognizable word or phrase. From this confusion emerged an uncomfortable point of law; if the insurers of the *Lews Castle* could prove my skipper to be incompetent I could be sued for a sum much larger than that covered by my insurance policy. It was with that cloud upon the horizon that I began the New Year.

On a more positive note, the prospects looked good on the oil price front in 1945, with the oil buyers quoting £50 per ton. One of those buyers, Gordon Davidson, a senior partner in a major Glasgow firm involved in all fisheries, wrote to him outlining his belief (perhaps as a consequence of his former experience with Anthony Watkins) that there were many commercial possibilities to be exploited from the shark, beyond the liver oil, that he would like to see investigated:

> He told me that he expected the flesh to be marketable either salted, fresh, or as fish-meal, that manure could be made from the refuse, that glue could probably be extracted from the membranes, that the skin could have a high market value, and that there must be many more possibilities at present entirely unexplored. He was enthusiastic and encouraging and offered his help in designing a small factory at Soay.

Sensibly, in the early stages Maxwell stuck to the basic idea of only including an oil extraction and fishmeal plant at the factory site, with a small laboratory for experimental work to be

carried out. The work began on the site at the beginning of 1945, and he hoped to have it working by June of that year to handle a few sharks caught during the latter part of the season.

The question of a suitable gun was proving problematical, for many of the same reasons that Watkins was encountering, post-war. Finally, a custom-made gun using a 20mm Oerlikon barrel was ordered from an arms expert, to be made in his private workshop. The project, on the face of it, was taking shape. He had a gun on order, a factory in build and two small vessels, the *Gannet* and the *Dove*.

As well as an office in Mallaig, through which poured a torrent of letters, permit forms, legal correspondence concerning the *Dove* and a hundred and one other matters that would need addressing concerning the new venture. Being on his own, with no secretary or assistant, this must have been an arduous and soul-destroying task, and Maxwell battled bravely to stem the tide of demands on his time that such an avalanche of paper entailed. Red tape proliferated in the aftermath of the war, as though it had been stored for use once hostilities had ceased, and then the floodgates had been opened. It seemed that everything that he would need to do now needed a permit. Even the construction of a tiny jetty on Soay of just a few square feet required approval from the County Council, HM Customs and Excise, the Admiralty, the Board of Trade, and the Ministry of War Transport.

Not all of it was without some potential value, though, and Maxwell (with apparently more lofty connections than Watkins before him) did try to gain some advantage from the unpopularity of the Basking Shark.

> The Secretary of State for Scotland continued to give me really valuable help and encouragement in every possible way short of actual financial assistance. The department was naturally interested in the project; it held the possibility of a new industry in one of Scotland's problem areas; it promised the production of oil and foodstuffs at a time when both were desperately short, and the reduction of a pest by which thousands of pounds' worth of herring nets were destroyed every year.
>
> With a price on the head of every shag, cormorant and seal, it seemed to me that the Herring Industry Board should have offered a small subsidy on the killing of each shark. They refused – I suspect because it must have been plain to them that no larger amount of sharks would be killed as a result of such a subsidy. No-one would go to the lengths of killing a Basking Shark and producing evidence of having done so merely for the five pounds for which I was asking.

Sometimes things did seem to be moving in the right direction. Maxwell was able to engage a young marine biologist, Gilbert Hartley, through his friendship with the well-known ecologist Frank Fraser Darling. Hartley was keen to do original work on the Basking Shark, and had contacted Fraser Darling seeking to find out if there were any opportunities that he knew of, and the connection was made. Maxwell and Hartley agreed to collaborate, and Hartley contributed a list of equipment needed to enable his studies to be carried out. It was a welcome small victory.

But it was a rare event. When the *Dove* finally arrived from Stornoway in February 1946:

> It became clear that we were still at the very beginning of our troubles. From the moment I set eyes on her I knew, and at the same time tried to conceal from myself, that I had made a really gigantic blunder. She was in roughly the condition

one might expect of Noah's ark were it thrown up by some giant subterranean upheaval, nor would the engines have made one marvel at Noah's mechanical genius.

But Maxwell's luck was about to take a decided turn for the better with the arrival of his very first employee at the shark fishery, Joseph 'Tex' Geddes.

Tex Geddes had served with me in Special Forces; he had spent most of his life in Newfoundland, and as a boy had been with the Newfoundland fishing fleet. He was in his late twenties at the end of the war, and it was difficult to fit into his years the variety of experience with which he was credited – lumber-jack, rum-runner, boxer, knife-thrower, Seaforth Highlander are only a few samples. He could handle a boat well and had a keenness for adventure which appealed to me; on the debit side were a rather violent temper and a periodic liking for drink.

Geddes brought a whole range of practical skills to the table, and was an experienced seaman who knew his boats, as was evidenced by his comments on the *Dove*:

'There's a year's work there. And she's as full of rats as a town is of people – black rats, and carrying the plague, I shouldn't wonder. And when you start scraping the filth in the galley there's no wood underneath it, just more filth. And the Stornoway crew brought her down on one engine   the other's all mucked up. You've been had for a sucker this time.'

As is so often the case with old wooden boats, the worst is concealed, and as successive layers of boards are removed, the vista only becomes more alarming. The *Dove* was no exception, and 'there was very little in the ship that did not need renewing.' She was a rotten as the proverbial pear, and only the massive structural timbers appeared to be sound. Three months of costly repairs in Mallaig only exposed more of her dreadful state, and the bills steadily mounted. As Gavin Maxwell had come to know, getting hold of, and then *keeping* hold of tradesmen in a place like Mallaig (during a herring fishing boom) was by no means easy:

The fishing fleets must come to Mallaig, the boats must be serviced – and serviced in a hurry, for in the middle of a good fishing every day in harbour may lose several hundred pounds. When the necessary repairs to some boat's engine should cost as little as twenty-five pounds, it might well be worth two hundred pounds to the boat's owner to get the work done at once and go back to sea.

If some fisherman was desperate enough to pay the engineer working on the *Dove* to down tools and move over to work on his boat – then money would talk.

But for the most part, one can sense Maxwell's fascination with Mallaig: 'It is a strange town, a boom town, a Klondike, a Dodge City; it is new and growing, but its newness and growth seem not of this century.'

By way of historical context, the coming of the railway in 1901 changed everything for Mallaig and the surrounding districts. The railway provided a natural conduit for equipment and supplies in and fish out. Suddenly Mallaig became the *de facto* fishing port for the Hebrides. As always in such 'gold rush' conditions, an army of chancers and wild men flocked to the area. Small contractors set up and became rich in only a few years, others lost their shirts in ill-judged schemes. Fishing vessels flocked from both coasts, and 'Chinatown', Mallaig's slum area full of girls who gutted the herring, grew up between the railway and the sea. New developments helped to establish a firmer presence; an ice plant and a kippering factory provided steadier employment and secured Mallaig's position. And whilst some local fishermen made fortunes in the good times of the herring fishery, those fortunes could just as easily be lost in a few months due to equipment breakdowns or prolonged spells of bad weather keeping a fishing boat tied up in harbour.

Then there were the crews; men used to huge cash payouts as reward for good fishing weeks might well move on to another boat if the one they were currently on was laid up for one reason or another. The Mallaig fishermen worked on a shares basis: if the boat wasn't working, they weren't earning (and Mallaig fishermen were not known for their patience). The best of the crewmen were much prized, and were in constant demand, so it was definitely in the interests of any skipper-owner to hold on to his crew, or be condemned to end up with whatever crew he could scrape together.

As Maxwell rightly identified, drink was a perennial problem with some of the men:

> Among those who often change boats, a frequent trouble is drink; they may be good enough workers otherwise – indeed, some of them are the best of all – but unreliable on this account. The euphemisms, and the precise degree of unreliability that they imply, are many. 'He takes one now and again'; 'He likes a drink'; 'He takes a good dram'; 'He's all right when he's off it'; 'He's aye fou' (roughly translated to perpetually drunk). It is to this last category that a skipper may have to turn to in the end, if his crew have become impatient with too much waiting in harbour. To the 'aye fou's', the more days in harbour the better.

But, by and large, Maxwell liked Mallaig and its inhabitants:

> The people have, in the main, a great friendliness and a natural generosity that is often unexpected; one and all, too, seem to share a common motto: 'I'll not see you stuck.' I was always on the verge of getting 'stuck', and I heard those words, and saw them proved, many times.
>
> To the tourist, Mallaig may be its poster self – the gateway to the Hebrides; to me and to many, it is a town of a different sort of romance: of herring scales and a million gulls, of energy and squalor and opportunity, of feud and fortune; the 'end of steel' – the railhead – beyond which all is gamble.

Which must surely beg the question: if he had to have a factory for his shark fishing enterprise, why on earth didn't he base it in or around Mallaig? Mallaig had everything he needed – cranes, slipways, fuel, support services like engineers and an available workforce who knew boats and were used to the hard work that fishing entailed. It also had a population sympathetic to such wild characters and mad projects as his own, and who knew that the smell of fish was also the smell of money.

Mallaig had the best road, rail and sea links in the whole area, making shipping materials in and products out as easy as could be managed in such a remote region. And Mallaig was an all-tide harbour, with no bar to negotiate as there was at Soay, meaning that no time need be wasted waiting for the tide to rise to get his boats in or out. All in all, the choice of Mallaig for the factory would, at one single move, have removed a huge cost and logistical burden from his nascent enterprise. That he apparently didn't even consider this is hard to comprehend.

The 1945 season

In late April 1945, the *Dove* was finally ready for her trial voyage. As Maxwell had guessed, with the ring net fleet already well into the new herring season, crew were hard to come by, and he set off with an unnamed temporary skipper, Foxy, Tex and two boys, one of whom was nominally ship's cook, but as Maxwell drily observed, 'he could be more correctly described as ship's tin-opener.'

The first voyage went smoothly enough, but was a shock to Maxwell, even though he was certainly used to basic living after his wartime experiences in London. The shattering racket from the engines made conversation nigh on impossible, and the all-pervading miasma of paraffin from the engines, the cooking stove and the Tilley lamps that lightened the gloom below decks infected everything. As he remarked contemptuously: 'The *Dove* was no luxury yacht; she had been one of that grim black procession that sails in line ahead from Stornoway to the distant fishing grounds, each skipper stepping out from his wheelhouse to spit as he passes the ring-netters from the south.'

Not his sort of craft, obviously. Nor had the rats been eradicated, as he found when he left the contents of his trouser pockets on a convenient ledge by his bunk. When he went to retrieve them in the morning his hand was covered in rat dung.

For once the *Dove* behaved, though, and the same week she began carrying materials from Mallaig to Soay, all of the essential items to build the factory such as bricks, cement, wood and machinery, including an eight-ton boiler. But wood was in desperately short supply after the war, and the *Dove* and her crew were soon out in the islands scavenging timber from the many windward facing bays where huge piles had accumulated after the wartime depredations of the U-Boat fleet on the convoys carrying supplies to Britain.

The west coast of Rum was a prime site for such timber, and the *Dove* set off there with the intention of towing back bulk loads of timber to Soay. The only beach on that wild coastline is at Harris Bay, where a half-mile of shingle beach collects all that floats in from the Atlantic. Above the beach, the bowl of Glen Harris rises up into the lonely and bleak hills – Askival, Ainshval, Hallival and others – that are seldom free from cloud. Only the bizarre sight of the Bullough mausoleum, a pink marble miniature version of the Acropolis, and a small croft house near it, give away any sign of human habitation since time began.

Here the crew rummaged through the debris of war and 'timber, timber, everywhere, like a fallen forest', working through the day, lashing the beams and planking into rafts to be towed back to Soay by the *Dove*. This stretch of the shore is a favoured spot for Basking Sharks, and a hail from the *Dove* caused them to swing around from their work to watch a black fin cruising between them and the shore. All afternoon it remained in sight, 'a reminder of the adventure ahead.'

The factory was slowly coming together, and Maxwell determined that the time was right to take a shark for experimental purposes. The gun maker told them that the gun and harpoon would soon be ready; he would come up and test them himself, and a date was agreed. But the good old *Dove* had other ideas and was once more back in the queue for repairs, and still laid up when the gun maker arrived in mid-July. The factory was way behind schedule, too, and was barely half completed, with much essential machinery still missing. Maxwell sensibly decided to give up all thoughts of trading in 1945, and spend his time on researching the catching of sharks instead.

Then the gun arrived. Although it looked impressive, and showed 'the undeniable engineering genius' of the gun maker, it proved to be basically useless. And as if that wasn't bad enough, the harpoons were hopeless. Maxwell had already told the gun maker that his harpoon design was fatally flawed, and that if he wanted to try this out, he must also bring some harpoons to Maxwell's own design – which he had failed to do. So they were stuck with 14 of the gun maker's barbless harpoons, and two of the hand harpoons that Maxwell had had made in Mallaig the previous year.

The usual business of cutting and shutting to fit any unusual item to a ship's deck meant that it was a whole week before the gun was satisfactorily fitted in the bows of the *Dove*. The gun trials out in the Sound of Sleat showed that the harpoons were prone to tumbling before impact, but on the basis of one successful shot, the team headed out across the Minch where they had heard sharks were numerous, working their way up from South Uist to Harris. Whilst they did find sharks, the results were desperately and predictably poor: 'Hope is long in dying; we fired fourteen shots at sharks and five at killer whales – one of these being the only fair chance I ever had at a big bull killer – and not one single harpoon struck home.'

There was nothing else for it but to rely on hand-held harpoons, but the only one aboard required modification, and the local blacksmith on North Uist didn't have the tools to do the job, so in a fog of black depression the crew set course back to Mallaig. With no sample specimen to assay for the potential new markets and the season half over, a sense of urgency gripped them and two new hand-held harpoons were soon made.

A few days later, reports of sharks on the west coast of Rum sent the *Gannet* off in pursuit. On board were Geddes, Maxwell and his new skipper, Bruce Watt:

> Bruce had joined us a week or so earlier, and remained my skipper until the middle of the 1947 season. He was a man of my own age, a former Merchant Navy Engineer Officer; a good seaman, solid, reliable, teetotal, and of native common sense, an absence of impetuosity, that sometimes made him seem slow in comparison with more volatile characters.

Once again, Gavin Maxwell had made a sound choice. Bruce Watt would prove to be a first class skipper, and a good foil for the more excitable members of the crew.

As Rum is only a few hours from Mallaig, the crew had taken no food or spare fuel with them. Equipped with the two hand-held harpoons, now with 15 feet of iron piping for shafts, whose weight would assist in the harpoons penetrating deep, their aim was to place both harpoons into the shark, with Geddes and Maxwell as the harpooneers. Each harpoon was attached by a wire trace to 80 fathoms of rope, to which was also fastened a canvas buoy to give some indication of the direction of the shark. Two hefty coil springs were attached near the outboard end of the rope to give some additional shock absorption and so hopefully avoid the rope parting.

About a mile off Harris Bay they sighted a large shark, with a huge fin standing high out of the water, and Watt swung the *Gannet* around to make the approach.

Geddes yelled 'Let him have it!' and the two men drove the harpoons home with all their might:

> Nothing happened; it was the anti-climax of bayoneting a sandbag. I leaned on the pipe and pushed as hard as I could; from the corner of my eye I could see Tex thrusting and shoving furiously. Then suddenly through the long shaft I held I felt a volcanic surge of strength as the tail of the sharks swung towards the boat in an effort to crash-dive. Everything was hidden in a great shower of water and spray. As

the spray cleared I saw that the rope was running out at a tremendous speed. The canvas net buoy, with its own six fathoms of rope, went overboard and submerged in the same instant. As the first rush began to slow a little we slipped the rope in a half-turn on the drum of the winch, and the gannet began to be towed slowly ahead in widening circles. We seemed to be attached to a shark at last.

Every now and again the buoy would appear on the surface for a few seconds. At first we used the tiller to follow it directly; then, as half an hour became an hour, and an hour lengthened to two, we realised that until we started to haul in the rope there was nothing useful that we could do. We waited for the shark to tire himself out – we might as well have waited for him to die of old age.

After four hours the shark was still in charge. The day was drawing to a close, the wind was steadily rising from the south-west, straight from the Atlantic, and they were now some three miles west of Rum in open water. Concerned that they lose this shark by trying to pull it to the surface too soon, before it had tired, they were determined if needs be to let the shark tow them through the night.

Easily said in a less exposed situation, or in a place where the weather is far more benign than the Hebrides. As dusk fell some herring-fishing boats came by, heading for the shelter of Canna Harbour, one pausing to shout across to them that they'd better make tracks for Canna, too, as there was a gale warning out on the wireless. To do that would be unconscionable though, as it would mean cutting the shark free, so they decided to tough it out.

Two lighthouses were visible to them. The small light on Sanday, the island off the south coast of Canna, and the Hyskeir, a tall light mounted on 'a lonely wind-beaten rock' 10 miles to the south-west of Canna. At that time the light on Hyskeir was some 10 miles to their north-west, and both lights were visible as Rum faded into the blackness.

Within an hour the wind began to rise still further and the night grew increasingly cold. The sparkling display of phosphorescence helped to alleviate the gloom as breaking seas swished by, but really there was nothing much to be happy about. The shark was speeding up, dragging them into a rising sea, but Maxwell's artistic senses could still see some of the beauty in their predicament:

> Except for the sound of the breaking water, the night was very quiet; the *Gannet* had no rigging for the wind to play tunes upon. We sailed a dream sea in the dark and the eerie phosphorescence, towed by the wounded shark far below us in the dark water.

But still their course continued west, and by midnight Hyskeir light bore due north of them. Still the wind rose, and the seas grew higher, and they were still heading out into bad water with little fuel and no food. Even unflappable Tex Geddes didn't much like it, opining that 'If we need to do this every time we catch a shark I'll be needing double pay.' Maxwell collapsed into sleep on the bottom of the boat, wrapped in his duffel coat.

I know from bitter experience that this is not a kindly piece of water on which to find yourself during bad weather. Local seafarers know that this sector of the Sea of the Hebrides is one of the worst places to be in a gale, as the tide runs hard between the west of the Small Isles (Muck, Rum and Canna) and the Outer Islands. And right in the middle lies a long tongue of shallow water that runs approximately 17 miles south-south-west of Hyskeir and 10 miles north up to Garrisdale Point on Canna. In a gale, with wind against tide, the sea breaks heavily all along that shallow bank, and worst of all in the area of the Mill Rocks, a really vicious plateau awash just two miles south-west of the islet upon which Hyskeir Light stands. This wouldn't have been just an unpleasant experience in a small boat – it could easily have proved fatal, especially for a small, open boat like the *Gannet*. They were being drawn inexorably into that maelstrom, and the more experienced members of the crew of the *Gannet* knew what they were in for.

> Through my sleep I heard voices once or twice – when I became conscious enough to understand. I realized that Hyskeir Light was bearing northeast and that Sanday Light was obscured. I roused myself, stupid with sleep and cold, to find that we were in a really heavy sea, the breaking phosphorescence stretching around on every side to the limit of vision. There was distant undercurrent of sound, deeper and heavier than the nearby breakers, which at first I could not place. Then through it came an unmistakable call, thin and buffeted by the wind but sweetly familiar, the calling of curlews – curlews that meant rock and reef.
>
> We trusted to Bruce, and Bruce gave the inevitable verdict that we were as unwilling to accept as he was to give. We must get free of the shark somehow and at once.

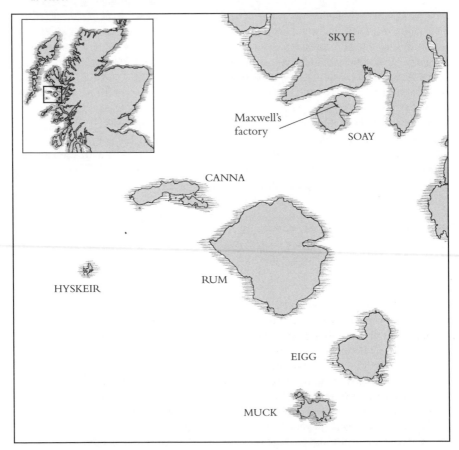

The Small Isles and Skye, Scotland.

The rope still stretched out ahead of them, at an acute angle – the shark must have been nearly at the surface. They decided to try and pull the harpoon out by starting the engine and going hard astern, but irony of ironies, for once the harpoon simply wouldn't pull out. There was nothing for it but to cut the rope.

Bruce chopped through it with an axe, seven and a half hours after I had planted the harpoon.

Dawn came as we passed between Rum and Canna, a bleak dawn with a heavy sea and lashing rain. The journey home took five hours; we were soaked and shivering and I felt seasick for the first time in my life.

Heartsick, too, I shouldn't doubt, and who could blame him? But giving up at this stage was not on the agenda, and pausing only to return to Mallaig to re-equip with two new hand-held harpoons, the doughty crew of the *Gannet* was soon once more ready for action.

Maxwell's friend John Lorne-Campbell of Canna sent him a telegram that simply said: 'Huge shark off Compass Hill'. Compass Hill is the highest point on the north shore of the island, and is so called because it creates a magnetic anomaly affecting ships' compasses in the vicinity. It forms a magnificent centrepiece to the otherwise rather bleak north coast, home as it is to thousands of auks and other seabirds, and with occasional eagles soaring above it all.

There had been an unexploded mine reported in the area, though, so all eyes were carefully scanning the sea not just for the shiny flag of a dorsal fin, but for anything else of an unusual nature that might just be a mine. Having found the former object and closed in for the kill, unbelievably Maxwell now found himself looking over the side of the *Gannet* at the latter object, at far too close a range for comfort. Meanwhile, the shark sailed serenely around the mine, and Maxwell was left to wonder whether 'he would go that inch too close that we ourselves had so nearly been.'

Stopping for a social call on Canna, the crew split up. As Maxwell returned down the path to the quay, he saw Tex Geddes waving wildly and shouting to him, pointing to the fin of a large shark close to the rocks right in the harbour. Bruce Watt was already getting the harpoon gear in place and they were underway in less than five minutes.

And so began a wild goose chase among the rocks, with the shark seemingly playing hide-and-seek with them, staying right on the shore then diving for minutes and coming up far away. The crew was losing patience when the shark suddenly surfaced right in front of the boat and Geddes, unwilling to wait for a better shot, drove his harpoon deep into the shark's back near the tail. Maxwell didn't get the greatest view of what happened next:

when the whole boat seemed to be struck by a high-explosive shell. From the corner of my eye I was aware of the shark's tail somewhere high in the air and apparently about to slam down right on top of me. I ducked instinctively, and I was almost on all fours when two objects hit me almost simultaneously. The first was a harpoon pipe, whose end struck me a tremendous blow in the groin. Then Tex, whose body felt like a ton of rocks, fell backwards on top of me, bounced off, and tumbled right back in the hold upon the coil of rope, which was streaking out at high speed. I realized dimly that Tex was in very great danger and that if any part of him got tangled in the rope he would have broken bones before he was dragged overboard. I fell back into the hold myself and was trying to as I did to lug Tex clear of the rope coil. He seemed unconscious and had a lot of blood coming from

his head. Bruce had jumped forward from the tiller with the same idea, but as he landed in the hold his foot caught in a snarl of rope, and his problem was added to ours.

The *Gannet's* bows were jerked violently round into the bay, and for a second or two we seemed to be travelling at enormous speed with no-one at the engine or the tiller. Then the harpoon pulled out, and the *Gannet* gradually lost way, while Bruce and I struggled with the situation in the hold.

Geddes was out for the count with a large gash in his head, Watt had twisted his ankle and Maxwell was feeling sick from the blow he had taken right where it hurts. Seemingly Geddes's harpoon had gone in too near the powerful tail and the harpoon pipe had been swung away from the boat the first time by the tail, before being driven back again, catching him a glancing blow on the head as he ducked to avoid it, before whacking Maxwell in the groin. They were lucky it wasn't far worse.

We took Tex up to Canna House, and a guest staying there, a trained nurse, examined his cut. It was long and gaping and obviously required stitches. Tex was given the best part of a tumbler of whisky, and in the warm kitchen, among the Siamese cats, she stitched it neatly and efficiently.

A change of luck

Maxwell had decided to remove the Oerlikon gun from the *Dove* on to the *Gannet*, partly due to the endless breakdowns of the former vessel, and also to lower the height of the gun in a bid to reduce the problems already encountered with deflection. The gun platform on the *Dove* was 10 feet above the water, whereas it was only three feet on the *Gannet*, allowing for a shot at almost point-blank range. However, the work required to make the transfer was far from easy, as the *Gannet* was a much smaller and therefore more lightly constructed boat, so much reinforcement of the bow area was required to cope with the recoil of the gun. Eventually the work was completed, and with the gun mounted well forward on the starboard side of the bow and two fresh harpoons, they were once more ready for action.

The fish telegraph was working well, and acting upon a report of a big shoal of sharks across from Mallaig near the Point of Sleat on Skye, the crew cast off and headed out. By the time they got to the shoal, though, most of the sharks had dropped below the surface and just a few were left as targets. At least, until they got closer, when even those last few went down. This made a big impression on Maxwell:

Then, and up to the very last shark I killed years later, this waiting for a shark to resurface, straining one's eyes for the faintest ripple or gliding bulk below the water, set my heart hammering savagely against my ribs, as though it were a sort of overture, a roll of drums leading up to the climax – the gun's roar and the flying rope, and the tail towering out of the water in a drench of white spray. This was the first time that it really happened.

At last a fin appeared, 50 yards ahead of them. The *Gannet* set off in pursuit, before Watt gave the order to come out of gear in the final approach. At the last moment the shark swung across their bows, and Maxwell was forced to take the shot:

The fish had been swimming very high; there were only a few inches of water over his back when I fired, and I felt quite certain that the harpoon was in him.

The tail behaved as usual, hiding everything with a storm of spray; then, when it had subsided, I saw the shark a fathom or two down in clear water, swimming fast on an opposite course. I could see the end of the harpoon shaft sticking a foot or so out of his side – below the point I had aimed for – and a dark plume of blood trailing from it in the water, like smoke from a chimney. Tex saw it too, and gave his war cry for the first time, a war cry that I came to associate with every kill, and which in a later season I remember hearing across half a mile of sea, following the boom of his gun in the summer dusk: 'He feels it! He feels it!'

Determined not to lose this shark, the crew let the animal take plenty of rope before slowing the rush to the bottom on the winch, then let the shark tow them sluggishly for two hours before they began to haul up. Slowly but surely they brought the shark to the surface. Finally, it came in sight, with the rhythmical sweep of the tail as the shark tried to bore downwards, being the first part of the animal on display before the rest of the creature took shape:

I could see part of the body beyond the tail now; the body of a dragon, six feet through and showing a glimmering white belly as he twisted and lunged. At the far end of the belly there seemed to be two gigantic flippers – I was unprepared for the size of these pectoral fins, which had been minimized in the drawings I had seen.

As soon as his tail came clear of the surface the power of that punkah action became apparent; at each lunge it exploded a fountain of water from the sea. Several times it struck the *Gannet's* stem, leaving gobs of black slime as it struggled free. We were busy with ropes now, and after several near misses succeeded in dropping a noose over the long upper half of the tail fin as it jammed momentarily against the bows. The next lunge carried the tail below the surface, and for a moment it looked as though the rope would be flung free, but as the tail rose again towards the boat we saw the other half of it slip through, and the whole fin was in the noose. We almost knocked each other overboard in our hurry to pull it tight, but we saw it close firmly on the narrow isthmus of body below the fin, and the shark was ours.

By now it was late afternoon, and the long, slow slog of the tow back to Mallaig was ahead of them – their factory on Soay was still a construction site. The shark was still alive, and towing it with the *Gannet* at half a knot by the tail was like 'towing a house'. It was only later that they learned to tow the sharks alongside the boat head first with the jaws tied closed to reduce drag. Finally, a Mallaig boat came up to them and agreed to take the animal in tow alongside the *Gannet* and with both their engines full ahead, three or four knots could be made. Mallaig was in sight now, and a festive atmosphere was building up on the quayside in advance of their arrival – the trusty fish telegraph was at work again:

It was near dusk when we reached Mallaig. It was the tourist season; Mallaig was performing its brief seasonal function of Gateway to the Hebrides, and the train had disgorged its cargo of holiday-makers, would-be mountaineers in huge climbing boots and swathes of rope, brightly tartaned Highlanders from Glasgow and the industrial cities, earnest hikers with gigantic rucksacks and skinny legs, and sad-looking elderly couples who seem to visit the Hebrides annually to lament the climate. Word had reached Mallaig that the *Gannet* was bringing in a shark,

and the attention our project had received from the daily press had thronged the piers. As we passed the big stone pier where the island steamers berth, and headed for the inner side of the fish pier where the herring catches are landed and bid for, I remember that parts of the crowd began to run back to intercept us before we reached our berth beside the *Dove*. There were something like fifteen hundred people crowded on that short pier by the time we churned laboriously round the end of it.

The shark was still very much alive, and transferring the rope it was attached to from the winch to one of the uprights of the pier nearly cost Bruce his arm as the shark vigorously took up the slack: 'The rope slammed taut with a noise like the plucked strings of a double bass, catching the boat's gunwale a hundredth of a second after Bruce had snatched his arm away.'

Several attempts were made to lift the shark on to the deck of the *Dove* by means of her steam capstan, but all failed when the full weight of the animal came on the line. As darkness fell, the disappointed crowd dispersed, but reconvened when they heard that the skipper of a boom ship with a huge crane had agreed to lift the shark. Manoeuvring the *Dove* alongside her wasn't easy, but the crowd simply grew larger as the excitement rose.

At last, the shark began to rise from the water, with the two-metre broad tail the first part of the body to break the surface. The shark, still struggling, sent ripples from side to side as it fought to escape its captors. The shock of the sheer scale of the animal was not lost on Gavin Maxwell:

> The narrow neck just below the tail seemed infinitely prolonged; then, very slowly, the girth of the body began to rise. Size always appears greater in the vertical than the horizontal, and by the time fifteen feet of the shark were clear of the water and the girth was still increasing, he appeared literally monstrous, a creature of saga or fantasy, a dragon being hauled from its lair. The darkness, the shifting yellow reflections of the harbour lights, and the white glare of the searchlights combined to give a stage effect of mystery and magnification. There was an excited gabble from the packed crowds on the pier, gasps and exclamations, and a group of women near the edge panicked and forced their way back into the press behind them.
> 'Oh, what a crayture!'
> 'Ye wouldna' believe it!'
> 'It canna be a fish!'

The strain on the gear of the crane was now very evident, whilst the monster grew ever longer:

> Then, with twenty feet of him standing clear of the water, there was a slight snapping sound, as a man makes when he steps on a rotten stick. The crane stopped at a quick order, and for a second there was utter silence – followed by a tremendous crack and a sickening, tearing sound as the great carcass plunged back into the water. For a moment the severed tail hung suspended in midair; then that, too, fell with a mighty smack into the oily black water of the harbour.
> After these six years I can still hear the noise from the watching crowds, feel again the almost unbearable disappointment of that moment, that last unbelievable frustration of Tanatalus.

96

I did not think that the shark was dead; even tailless he would probably wriggle or drift with the tide into deeper water outside the harbour. I felt quite certain that we should not see him again and not even the tail remained to prove that we had at last caught a shark and brought him to harbour.

But he was wrong. The following morning a lobster boat spotted the shark on the bottom outside the harbour and soon, with the aid of a grapnel, they had their shark attached to the *Gannet* once more. By now gases had formed inside the belly of the shark, adding buoyancy to the great body, and within 10 minutes the shark was at the surface. Even the tail was recovered, and after securing permission to beach the shark at Henderson's slipway, 'with the huge tail shining like patent leather, our dragon was complete when Mallaig woke.'

With the shark hauled up on the slipway, this was an ideal time to conduct some basic dissection, and a telegram was sent to Hartley, inviting the marine biologist to come and carry out some more in-depth work, too. The shark was a sizeable female (so much for all the talk of 'he'), and at 7.5 metres overall length (and 5.5 metres in circumference) gave them their first experience of cutting up such a behemoth, which must have given them some alarming insights into the sheer difficulty of dismembering a beast of this scale. Referring to a textbook on skinning sharks, when Maxwell reached the words 'now turn the shark over', he knew that they were out on their own with the Basking Shark. At the same time, they noticed that the fearsome 'dragon' they had slain was only equipped with rows of minute teeth, 'as small and needle-pointed as a kitten's'

While the necessary samples were taken of the skin and flesh for technical purposes, a less scientific form of analysis was conducted by the local population who helped themselves to flesh for fish suppers. Even the local hotels served shark steaks without complaint. By the time Hartley arrived to conduct the more scientific examination, the crew had returned to sea to look for more sharks. As it was by now late in the season, Maxwell feared that they would find no more, but in the weeks that followed he saw Basking Sharks down as far as the Treshnish Isles off Mull, and finally near Soay, killing two more the following week. A few small sharks remained in the area until mid-October, giving them the chance to enlarge their pitifully small repertoire of shark-hunting skills before winter closed in.

Their luck seemed to have changed for the better at last. A good thing – they were going to need all the luck they could get in the following season.

Chapter 7
Reality Bites

At the beginning of 1946 all that Gavin Maxwell's new firm knew of hunting and marketing the Basking Shark came from a few sample specimens that they had caught, and some very basic assays that had established potential markets for most parts of the shark. And he was honest enough to acknowledge some serious doubts about the wisdom of adhering to a policy that took the latter point seriously.

He quite rightly realised that a 'gigantic' problem of transporting, handling and converting the giant carcasses into so many products lay before them, requiring serious investment in plant and staff – and so far he had not even begun to address those needs. He also knew enough to suspect that to do so would be to enter a blind alley: 'From the time the project had first been conceived, my instinct had been to confine ourselves to the marketing of the liver oil, the value of which had been high in peacetime and had now practically doubled.'

But his advisers insisted that the only way to succeed would be through 'using every part of the fish. Nothing must be wasted, no possibility unexplored.' Easily said if the fish before you is a herring, but with a creature in which the head alone could weigh one ton, a totally unrealistic strategy unless massive amounts of capital could be deployed to finance an operation on a far larger and more mechanised scale.

> I see it now as I saw it at first: an ivory hunter in the deep Congo jungle,
> standing by the mountainous carcass from which he has cut out the tusks
> and pondering how he may capitalize the tons of flesh, the hide, the bones –
> all the apparent and gigantic waste. But so insistent were my advisers, and so
> apparently experienced in the mass handling of all that comes out of the sea,
> that I was won over completely to their point of view. The factory, not the
> ships, was to be the nerve centre – a nerve centre doomed from its inception
> to starvation from blocked arteries. I know now that the shark's liver is the
> elephant's ivory.

But the die had been cast, and the catastrophic decision had been made to either market or explore no less than 10 different products during the 1946 season: 'Liver oil, liver residue,

glue from membranes, frozen flesh, salted flesh, fish meal, dried fins, bone manure, plankton stomach contents, and glandular products.'

And all of this from a tiny factory on the remote island of Soay, in the wildest part of the entire British Isles.

Maxwell recognised that this was a major stumbling block, and that if the factory were to function at all, some of the processing would have to be carried out at sea, and he proposed the purchase of a Tank Landing Craft Mk IV to act as a parent or backup factory to the shore base. But to do so would cost more money, and that was by now in short supply. And before that, something would have to be done to replace the ailing, rotten *Dove*.

To take into account the changes envisaged, the Isle of Soay Shark Fisheries would have to commit to spending something like £12,000 (around £450,000 in 2017) to reach the starting gate in 1946. Maxwell had been banking on securing further funding from the Scottish Development Council, only to eventually find that they had no funds whatsoever to disburse. They in turn put him in touch with the Industrial and Commercial Finance Corporation 'whose terms proved unattractive.' Maybe they looked too closely at the business plan. Whatever the case, Maxwell acknowledged that 'It was clear that the programme would require drastic modification.' He even approached a number of Captains of Industry for their view:

> Their attitude was in each case the same, expressed with the greatest clarity by the kindest and most human of them in the words, "My dear boy, if we were going in for a new industry like this, we should write off fifty thousand pounds and five years to experiment – you are expecting to make a profit on twelve thousand pounds and one year's experiment." This was wisdom, though I did not recognize it – in fact, it took three years and the partial misapplication of twenty thousand pounds to bring us tantalisingly within sight of success.

A revised scheme without the factory vessel and concentrating on the bare minimum for continuation with a capital budget of £7,500 was prepared and circulated to many of Maxwell's friends who he thought might be interested. Nine of them came forward with loans of £500 each to keep the business afloat. This might have been an apposite time to ditch the idea of so many products and concentrate solely on the oil, or perhaps just the oil and fishmeal, not least because the factory that should have been finished in October 1945. It was now long overdue for completion, largely due to the need to install additional plant to cope with the myriad products. In fact, it would be a full year more before the factory was completed. And still the *Dove* did her best to bankrupt the little enterprise, one of her engines throwing a rod through its crankcase during a transport voyage to Soay.

But fortune still smiled on Maxwell occasionally, and Christmas brought a present in the form of a man 'who had stepped straight from a Wild West film', who bought the *Dove* on behalf of a firm that aimed to start an Arctic fishing fleet. Maxwell 'pocketed his cheque and said goodbye without sentiment to the *Dove*, thankful that she had found no resting place with me. … She got no farther than Loch Fyne, via the Crinan Canal, and in the spring was still lying at Ardrishaig with engine trouble. Later in the year I was told that she had struck a mine and sunk off Kintyre.'

Some ships are just plain jinxed – ask any sailor! But with the *Dove* safely off his hands and money in his pocket, Maxwell had a perfect chance to make a major step forward on a number of fronts, including the purchase of a practical, tough, multi-purpose craft as his main working vessel. But that would have meant another fishing vessel, and perhaps after his experience with the *Dove*, Maxwell was understandably jaundiced in his view of second-hand fishing vessels. In his employ he now had good men like Tex and Bruce who knew boats inside out, and who could undoubtedly have found him a suitable boat. But for Maxwell, being the man he was, a mundane craft like a functional fishing vessel was probably anathema in any case, as it wouldn't fit in with his self-image. So instead he decided upon an ex-naval Harbour Defence Motor Launch (HDML) which, in keeping with his desire to create a warrior-like air about his shark hunting enterprise, he would rename the *Sea Leopard*.

Maxwell creates an image of joint agreement in his explanation of this surprising choice: 'It was clear, after much discussion and detailed comparison, that the main catching craft should be an H.D.M.L.' It's tempting to ask who else was involved in these discussions and comparisons. Nobody in the history of Basking Shark hunting ever chose a vessel anything remotely like an HDML as a shark catcher, and it would be very hard indeed to imagine pragmatists like Tex Geddes and Bruce Watt going along with such a perverse selection. In fact, Tex was later to remark that during his time with Maxwell:

> We improved our technique through experience and a great deal of trial and error, until we arrived at what we considered to be the perfect gun and harpoons, but not the perfect boat, for we were fishing from a seventy-five foot, ex-Admiralty, Harbour Defence Motor Launch, powered by a pair of 160 h.p. diesels.

What an HDML did have was style, even if its practicality didn't rate highly; and as Gavin Maxwell showed throughout his life in his choice of cars and boats, style outdid substance every time. HDMLs are undeniably handsome craft, with a lean, rakish, purposeful profile, unmistakably a military vessel. Superficially they resemble their far more potent relative the Motor Torpedo Boat (MTB), but with a round bilge hull and twin diesels of around 150 horsepower each, versus the stepped hull and three huge Packard petrol engines (4,200 horsepower in total) of an MTB. At 22 metres length overall and just under five metres beam, the HDMLs were typical military vessels; long and narrow in the pursuit of speed rather than comfort. The draft of 1.5 metres made them a practical choice for use in inshore waters and for their more-typical assigned work, protecting ports and harbour installations. On the positive side, the top speed of around 11.5 knots would prove useful for reaching the hunting grounds quickly, and the power would enable faster towing speeds with the sharks lashed alongside, but that's about the best you can say for such a bizarre choice of craft.

If lacking the practical characteristics necessary for shark-hunting was not enough, it was entirely the wrong boat given the state of finances of the company at this stage, where only one substantial vessel could be afforded. The replacement for the *Dove* should have been a multi-purpose vessel, probably a fishing vessel similar in size and style to Watkins's catchers, capable of carrying bricks to the factory as much as hunting Basking Sharks. It needed to be economical to operate, maintain and crew, given that Maxwell already had concerns over running costs; even with a vessel as simple as the *Dove*, he'd remarked that 'it was difficult to justify the high running costs of the *Dove* on her carrying work only'. The running costs of an HDML with twin 160 horsepower diesel engines were bound to be considerably higher than those of a small fishing vessel, and an HDML also lacked all of the practical load-carrying advantages of a fishing vessel, like a steam winch, derricks and a large hold. Maxwell claimed that 'the running costs of the H.D.M.L. were comparatively low.' When compared with what – an MTB?

I asked a good friend of mine, Ronnie Dyer, for his view of the choice of the *Sea Leopard* as a suitable platform for shark hunting. As for many years Ronnie skippered an ex-HDML (the M/V Etive Shearwater) that had been converted to a passenger ferry to run between Arisaig and the Small Isles of Muck, Eigg, Rum, Canna and Soay, I felt he might be best placed to give a coherent opinion: 'Madness. The HDML was a great boat, but far too light, very rolly and prone to rot. A fishing boat would have been a much better choice.'

But the choice had been made, for good or ill. Frustrated at the difficulties involved in buying an ex-service HDML at sealed bid auction, Maxwell sought the help of the Scottish Home Department, who helped him arrange to buy an HDML by private treaty from the Admiralty, securing the boat for nearly £4,000 (c. £160,000 in 2017). It is perhaps worth considering that £4,000 would have bought three practical ring-netters like the ones that Watkins was successfully using, and put guns on all of them. Simple, tough, low-running-cost, easily repairable boats that could do everything he needed, and could be converted back into profitable herring fishing vessels in a few days to allow them to work in winter or when sharks were in short supply in summer. It is tempting to ask how differently things might have panned out for Maxwell if he had made that choice.

The work at the Soay factory continued, during which time Maxwell explored many avenues in a bid to find some further profitable enterprises for the factory and staff to be engaged in during the off season. Many potential options were examined, such as peat drying, seaweed drying, a lobster pond and stone quarrying, but all of these were discarded as being too capital intensive or unprofitable, no doubt due in part to the remote nature of the island. Yet no mention is made of the one obvious alternative source of income that most of the other hunters turned to or never completely relinquished – herring fishing – which a more adaptable choice of catcher boat could have easily accommodated. But slowly the construction project began to take shape.

Then the weather gods decided to intervene, sending down a cyclone to obliterate the part-built factory. Maxwell, who had been down in London's docklands inspecting HDMLs, rushed back to a scene of utter devastation. The cyclone which had reached 120 miles per hour nearby had 'handled much of it like a card castle. The spot estimate for repairs left me silent; moreover, we now had no boat to act as carrier.'

This must have been a terrible time for Maxwell, with everything either being repaired or awaiting delivery: boats, guns, factory equipment. Yet on paper all was in order. The *Sea Leopard* was to carry the Oerlikon gun, and the *Gannet* was to be fitted with a muzzle-loading whaling gun. Harpoons, ropes and all other gear were being selected to reflect the lessons learned to date, and special permits had been obtained for supplies of vital equipment such as expensive yacht manila rope, to cope with the enormous forces of a struggling shark.

But nothing had been tried in practice, and they were woefully prepared for the simple, practical difficulties that would present themselves on a daily basis when it came to both hunting and handling sharks.

The first of these problems presented itself once the first shark had been towed to Soay Harbour. If it was low tide even the shallow-draft *Sea Leopard* would have been unable to cross the boulder bar that blocks the entrance. To get around this, Maxwell had bought a 4.5-metre-long power-driven boat called the *Button*, to tow the carcasses over the shallow bar and up to the slipway at the factory. There they would be floated onto a railway bogie-truck on rails to be hauled up by steam power onto the concrete cutting-up 'stance' for dismemberment into the various raw materials for processing, starting with skinning the animal:

Next the liver would be removed, cut up, and put into the barrels of the oil-extraction plant, to each of which led a steam pipe from the boiler. Then the fins and tail would be removed and placed in tanks for the extraction of glue liquor. The vertebrae would be set aside to dry for later shipment in bulk for manure. The flesh would be cut up and put into the icehouse, of which the concrete cutting-up stance was the roof. All the suitable residue would then go through a plant for conversion to fish meal. This plant, which, like most of the factory components, never fulfilled its function adequately, consisted of a mincer, a press, and an eighty-foot tunnel filled with trays moving on rails, through which a fan blasted hot air from the boiler furnace.

Here was the ivory hunter commercializing the carcass in the jungle. I was a convert, and, like most converts, I was beyond reason. We had no separate carrying ship, and if any of these components were to break down, the catchers must leave their work and sail to the mainland for spare parts and replacements.

Finally, the *Sea Leopard* arrived from the Clyde in late April. Maxwell's true reasons for choosing such a craft may be gleaned from his remarks on her arrival:

I went on board her in Mallaig harbour with very different feelings from those with which I had greeted the *Dove* a year before. She had cost nearly four thousand pounds, and she seemed to be worth every penny of it. Lying among those fishing boats, she was like a greyhound among bulldogs, seventy feet long, sleek and graceful, and with Admiralty written in every line of her. The engine room was amidships, below bridge and wheelhouse; it was twenty feet long, with a high deckhead, and the twin hundred-and-sixty-horsepower Gleniffer-Diesel engines gleamed with copper, chromium and fresh paint. The crew's quarters were for'ard, and the afterpart of the ship contained the officers wardroom, minute but with all the comfort and fine fittings of an expensive yacht. I went to live on board her at once, and for two years I regarded that cabin, ten feet by ten, as both my office and my home.

Ominously, Maxwell makes no mention of a thorough inspection or professional survey of the *Sea Leopard*, although he could have had one carried out during his time down in London at the beginning of the year. Perhaps she was surveyed, or maybe he didn't think it necessary for a craft that was less than five years old and Admiralty maintained. But given that her purchase represented a large part of the money that he had just raised as capital, one would hope that she was thoroughly inspected and approved as fit for service – and not just bought sight unseen, again. Time would tell.

With things finally gaining momentum, the *Sea Leopard* receiving new paint and the *Gannet* on the slip being overhauled, Maxwell assembled his new crew, including Skye man Dan MacGillivray as mate. Dan was Bruce's brother-in-law, and 'a really efficient seaman, quick in decision and imperturbable, one of those men whose very presence inspires confidence.' Another excellent choice of crew member. The temporary engineer stayed only long enough to nearly wreck one of the *Sea Leopard's* engines by allowing it to run dry of oil, and raid Maxwell's drink cabinet. Tex and a young Soay man, Neil Cameron, made up the remaining crew of the *Gannet* – the fleet was nearly ready to go.

Then Watkins arrived in Mallaig, and (as described in Chapter 5) put the cat amongst the pigeons when the two hunters met. We have already heard Watkins's version of the meeting, and unsurprisingly, Maxwell's take was little different – he simply didn't want competition in what he viewed as his 'patch'. This was especially the case as: 'I knew enough by that time to be sure that the sharks that visited the Hebrides had certain favoured bays, and that if Watkins and I had equally good judgment and information service, our catchers would be in direct competition on the same fishing grounds.'

And sharks were already being reported in Loch Bracadale and Moonen Bay on the west coast of Skye, and Maxwell and his crew weren't ready. They now had an enemy, and he was going to beat them to it – or so it seemed.

Then Geddes and Watkins locked horns after Watkins appropriated the engineer working on the *Gannet* to work on reinforcing their harpoons. Tex Geddes was understandably livid:

'He's got wooden harpoon sticks, and he wanted them bound with metal at the bottom, the same as ours, so the wood won't get blown to bits. He knows there's only one man in Mallaig could do it, and he was on the *Gannet's* engines. Watkins said money was no object, and the b------- just dropped his tools and away back to his workshop. That's business.'
'Did Watkins say anything else?'
'He asked in that sneering voice of his if we thought we'd catch a shark in that silly little boat. I told him that if I ran across him on the fishing grounds I'd let him have a harpoon through the boat's side.'
Tex did not like Watkins, who later became my good friend.

Watkins and his fleet got away first, but reports filtered back that they lay every night in Loch Bracadale, seemingly not working. Maxwell finally made it out after him on 7 May, and in the manner of these things kept the feud on the boil:

Our hackles did not go down that season. Neither trusted the other an inch, and each crew vied with the other in the magnificence of their lies, forgotten failures, and magnified successes, though Watkins and I became more frank with each other. When the crews met throughout the summer, they would ask the other casually where the other was sailing for the next day, and each would be immediately sure that there could be no sharks there. One Sunday late in May I met Watkins' skipper in Mallaig, and he asked where I was going on the morrow.
'Moonen Bay and the Skye shore,' I replied, though we were sailing at midnight for Barra. 'They're thick there.'
'Aye, I've heard that. That's where we're going ourselves.'
Very early the next morning we were stealing through a cold grey sea mist off the Barra coast when I made out the indistinct silhouette of the Perseverance's bows and gun, running on a parallel course two hundred yards away.

Watkins might have got away first, but he immediately ran into a sea of trouble. Proud as he was of his little fleet, he knew only too well that they were untried in many aspects of their operation:

I suffered a momentary qualm when I thought of all the untested equipment we were carrying. The guns, the wooden harpoon shafts, the lifting gear of the *Gloamin'*, and oil plant, all could be described as experimental. Even the men,

though all experienced fishermen, were mostly new to this particular job. None of
the *Gloamin'* crew had ever seen a Basking Shark, and even in the catchers, John
Paterson, myself and Bob Ritchie were the only ones with actual experience of
shark-fishing. There were bound to be teething troubles.

He was quite right. Despite the basic experience that he and the core of his team had
accumulated, they were in many ways just novices. A well-oiled machine they might have
been in the pre-war days, but now they were right back where they had started.

Leaving Mallaig, their plan was to head down to Barra Head, and then steadily work their
way north up the inside of the Outer Hebrides. And in a flat calm sea the fleet had made their
way up that wild and familiar coastline as far as Castlebay, Barra, when a shoal of fins broke
the surface. Watkins felt a surge of excitement, as he noted that there were several shoals in
sight, and all within 12 hours of leaving Mallaig! Looking around, he saw the other catchers
alter course towards the *Paragon*, and noted with satisfaction the bustle of activity on the deck
of the *Gloamin'* as her crew readied the lifting gear in anticipation of the first sharks. It was a
moment of the highest drama: 'The boy of the *Paragon* – each of the catchers carried a boy
of fourteen or fifteen as general dogsbody – a keen student of cinema, said: "This is it."'

Running up on the first shark, Watkins got in a perfect shot, despite finding the gun
clumsy and awkward to handle. Not for the last time would he miss the purpose-designed
and built Kongsberg gun. But after 20 minutes the shark was up at the surface and lashed
alongside, 'the most peaceful shark I had ever caught. "It's a piece of cake," I remarked to the
boy to show that I, too, was a keen film-goer.'

It was a good big shark, a 10-barrel liver, he thought with satisfaction. Then his eye
caught sight of the harpoon sticking out of its back. The shaft had sheared off as expected,
but the head of the harpoon had only partly penetrated the shark at an oblique angle – it
was only just attached by a thin strip of sinew. A less quiescent shark would have broken free
in an instant. This didn't bode well for the new guns and harpoons. Still, that would have
to wait and be proven. They were now approaching the *Gloamin'* to commence the next
experimental phase, lifting the shark aboard. And the previously somnolent shark was now
rapidly waking up to its predicament, fighting vigorously against the ropes that constrained
it, alarming the newcomers aboard the factory ship who had never seen a creature of this
scale before.

'We'd best kill it before we try to lift it aboard,' suggested the skipper, clearly
voicing the general opinion. 'We canna have a great fish like that flapping around
the deck. It'd smash the ship up.'
'You can't kill a shark,' I said. 'But it's quite alright. They're like whales, they're too
big to move when they're out of the water. It won't be able to lift its tail. Come on,
let's get to work.'

Passing two strong wires around the shark's body to act as a sling, and then attaching them
to a block on the derrick, the lifting began. The stays of the derrick grew bar taut and the
woodwork creaked and groaned as the struggling winch lifted the animal slowly aboard. The
strain on the gear was alarming, and Watkins remembered the men in Peterhead who had
assured him that such weights were way beyond the safe capacity of the gear of a drifter: 'The
crew, I noticed, had the same feeling, and unobtrusively they had all stationed themselves
where it would be relatively safe if something broke.'

Steadily the shark came aboard, the struggle all but over, with only the mouth still active, vigorously gasping in a final, pitiful sign of life. 'Even the hard-boiled *Gloamin'* crew, who had seen millions of small fish die in their time, were moved by it. "Glad I'm not a ruddy shark," commented one of them as he turned on the capstan and the dying monster swung slowly in-board and came to rest on the gutting deck.'

Gutting the shark and handling the two giant lobes of the liver into the boiling-tanks took moments, the liver all but filling the 250 gallon tank. The rest of the huge carcass now lay sprawled across the deck 'like a giant kipper', and the crew set to lifting the remnants to dump them overboard. But despite the removal of the heavy livers, this was still four tons of fish, and it was about to bite back, as the body, now clear of the deck, hanging from the derrick, began to move:

> 'Lower away' I bellowed at Jim Smith, but it was too late. With an awful and majestic deliberation, the great carcass wheeled over, slipped out of the sling and crashed through the gutting-deck, smashing through it with a splintering crack and coming to rest on the main deck below amid a jumble of shattered planks and cross-beams. A cloud of dust arose from the debris in the still air as though from an explosion.

Watkins was relieved that no one was hurt, but knew that this was a colossal setback to his plans, as the damage would likely put the *Gloamin'* out of action for many weeks, and the cost of repair would be high, perhaps as much as £200 he thought. And without the factory ship, some new way of handling the livers would have to be devised. Right at the start of their first working season with the new fleet, and with sharks all around them in perfect weather conditions (and an unexpected competitor hot on his heels), this was a serious blow indeed. Watkins decided to put a brave face on it, if only for the sake of the crew:

> 'Obviously,' I said, in as matter-of fact voice as I could command, 'that's not quite the right way of getting the carcass over the side. We'll try sliding it off the next time with a sort of horizontal Spanish windlass. In the meantime, skipper, you'll have to get this blasted fish cut up and dumped over piecemeal. We don't want to go into Castlebay looking as if we've been hit with a six-inch shell.'

And this was not the last of the setbacks. Looking around the horizon Watkins could see all of his catchers, and even saw the puff of smoke from the gun of the *Paragon* that indicated that the hunt was still on. But the *Dusky Maid* was heading towards them at full speed, far too fast for a shark to be tied alongside – what was going on? The *Paragon* swung alongside, and John Paterson eyed the wreckage of the gutting deck:

> 'We had a bit of an accident here,' I said. 'How have you got on?'
> 'Och,' said John in disgust, giving that word a meaning and expression that only a Scotsman can produce. 'There's plenty of sharks about, but the harpoons are damn useless. They won't go in at all. We've fired at four sharks. And the *Perseverance* and *Paragon* are just the same.'

The wooden shafts were too light to hold the harpoons straight when they hit the water. Watkins knew that his own catch was just a fluke. So with fish all around them, but no effective way of catching or handling them once caught, he gave the order to set sail for

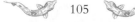

Castlebay, to lick their collective wounds. Not much more than 12 hours into the new season, Anthony Watkins had lost all pride in his fleet, knowing that this complete failure was also a financial disaster in the making, as his running costs at this stage were around £200 per week, and with no chance of any income things were looking very bleak indeed. Looking around at this sorry set of affairs, he had a sudden thought: 'I wondered why I had told the skipper to head for Barra. What on earth could we do there? I walked forward to and told him to alter course south.'

And to add insult to injury, on their way home they passed the *Sea Leopard*, heading for Skye with a shark alongside – the final ignominy.

Chapter 8
In Action at Last

A nthony Watkins might have been down, but he was far from out. By the time the fleet were back in home waters and safely tied up in Tarbert, Loch Fyne, the expert gambler had concocted a sensible defence. He gathered the entire team on the deck of the *Gloamin'* and announced that the *Dusky Maid* and the *Perseverance* would go over to herring fishing, while the *Paragon* and the *Gloamin'* would continue shark fishing with a view to perfecting techniques and equipment.

Slowly and systematically Watkins set about improving all aspects of their shark fishing equipment, starting with the wooden-shafted harpoons that had proved so useless. Harpoon heads of different weights were made, and metal shafts developed. These were tried out not only on live sharks but also dead ones anchored at varying depths in the water. With further modification to the balance and a more streamlined head, he finally declared that these harpoons were 'infinitely superior' to the Norwegian harpoons he had used pre-war, with far greater range and penetration.

Trials were also conducted to find the heaviest line that could be employed without affecting the penetration of the harpoon. By hauling the shark in immediately with the steam winch, the breaking strain of the line could be tested, although some sharks were lost in this way. Eventually a general rule was established that a shark should be played for 15 minutes before any attempt was made to winch it to the surface, a significant improvement on the pre-war time of one hour.

By improving the mounting of the two-pounder gun, it eventually became a capable weapon, although this was made redundant by the long-anticipated arrival of the Kongsberg guns in August. As Watkins well knew, there was no further point in trying to develop a compromise gun when purpose-made weapons were to hand.

Efforts continued on the *Gloamin'* to arrive at a safe and speedy means to get the shark aboard, gutted and back into the sea. The difficulty of getting the slings around the shark to lift the body was eventually solved by having two crew members climb over the side onto the struggling shark to do the job, although Watkins doesn't mention their thoughts on this development, except to mention that 'this could be a difficult and hazardous business.' One wonders what today's Health and Safety officials would have made of this practice....

A simple means was devised of sliding the carcass back over the side after gutting. Not that these improvements made things easy:

> We had two or three hair-raising experiences lifting sharks in rough weather. They swayed about ominously when coming aboard and were apt to hit the deck with a crash which made the whole boat shudder, no matter how carefully the winch-man controlled the steam cock. Sooner or later we should have had a nasty accident, with possibly fatal results to one or more of the crew, and I quickly reached the decision that it was impracticable to use the gutting-deck at all in bad weather. Fortunately I found that a compromise of gutting the shark as it hung in slings over the side was a perfectly feasible operation, though it took rather longer.

The oil plant on the *Gloamin'* worked better than expected, the large triple expansion steam boiler of the drifter providing far greater power than the old shore factory set-up. With a few modifications, the *Gloamin'*'s arrangement could produce as much oil in one hour as the old factory could in one day.

Watkins also commenced a training regime for the new harpoon gunners, keeping them practising until he could be confident that a missed shark in the next season would be a rarity. Meanwhile, the *Dusky Maid* and the *Perseverance* were still gainfully employed herring fishing, a less dramatic but still profitable trade for them.

One evening, somewhere between Arran and the Ayrshire coast, Watkins and his shark fishing team noticed signs of substantial shoals of herring: 'My companions' eyes gleamed with excitement, for we were looking at shoals of herring as had not invaded the Clyde for years. Shoals is a feeble word to describe the massed armies of fish which were manoeuvring in the area.'

Gannets and dolphins were already on the scene, harrying the immense shoals, but there was no sign of the greatest hunter of them all – man – a situation that Watkins soon put right with a phone call to Upper Loch Fyne, where the *Perseverance* and the *Dusky Maid* were making indifferent catches. They wasted no time making their way south and 'wading into these stupendous shoals, and in a few days wiped out my financial losses on sharks up north.'

Naturally, the crew of the *Paragon* wanted to join in on this bonanza, as during the experimental regime for the shark fishing they were on a fixed wage, with only a small bonus for each shark caught, and so far it had been a poor season for shark fishing. If the *Paragon* were to be put back into herring fishing mode the crew would revert to the traditional shares payment method, with one share apiece, which could be worth £20 or more in a good week. But Watkins was determined to see the research programme through until he felt that there was nothing further to be learned. Once he was satisfied that the technique and equipment was as good as it could be, he gave the order to switch over to the herring. As he had designed the shark-fishing gear to be easily removed, this took only a matter of days, and with the exception of the *Gloamin'*, all his fleet were soon busily cashing in on the massive shoals of herring.

Converting the *Gloamin'* back into a herring drifter was a far more time-consuming proposition, so Watkins decided to take a more oblique route to making her profitable and converted her into a herring *carrier*. As he well knew, the Outer Hebrides was a highly profitable fishing ground for herring, with one major drawback that the only market for the fish was in Mallaig, back eastwards across the Minch to the mainland. So, all of the fishing vessels had to stop fishing once they had filled their hold, and set off across 'one of the wildest and most treacherous seas in existence' to sell their catch:

For the small ring-netter this journey in the winter was at best a waste of time and money; at worst it was a dangerous ordeal. After a successful night's fishing, I have seen heavily-laden boats setting out so low in the water that they made my hair stand on end.

Having had the foresight to make the oil-extraction plant tanks easily removable and the fish holds simple to reinstate, the conversion was relatively quickly carried out. Watkins, with his largely Clyde-based crews and good connections in the area, soon found ready takers for their herring-carrying service, and the fleet set off for Mallaig in November. Arriving there, they found the place 'bubbling over with excitement' at the prospect of the winter's herring fishery. As the waters had been very little fished during the war, the herring stocks had rebounded (as they had in WWI), prices were excellent and demand high. A huge flotilla of fishing vessels had gathered in the port from the Clyde and the Scottish east coast to join the local fleet in the hope of making a killing. Fish buyers were everywhere and lorries awaiting the catch blocked every available parking space. The pubs were chock-a-block with fishermen, even the 'aye fous', who in normal times would never have been advanced credit.

Watkins's group of six ring-netters set sail for the Outer Islands in advance of the *Gloamin'*, as she had to take on coal and a stock of fish-boxes, and he arranged with them that she would be in Lochboisdale by midnight. There she would await the ringers and take on their catch before making the return run to Mallaig. Arriving in Lochboisdale before 11 o'clock, she went to anchor, and the crew settled down to smoking and playing cards in anticipation of the first arrival.

Within an hour, the bump of a vessel coming alongside alerted them that the *Mairi Bhan*, one of Watkins's fleet skippered by the former harpoon gunner of the *Dusky Maid* had arrived.

'Good heavens, Alec!' I exclaimed. 'We didn't expect anyone for hours. You filled up damn quickly?'
'Aye, be God, there's all the herring in the world out there. The others'll be along in no time. We'll be out again and get another boat-load before morning.'

By two o'clock the *Gloamin'* left for Mallaig with no space left for even a single extra herring, and with her decks almost awash. Having the carrying vessel had made it possible to double the catch the six ringers could have made without her. And this was just the start:

The rest of the week was a repetition of the first night. The group caught a fabulous amount of fish, and, working out the profits of the *Gloamin'* alone, I found she had cleared £700 (£28,000 in 2017). I was frankly awed by this figure. I had naturally expected the ship to earn some profit during the winter – after all that is what a ship is for – but £700 in a week, that was beyond my wildest hopes. The amazing thing was that there was no reason why it should not continue on that scale for most of the winter. For the first time in my life I felt rich. It was a splendid feeling.

And that was only the profit made by the *Gloamin'*. Undoubtedly the *Paragon, Perseverance* and *Dusky Maid* would have made colossal profits too in this herring gold rush from the sale of their catches. No wonder Anthony Watkins felt rich.

Lochboisdale was to be their hub for the winter. In those days it was a fairly bleak, small village of rough cottages centred around a large upmarket hotel catering to the tourist trade and the salmon fishing fraternity. Watkins had recently married and now decided to bring his wife up from Carradale and install her at the Lochboisdale Hotel. After all, he could now well afford to do so. So he arranged to meet her in Mallaig when the *Gloamin'* came across to discharge fish and coal up.

They set off the same evening for Lochboisdale, straight into the teeth of a gale.

Even though the *Gloamin'* was a very stable boat, she rolled sickeningly. Down below, the crew had organised a sweepstake to see how long it would be before the new Mrs Watkins was seasick, but that was the limit of the entertainment on offer. The visibility was terrible and Watkins was very concerned that they might not be able to pick up the Calvay light that marked the entrance to the narrow channel leading into Lochboisdale. This was no coast to err in your navigation when making a landfall, and he well knew it.

Solid green seas cascaded over the deck and the men in the wheelhouse peered intently into the blackest of nights:

> 'Bit of a nuisance if we don't pick up the light soon,' I remarked.
> 'Aye' said the skipper, and the two hands threw in a few 'ayes' for their own good measure. You never get a Scotsman to pass up the chance of getting in an 'aye', and the more slowly and despairingly he can say it the better.
> We opened the windows, and two pairs of glasses were focused ahead. Icy blasts of wind and rain swept through the wheel-house, destroying in a second the snug atmosphere which we had created during the last few hours.
> 'There she is, there's the light,' said the skipper suddenly, unable to keep a note of relief out of his voice. I peered through my glasses, and now I saw a winking so faint that it was some time before I was certain it was not imagination. I felt like slapping him on the back, but restrained myself in time and merely murmured:
> 'Well done, Skipper.'

As all skippers know, the collective relief at such moments is almost overwhelming, and so it's terribly easy to drop your guard with safety seemingly so close at hand. But making the safe landfall is only the first step – you still have to get in, and so you have to stay on full alert until the anchor is down.

Entering the channel, the raging seas went down and 'we sailed as serenely as though in a canal. It was still pitch dark, except that the faint glimmer of the pier light could be seen in the distance to guide us in.'

Going forward to let the crew in the cabin know they would soon be in, Watkins must have felt supremely relieved, but:

> I was just descending the companionway when there was a terrible, grinding crunch, and the ship lurched to a standstill so suddenly that I fell in a heap on the lower deck. I picked myself up, ignored the sea of startled faces in the cabin, grabbed my torch from the table and hastened back on deck.

Shining the torch over the side Watkins could see jagged rocks so close that he could have jumped onto them. The *Gloamin'* was bolt upright and obviously hard aground. By now the rest of the crew were up on deck, and with good reason:

When a ship hits something good and hard, everyone likes to get out on deck as soon as possible. Drifter men in particular move smartly on these occasions, for drifters are very heavy ships for their size, and if they sink they go down like a stone. Without a word, like a well-trained body of soldiers, the crew doubled aft and began to swing out the lifeboat. It was pitch dark, and for all they knew we might be a long way from land.

'There's no particular hurry,' I called out. 'We're in no danger of sinking. We're already on the blasted bottom.'

They were safe for now, but the question was: how to get her off? And that depended entirely on whether she was holed or not. If she wasn't holed, with luck she could be floated off and repaired. But suddenly an awful thought struck Watkins – it was low water when they left Mallaig, so it should be around high water now. And whether the tide was still rising might make all the difference in the world to the survival of the *Gloamin'*. If it was still rising, and she was undamaged below the waterline, then she might float off at high water. But if the tide was falling, then as long as the wind stayed where it was from the south then the channel should stay calm and she would just settle on her side without further damage and float off at the next high tide. So many ifs …

Checking the *Nautical Almanack* brought no comfort at all; in fact it delivered the worst possible news. The *Gloamin'* had struck at 3.15 a.m. and it was now 3.20 a.m. She had gone aground at exactly high water. There would be no extra water to speak of at the next high tide. Floating her off was not going to be easy. A check of the bilges revealed that at least she wasn't holed. If the channel stayed calm, there might be hope yet.

The lights of a fishing vessel appeared in the channel, and they signalled to her. It was the *Mairi Bhan*, and Watkins related the grim tale to her skipper, and asked him to gather the other ringers to help pull the *Gloamin'* off before the next high water. Taking Watkins's wife on board, the *Mairi Bhan* set off for the village. Watkins and the crew set about lightening the ship, dumping the coal over the side and draining off the water in the boiler; everything to give her more buoyancy when the next flood tide would lift her. As she settled on her side, the cracks and groans as the ship took up the unnatural stresses and strains drove them from the cabin into the lifeboat – just like a shark out of water, a boat does not usually come off well in the war with gravity.

As dawn finally broke, the poor *Gloamin'* 'lay on her side in a few feet of water, helpless and pathetic as a beached whale.' But from what could be seen from the lifeboat, she was largely undamaged except for some iron plates missing from the bottom of the keel. So far the channel had remained calm, although the wind was rising, and they hoped that as the tide rose she might be towed off backwards. As the tide started to flood, the crew moved back aboard the vessel as she began to rise.

In the cabin, the mood must have been strained, to say the least. They would need all of their luck to get the *Gloamin'* off. And one thing they would need above all was for the sea to stay calm. Some hope in the Outer Hebrides.

'The glass is dropping,' said the skipper. It was an unnecessary remark, for the barometer was the first thing any of us had looked at as we came in.

'Yes,' I said, 'I thought we were in for a breeze. I hope to God the wind stays south. The channel'll stay calm then. But if the wind backs to the east we're finished.'

'Aye', said the skipper heavily. 'We'll get the weather report soon. That'll ease our minds.'

I switched on the set, and we listened to the end of a cheerful programme of dance music. It faded, and then came the familiar, well-bred voice of an announcer, maddeningly deliberate and expressionless as he dealt systematically with each sea-forecast area. Finally, he came to us.

'The Minches – forecast for the next twenty-four hours – strong southerly winds increasing to gale force later and backing to east – visibility moderate.'

I switched off, and for a moment there was silence in the cabin. Then the engineer said in a consoling sort of voice: 'Ah well, it might be worse. It said "later". She should at least be up on her keel by then. We may even have her off.'

'Yes' I said, trying to make my voice as unemotional as the announcer's, 'but if we don't get her off this tide, we won't get another chance. That's quite certain.'

During the afternoon the crew ran the anchor out over the stern attached to 70 fathoms of chain in the direction of their planned escape route. The capstan would have to be worked by hand in the absence of steam power, but with the mechanical advantage of the capstan and the combined engine power of the six ringers, that would have to do. Watkins began to feel more hopeful.

Shortly before high water, the ringers assembled and each one passed a heavy hawser aboard the *Gloamin'*. The slack was taken up and the capstan manned, and at six minutes before high water, Watkins gave the signal for full power from the ringers. It was now or never. The water boiled under the sterns of the ringers and the capstan crew gave it their all. Twenty feet was all that was needed.

A small wave lifted the stern an inch or two, and a few links of chain came in. Watkins called for the capstan crew to double their efforts. Another wavelet came in, and this time a few feet were gained. Then six feet. Maybe they were winning? Then a smaller gain of a few feet, then a few inches, and – it was over. The *Gloamin'* was stuck fast, and Watkins's watch sounded the death knell. Seven minutes past four – the ebb had begun and the *Gloamin'* was lost.

For a few minutes I let the ringers and the capstan party sweat away, although I knew it was useless. The *Gloamin'* looked, and felt, safe and secure, but she was doomed just as inevitably as if she was ablaze from stem to stern. As the wind veered and increased, the ripples would turn into waves, and the waves would thunder unchecked through the channel. As the *Gloamin'* heeled over, she would be pounded mercilessly against the rocks. Her ribs would be smashed in first; then her back would break. This would be followed by the collapse of the superstructure, funnel and mast. In a few hours she would be an unrecognizable tangle of wreckage.

Watkins gave the signal for the ringers to cease trying and called the *Mairi Bhan* over to take the crew off with their personal gear and some valuable items from the *Gloamin'*; he and the crew silently made the short journey to Lochboisdale. There was nothing more to be said at this time, before the shock of the loss would sink in.

And that didn't take long. The devastating change of fortunes that had struck his small enterprise in the last 24 hours hit him with paralysing force, as he pondered his loss. The fine ship he had converted for her dual-purpose role with such care and cost was wrecked, and although insured, the insurance would only cover the value of the ship, not the lost profit or the cost of the conversion of her eventual replacement. Whilst he had money in the bank, he knew it wasn't enough to meet those costs. While only hours before 'the future had looked bright, glitteringly bright', Watkins suspected that 'the wreck of the *Gloamin'* might be followed by the wreck of my business.'

So far anger hadn't entered the equation. Up until the abandonment of the rescue, all had been concentrating on the possible salvation of the ship. There had been no time for recrimination. Now he felt entirely within his rights to give voice to his thoughts about the loss, including where the blame should be laid – at the feet of the skipper:

> There was no possible excuse for straying out of the channel. He had been into Lochboisdale before. It was a very dark night, certainly, but if he had been unsure of his position, he should have slowed down, or posted a lookout in the bows or both. All he could do was plead that old excuse 'error of judgment'. But you don't pay skippers to have errors of judgment. Or if they do, well, you are at least entitled to say a few strong words on the subject. Especially if that error of judgment has cost you thousands of pounds, and possibly your livelihood.

Mentally rehearsing a range of bitter, condemnatory phrases, he called the skipper back to him. Even in his anger, Watkins felt no desire to humiliate the man in front of the crew. The skipper waited for him to catch up, undoubtedly suspecting what was to come:

> I drew in my breath and opened my mouth to speak – and then I saw his face. His eyes were red and sunken, he had aged twenty years in the last twenty hours; he looked crushed and bewildered.
>
> For the first time I thought what it meant to him. For a skipper to lose his ship through his own error is the public proclamation of his incompetence, and his professional suicide. He would never again stand on the bridge of his own ship. Never again would he be addressed as 'Skipper', For the rest of his life he would be known as the 'man who lost the *Gloamin'*.'
>
> 'Oh, er, Skipper,' I said. 'When you've got your boys settled in, come up to the hotel for a drink. I think we all need one.'
>
> A look of relief spread over his face; he could not speak but he forced a smile of thanks, and I walked on by myself to the hotel.

Arriving at the Hotel, his wife took one look at his face and guessed the worst. '"What's going to happen to us?" she whispered as she put her arms around my neck.'

Suddenly full perspective was gained again. Boats were lost all the time in the world of fishing. This was a setback, nothing more, and if he pulled himself together he would make something of this worst of throws yet. The ringers were still making money, and he could mortgage them to cover the cost of the replacement boat's conversion. They would be back next season. Time to get a grip, and start the fight back.

> 'Good heavens,' I said as cheerfully as I could, 'nothing worse than another visit to Peterhead. And now let me have a bath and a change, and then we'll have some champagne for dinner. You have champagne when you launch a ship, so I don't see why you shouldn't have it when you wreck one.'
>
> An hour later, in the middle of an excellent dinner, with a half-empty bottle of Krug '28 in front of us, we felt much brighter.

After dinner they were directed to a separate room in the hotel to meet the crew of the

Gloamin', and settled down to neat whisky with beer chasers. Anthony Watkins's essentially decent and respectful gesture towards his fellow men worked its healing magic, and soon all of their troubles were put into perspective. 'Just before I switched out the light that night my wife murmured:"My God, what happens if you ever make money? The celebrations'll kill us."'

Maxwell's luck

After the encounter with Watkins in Mallaig, celebrations were far from the mind of Gavin Maxwell, who was now back at Soay. No more than two weeks into the new season he wrote in his diary:

> Lord ------ was right. Five years and £50,000 to experiment. I can afford neither. We have been at work ten days, the first ten days of what was to have been our first trading season, and every day has revealed more of our ignorance and inadequacy of our equipment. We should be at school to teach this new trade, not trying to earn money. Every day, almost every hour, teaches us a new lesson, but at a cost we cannot pay. I do not think our capital can carry us through this season. We must capitalize in experience every failure and every disappointment – nothing now can prevent this summer being full of both.

The season had begun on the evening of 6 May, when the *Sea Leopard* crossed the harbour bar at Soay with the *Gannet* in tow. She turned north up the coast of Skye past the grim, sheer cliffs, across the entrance to Loch Bracadale and on past Macleod's Maidens and the breaking ledge offshore at An Dhusgeir. In the early morning light they entered Moonen Bay and almost immediately encountered a shark at the surface.

Everything was ready to go, and Maxwell took over at the gun, with the shark's shining fin 'a stone's throw ahead'. It was an easy shot, but the harpoon flew over the shark's back, toppling in the air in the same way they had seen the year before. Re-loading the gun, they approached again for a second try – and missed again.

Maddeningly, they were in the middle of a big shoal, with Basking Sharks breaching all around them. The 'boom' as they crashed back into the water echoed around the bay and made it difficult to think. The bay was full of sharks.

Bruce Watt, Tex Geddes and Neil Cameron jumped aboard the *Gannet* to try a shot at another shark with the whaling gun. Watching from the *Sea Leopard*, Maxwell and the crew saw the little craft close with a big shark, which dived with a great flourish of the tail, but no sign or sound of a shot – perhaps they had run up too close and spooked the fish. But there was obviously something else at work, as the men on board the *Gannet* were struggling with the gun, and after a few minutes returned to the *Sea Leopard*. They hadn't been able to get the whaling gun to fire at all. The firing caps had gone off, but had failed to ignite the main charge of powder, and despite completely removing the old charge and replacing it with fresh powder, Maxwell couldn't get the gun to fire, 'for all the result I could get the black powder might as well have been coal dust.'

In desperation, they rigged two of the harpoons onto pipes, intending to use them as hand-propelled devices, without much hope of success. These harpoons were big and heavy, designed to be fired from a gun, not driven in by hand, and even a titan like Foxy would have struggled with one of them. And although Watt was stocky and enormously strong, he was no Foxy Gillies.

Eight times Watt struck home with a harpoon, and each time the shark broke free within half an hour. Eventually Maxwell decided to leave them to it, and move north around Neist Point to check out how extensive the shoal was – they might as well get some information, he reasoned, to add to their scant sum of shark lore. But the sea beyond the point was wild and white and the *Sea Leopard* 'bucked and jolted sickeningly', and searched around under the 900-foot cliffs of Poolteil without seeing a shark, before turning south again for the comparative shelter of Moonen Bay.

When he returned, Maxwell could see the *Gannet* being towed slowly around in the tiderip off Neist Point, and it looked as if they had at last got a harpoon to hold. But moments later even that hope was dashed as the rope went slack and a harpoon was drawn aboard 'whose barbs had bent backward like an umbrella.'

A freak gale then blew up, and they were forced to shelter beneath the cliffs for the rest of the day, giving them time to take stock of the day's failures. The most important of these was the whaling gun's failure to fire at all, followed by the bent barbs on the harpoons. Bruce Watt had managed to get the harpoons in far enough for the barbs to open, but they had simply bent – the steel was too soft. This was serious indeed, as it would take a long time for new barbs to be fabricated.

The gun was a puzzle, although perhaps it shouldn't have been – anyone familiar with the old saw about 'keeping your powder dry' might guess at just how difficult it would be to keep the black powder charge dry in such conditions on an open boat with the gun only a few feet above the water. They decided to test the Oerlikon gun on the *Sea Leopard* in ideal conditions to see whether the problem with deflection was consistent, and if so, to discontinue using it altogether. Back at Soay they fired 20 shots with the Oerlikon gun at floating targets, where they found that the water was deflecting the harpoons every time, and in any case, after a few shots the tubular shanks of the harpoons were bulging from the pressure and sticking in the barrel. There was nothing for it but to abandon using that gun altogether.

Maxwell was livid, and blasted his anger down the telephone line to the gunmaker in Birmingham, raging about the faulty equipment and the £160 a week running costs he was incurring with no means of catching a shark to meet the bill. He ordered another whaling gun to replace the gun on the *Sea Leopard* and some new harpoons with nickel-chrome steel barbs that (hopefully) would not bend. But the gunmaker could not promise to supply these before the end of the season …

Back aboard the *Gannet* to test the whaling gun, Maxwell was prepared to try everything to get it to fire. Muzzle-loading weapons are essentially very simple devices, but need to kept clean, and the powder and caps totally dry in order to function reliably. After two hours of cleaning the entire mechanism and probing the channels that connect the firing caps to the charge, they decided that there was nothing left to do but test it: 'Let's wake Mallaig up,' said Tex. 'If we're not firing a harpoon we can put in a hell of a great charge of powder without any danger.'

The gun had been proofed for a charge of seven drams of black powder maximum, but they decided to load it with around 20 drams, to clear it out, perhaps. The gun went off like a cannon, to cheers from the watching crew of the *Sea Leopard* and curious looks from the startled fishermen around them. The gun worked; at least, it worked some of the time, and that would have to be better than none of the time. Another trial shot with a harpoon stick and a less generous charge of powder blew the stick about a mile. Maxwell felt relieved. 'Success would begin tomorrow.'

The first catch

The *Sea Leopard* left for Soay and spent a few night hours alongside at the factory before setting off once more for Moonen Bay, with the *Gannet* in tow. Despite the lack of success so far, the crew weren't lacking in spirit and Geddes, in particular, was beginning to get ready for action. Tex stayed aboard the Gannet, declining all offers of breakfast until he had fired many practice shots with the whaling gun, which 'for some personal reason of his own he had now named "Sugan" and which boomed again and again in our wake.' Eventually he was satisfied and they took him aboard.

Everyone was keyed up, and excitedly downed successive breakfasts, thinking there would be no more time to eat later when they got into the sharks. But their high spirits took a blow when the weather forecast came through, as it wasn't promising, predicting a moderate to fresh north-west wind, backing west, strong to gale force later. They knew what that would mean on the west-facing, exposed shore of Skye, and decided to go straight to Moonen Bay to maximise their time there before it became a cauldron.

With Tex and Neil aboard the *Gannet*, the first sharks appeared just as they passed the Maidens, and the *Gannet* was cast off to go straight into action. The first shark was skittish, and Neil Cameron at the helm was having a hard time following it. Tex was 'yelling a steady and stentorian stream of blasphemy to poor Neil at the tiller' and after 20 minutes without getting a shot in, demanded a new helmsman. Maxwell, disappointed not to be behind the gun himself, took the second best option and jumped into the *Gannet*; after five minutes' pursuit and at a distance of only three feet, 'Sugan' blasted the harpoon squarely into the shark's side. Once more, Tex Geddes let out his war cry: "'He feels it!" yelled Tex. "He feels it!"'

Handling the shark as gently as possible to avoid a repetition of the bent barbs, the *Gannet* passed the rope over to the *Sea Leopard*. Glancing to the south, Maxwell saw a small flotilla of vessels heading towards them, and didn't need binoculars to know that a drifter and three ring netters together could only mean one thing – Watkins.

Maxwell swapped places with Watt and told him to get going after another shark as quickly as possible, before Watkins and his catchers could beat them to it. This was easier said than done, as the *Paragon*, *Perseverance* and the *Dusky Maid* all had the legs of the little *Gannet*, and to their astonishment, soon overhauled her, passing the *Gannet* on either side as they charged up to the shark that Watt was after, right under the cliffs. For a moment Maxwell feared Geddes's temper would get the better of him and create an ugly incident, but the *Gannet* simply sheered away. The *Perseverance* ran right up the shark at high speed, fired her gun, and all awaited the result. Which was a miss, right under their noses. Aboard the *Gannet* Geddes did a little war dance, and slapped Watt on the back, as a miss by the opposition was a good second best to scoring a hit themselves. The Watkins fleet, feathers ruffled, moved on.

Aboard the *Sea Leopard*, the delicate task of securing the previously caught shark alongside took the combined efforts of both boats. By now the wind had backed into the west, and it was time to make tracks for Soay. Maxwell shot the shark through the brain with an eight-bore shotgun in an attempt to kill it and they set off very cautiously for home, determined not to lose this first catch of the season. On the way they learned to tow a shark nose first with ropes around the jaws (to minimise drag), the gills and the tail. Partly this was due to concerns over the durability of the *Sea Leopard*, as Maxwell now understood how lightly she was constructed, but had no idea how much strain the thin hull could take. Under the circumstances they did well to make five knots, but that still meant six hours' steaming time to get to Soay.

By the time they were outside the harbour it was dead low water, so there was no chance of entry for at least four hours. Even the *Button* couldn't get out, as the water inside the harbour was several feet higher than that outside. To kill time, they experimented with inflating the shark with compressed air from the engine starting system on the *Sea Leopard* to make the carcass float and so be easier to handle. Eventually they were successful and the shark rolled belly up, with just a few inches of the white belly above the water; another tiny piece of the jigsaw to make the factory successful was in place.

Their first shark was a female, which over time would prove to be the general case rather than the exception. In the four years that Maxwell hunted sharks, the kill ratio was seven females to one male, a simple fact that a highly intelligent naturalist like Maxwell couldn't have failed to recognise might have serious implications for the long-term sustainability of his project. If you kill all the females, then where will your future generations come from?

Concerns for the long-term were far from his mind at this stage, though. What mattered now was the immediate future, and all in all, he felt that things could have been worse. They had overcome many basic technical problems, caught a shark and humiliated the opposition. 'We went to bed feeling we had done a good day's work.'

The Soay factory

Maxwell loved the chance to be 'in action', but sensibly, he decided that at this stage his time might be better spent watching the initial workings of the factory. So the next morning he directed the *Sea Leopard* and the *Gannet* to head off without him, and went ashore on Soay to examine first-hand all of their problems. Which were immediately evident when he watched from the raised site of the entrance as the factory foreman and a boy set off in the *Button* to tow the shark in to the slipway:

> It seemed a very incompetent piece of work. The boat was towing a carcass twice as big as herself, and they got into every possible kind of difficulty. The shark fouled the shore, the propeller and the bar; the boat wouldn't steer and went round in circles. There was a wonderful flow of entirely unacrimonious bad language – they had to shout to each other above the noise of the *Button's* engine, but to us on the shore the engine was practically inaudible and their voices stentorian. The boy was only fifteen, but he could have given a bargee points and left him flattened. All the monosyllables were used in turn, as nouns, adjectives, and adverbs in every conceivable permutation. It was nearly an hour before they had manoeuvred the shark to the end of the slipway and secured him there to wait for the tide.

And then he noticed that the platform and the rails for the bogie-truck to lift the carcass up to the cutting-up stance were too short. They would only be usable on a high spring tide. Maxwell suspected that this had been obvious to the factory staff for some time, but no one had bothered to inform him.

After three hours of struggling to float the shark on to the bogie, all involved were exhausted. But the shark was at last balanced precariously on the bogie and the order was given to start the steam winch and pull the shark up. Due to a further design oversight, the winch operator could not see the bogie, and so had to rely on hand signals from a middleman. The bogie hadn't gone a yard when it became clear that it was going to overbalance and the carcass began to slide. This information arrived with the winch man a second too late, and so the bogie toppled over and tipped five tons of shark into the rocky shallows. Cutting the shark up there was now the only option, and all the staff were working well into the evening, by which time the men were standing in water up to their waists.

Their spirits were buoyed a little by the arrival of Bruce Watt in the *Gannet* with another big shark in tow, one of two that Tex Geddes had shot, the other getting away, but not before inflicting some damage on the *Gannet*. But this new shark was only partial good news, if their experience with the first one was anything to go by:

> We kept on working till midnight on the first carcass. We got the liver into the
> barrels of the oil-extraction plant, the head into the experimental glue tanks, and a
> ton of flesh washed and iced in boxes. When at last I went aboard the *Sea Leopard*
> I had had no food for more than thirty hours, and having struggled all day in that
> mountain of soft cold flesh and entrails, I felt that I never wanted to touch any again.

This was like prison work, and although in times to come Gavin Maxwell may have had reason to complain about the workforce at the Soay factory, any sane observer would have to ask: who could blame them for disliking the work?

The problems with the slipway meant that there was no point in catching more sha.ks, as the factory simply couldn't handle them. The timber required to rebuild the slip was not to be had anywhere by conventional means, so they would have to go back to scavenging the beaches of Rum once more. And all the while, the shoal was out there ...

The rebuilding of the slip took all week, and none of it went smoothly:

> There was dissatisfaction among the factory staff, some of whom said they were
> employed only to cut up sharks; there were constant lists of new equipment to be
> fetched from Mallaig, fifteen miles away; there was a joiner who left on the steamer
> for more tools and did not return, and an engineer's mate who never turned up
> at all. An expert technician, sent out to produce an experimental sample of glue,
> spread poison around him like a upas tree. His job, he said in effect, was to advise,
> not to work; also, he had a sore back, and wouldn't be a labourer for any man. The
> men we needed to rebuild the slip must be taken off to do the manual labour of
> his experiments. He was a nagger, a grumbler, and a hyper-trades-unionist, an arch
> enemy of development and experiment. I could barely keep my hands off his smug
> face as he prophesied our total failure.
>
> I replied that it wasn't my crew's job to haul wood and build slips, nor mine, but
> no-one made a fuss; all were, on the contrary, anxious to help; that I had synovitis
> of the right ankle, a duodenal ulcer, and an enlarged heart, and that he was likely to
> survive me.

If Maxwell hadn't before fully recognised the difficulties he would encounter building a factory on such a remote island, he certainly did now. And it wasn't simply the case that all of the Soay men were lazy and/or incompetent. Maxwell knew that the best men were already employed either at Powrie's salmon fishery or as independent lobster fishermen. A few honourable exceptions apart, he was simply left with the rejects, and the men sent over from Skye to work 'only swelled the numbers as dummy figures upon castle battlements did in the Middle Ages.' He realised that what he needed most was a foreman/manager to oversee the day-to-day running of the factory, and keep the men at work. But in their current financial straits, that post would have to remain unfilled.

Meanwhile, the sharks were still out there, and once the slip was back in action there was nothing to stop the crews from catching them. But it was a desperately cold season, and conditions were difficult, windy and rough. Returning to Moonen Bay, in wild weather, Geddes got in a lucky shot that all but killed the shark outright. Nevertheless, there was still so much to learn and there were still problems with the gear, the misfiring gun and the unpredictable nature of the harpoons foremost amongst them. However, they were learning, and by the next day they had three more sharks to tow to the factory. They were making progress, on that front at least.

Consolidation

Over the next two weeks, they killed another 20 sharks before moving to new fishing grounds. Clearly the basic unit – Geddes, the tough little *Gannet* and even the misfiring muzzle loader – could be a force to be reckoned with. And Gavin Maxwell was at last able to fulfil his evident need for action; the glorious nature of his surroundings and the rhythm of the chase filled his thoughts, pointing to a far better (and infinitely more profitable) use of his real talents in years to come. Vivid descriptions of the sea and its inhabitants and reminiscences of shipboard life suggest that perhaps for him these were the best of days.

Until the weekends, that is, when the *Sea Leopard* returned to Mallaig, to be confronted with a mountain of mail, usually of at least 100 business letters. The most urgent would have to be answered before the vessel left on Sunday night, and it soon became obvious that this could not continue, so a resident shorthand typist was taken on board the *Sea Leopard*.

A sample of one weekend's mail shows that nearly half of these missives were to do with the potential or actual uses of the shark products extracted from the carcass, including a request for samples from a Chinese gentleman in Cairo interested in the dried fibres from the fins, and offering £1,000 (around £40,000 in 2017) per ton for export to the Far East, a portent of things to come. Then there were the myriad official applications and government permits required for rationed items essential to the operation of the factory, before the miscellaneous notes from journalists, people seeking employment and the halt and the lame looking for shark oil as balm for their ailments. And there were, of course, highly unwelcome letters such as the one from the building contractor, confirming that the icehouse was useless, with the suggestion that it be converted into a salting tank to pickle the flesh. As Maxwell simply noted, 'This was a serious blow.'

It was scarcely surprising that he had an ulcer. The level of responsibility involved in any business venture, particularly such a precariously funded one, is massive. And if the struggle is not just financial but also physical, then the stress that accompanies the responsibility has floored many a fitter man than Gavin Maxwell. His duodenal ulcer caused him serious pain and required a special diet that was hardly to his taste either. As Tex Geddes recalled to Douglas Botting in his biography *Gavin Maxwell a life*:

> He was bad-tempered sometimes, well, we were all bad-tempered. But you see he
> had a hell of a bad gut – duodenal ulcer – and I'd have to feed him on sago and
> cut out bacon and eggs and things like that, and he'd get bad-tempered over that,
> though you could hardly blame him. Sometimes when the buzzer went on the boat
> to signal a shark sighting, he found he couldn't straighten up, he'd be doubled up
> with pain, so he'd pour half a bottle of whisky into himself and *then* he'd
> straighten up!

But then there were the days when the sun shone and the sea sparkled, and the myriad seabirds of the west coast delighted Maxwell's painterly eye, and shone a little light into the gloom. Sometimes even the sharks turned up to throw a tentative lifeline to the ailing owner and his struggling enterprise.

Despite the information given to him by his earliest informants, the Sound of Soay had never proved to be a favoured locality for Basking Sharks. Leaving Soay Harbour for Moonen Bay as usual, no one was looking astern down the Sound as a result, but Geddes was in the *Gannet*, being towed as usual behind the *Sea Leopard*. Geddes had taken to using the east coast fishermen's name for the Basking Shark, 'muldoan' and Maxwell was suddenly aware of Geddes yelling from the *Gannet*, 'muldoan!' and pointing behind him at a sizeable group of sharks well up on the surface.

Jumping into the *Gannet* to steer for Geddes, Maxwell approached the shoal with caution, unwilling to risk spooking one shark for fear it would cause the others to dive, too. The shark he steered for swung across his bows at the last minute, and as Maxwell took evasive action to avoid a collision, he looked over the side to see: 'Down there in the clear water they were packed as tight as sardines, each barely allowing swimming room to the next, layer upon layer of them, huge grey shapes like a herd of submerged elephants, the farthest down dim and indistinct in the sea's dusk.'

With so many fish to choose from, how could they fail to make up for previous failures? The answer to that was 'Sugan', which repeatedly misfired, despite firing perfectly when not aimed at a shark. By late afternoon the last of the huge shoal had slipped below the surface, and the *Gannet* had taken four sharks, a successful day by their early standards, but with properly functioning equipment and improved technique, Maxwell knew they could have taken 20.

However, even when they'd towed the sharks the short distance into Soay Harbour, the battle wasn't over, and the first shark they had harpooned nearly got away, 'waking up' just as they were passing the tail sling ashore. After a brief but successful skirmish they managed to haul the shark to the shore, where the factory workers recounted even the next morning that shark was still alive and kicking.

As Maxwell well knew, though, 'alive' in a Basking Shark couldn't be related to what it would mean in another species:

> We found later that there was no way in which local muscular movement could
> be stopped quickly. If the brain was blown out with a shotgun it had no apparent
> effect; even the severing of the entire forepart of the head, from a point several
> inches behind the brain, sometimes produced no change for several hours. One
> Billingsgate firm to whom we sent samples of flesh wrote jocularly to say that they
> had asked for the sample to be fresh, not alive. The blocks of flesh, each no bigger
> than a large book, had been twitching in a disgusting way when the cases were
> opened in London, and had continued to do so for half a day afterwards.

But if the big shoal in the Sound of Soay had given them a glimpse of how bountiful the shark fishing could be, it was to be only a fleeting impression. The simple truth was that the shoals never stayed static, but moved on with the tides and the plankton, or set off to forage elsewhere once the 'patch' of plankton was exhausted. So unless you had the capability to stay with the shoal over a period of days, and the sharks had the courtesy to remain at the surface, then you were almost bound to lose the shoal. In the early northern summers, the hours of darkness were very few indeed, so the likelihood of losing a shoal due to lack of light was much reduced. But any need to return to shore, or the arrival of bad weather, would wreck your chances, and it could be weeks before another shoal was located – if you were lucky.

But as we've seen, Maxwell was anything but lucky, and after a few more days fishing in Moonen Bay and around the Dhusgeir, an engine problem sent them back to Mallaig for repairs; by the time they were back in action the weather had taken its usual Hebridean turn for the worse. The very few sharks that were about were nervous at the surface and dived before they could get anything like close enough to take a shot. Back in Mallaig, this time to sit out a southerly gale, they heard that sharks had been sighted in Loch Hourn, a remote and gloomy sea loch just to the north up the Sound of Sleat, and so set off the following morning. Inside the Loch all was quiet and still, as if nothing was happening outside, but no sharks were to be found, so they decided to return to Mallaig.

This was easier said than done. Towing the *Gannet*, as she wouldn't have been able to make headway under her own power, the *Sea Leopard* was already hampered, but as she turned south down the Sound of Sleat she met the full force of the wind and weather. With the wind against tide, the whole sea surface was white, and with short, steep breaking waves of around 15 feet they soon encountered problems with the tow, as the two boats alternately saw the rope go slack or come up taut due to the steep wave train.

The towrope was made fast to a hand winch in the *Gannet's* stem, and as the shock loading came on it in the troughs of the waves, it began to pull the foredeck apart. Geddes had shinned out on the deck with a knife in his teeth, ready to cut the rope so that the *Gannet* could turn back for shelter, when the winch pulled out of the deck and Geddes was catapulted into the sea. Now free of the tow, the *Gannet* surged ahead, with Geddes clinging on for dear life. Astonishingly, he had just about pulled himself back aboard over the bow when the *Gannet* was picked up on the crest of a wave and shot forward straight towards the stern of the *Sea Leopard*. Dan MacGillivray in the stern screamed at Geddes to let go, which thankfully he did, just before the *Gannet* slammed square into the transom of the *Sea Leopard* with a wood-splintering crash.

Neil Cameron in the *Gannet* had struck his head on the paraffin tank in the impact and came round to see Geddes in the water, so he immediately started to strip off his oilskins prior to jumping in, thinking the *Gannet* was on her way to the bottom. Finally shouts from the *Sea Leopard* brought him to his senses and he jumped back to the helm to try and steer the stricken craft back to Geddes, only to find the engine was jammed in astern gear. He then tried to unship the mast to throw it to Geddes to give him something to cling on to, but it wouldn't budge. Meanwhile, Geddes had managed to kick off his seaboots and tried to swim to the *Sea Leopard*, but the breaking waves kept driving him back and under. Driven under by a huge wave: 'I was quite sure I was going to drown then – it wasn't frightening, and I could hear and see, only I was seeing big trees in a wind and hearing music and bells, bells forevermore.'

Bruce Watt had started to swing the *Sea Leopard* round, desperately trying to anticipate where Geddes would come up next, but Tex had gone down again, and it looked like he was lost. After what seemed like an age, he reappeared and Watt, exercising excellent judgment, put the *Sea Leopard* right alongside him, and he was soon on board, saved from the jaws of death. Not that he was entirely happy about the manner of his salvation:

'You may think I was unconscious, Major, but I tell you I was madder than a wet hen. First I had it in my mind that it was all somebody's fault, and I wanted to find out who it was and knock the guts out of him. Then I was tired out, had it, finished, and you and Dan were sitting on top of me pumping the water out of me like a fountain, when all I wanted was to be left alone and go to sleep. I could have killed the lot of you. And then someone was trying to make me drink whisky, and I didn't want the damn stuff. I knew it would make me sick, and it did.'

We got Tex stripped and rubbed down and wrapped in blankets in the wheelhouse, and an hour later, when we were back in the shelter of Loch Hourn, the strains of his signature tune, played on the bagpipe chanter, were coming from the wheelhouse. Tex was indestructible.

To the Outer Islands

Repairs effected, the fleet set sail once more. The atmosphere aboard must have been sombre, for the time wasted and the cost of repairs had seen the inadequate capital of the company shrink still further, to the tune of £500. Moonen Bay and An Dhusgeir were devoid of sharks, but a telegram from his friend Davidson (the chief lighthouse keeper at the Uishenish Lighthouse on South Uist) told him that sharks were there, so they shifted their fishing grounds to the Outer Islands, where they stayed for the rest of the 1946 season. In fact, they never returned to the inner shores again, even in the 1947 season, and hunted almost exclusively in three places in the Outer Hebrides: Uishenish, Barra Sound and Scalpay Island in Harris. Maxwell commented:

> If there were sharks about at all, they were always to be found in one of those three areas. Uishenish, and the bay of which the lighthouse rock forms the south headland, locally called Shepherd's Bight, was the favourite of them all, and we hunted sharks there days without number, so that it is difficult for me now to separate one day from another.

However, when they got there, no sharks were to be seen; as these were the days before VHF radio was widely available, it was necessary to send the *Gannet* in to pick up the lighthouse keeper, Mr Davidson, to learn more. He told them that the sharks would surface in the evening, and that all they needed to do was wait and they would see all the sharks they could wish for. He was right, because at 7.20 that evening the first shark was spotted in glass-calm water just inside Shepherd's Bight, and by the time they had fired the first shot, 20 were in sight.

'Sugan' performed all of her usual tricks, and a few harpoons failed due to bent barbs, but they had five sharks alongside by the time they went to anchor in Lock Skipport. Surely they would now begin to turn the financial tide that was threatening to engulf the business? Not so, as the next day proved:

> Early the next morning we sailed for Soay, forty miles away, towing the five sharks to the factory. It was our first clear object lesson of the folly of a shore-based factory to which every kill, no matter how distant, must be towed. To make matters worse, we could not leave the Gannet to go on fishing in Loch Skipport, for neither boat was independent of the other. The Sea Leopard had no working gun, and the Gannet had no winch, and to kill a shark needed the cooperation of both. It was Friday, and it would be Monday before we could be back on the fishing grounds.

The season progressed in much the same way. Brief forays out into the Minch, catching a few sharks, before retreating to one of the established hiding places like Lochboisdale, Lochmaddy or Barra to sit out an endless succession of gales from all around the compass. Not an untypical Hebridean summer, in fact. Progress was made when the gun for the *Sea Leopard* finally turned up, although it had the same disastrous firing mechanism as the capricious 'Sugan', a failing that would not be righted until two years later. A successful adaptation was made by attaching a 40-gallon steel paraffin drum to the end of the rope instead of making it fast to the deck of the catcher. As the shark sounded it would pull out the rope, and at the last minute one of the crew would throw the barrel after it. The barrel would float at the surface, to allow the crew to recover it (and the shark) later, while the catcher would now be freed to chase other sharks. It worked very well. Maxwell and Geddes tried increasing the powder charges, presumably to aid penetration of the harpoons, with less success; Geddes typically decided on putting in a charge of almost three times the amount the gun was proofed for, which blew the gun clean off its mounting into his chest with predictable results. Back at the factory, despite the failure of the icehouse, a market for the salted flesh had been identified for the starving population of Germany via the United Nations Relief and Rehabilitation Agency. Things could have been worse.

But the weather sent the crew mad with 'cabin fever' during the seemingly endless days tied up in a range of grey, wet harbours, waiting for it to improve. An interesting insight into the whole operation at this time came from a journalist from *Picture Post,* John Hillaby, who visited the crew during July 1946 with photographer Raymond Kleboe, and between them they produced a fascinating and evocative report of their visit. John Hillaby was utterly seduced and fascinated by the whole affair from the moment he arrived. Maxwell had sent the *Gannet* across to Mallaig to pick them up, just as a storm blew up, confining them to harbour for the morning until it blew out, with only the company of a bottle of whisky to keep them occupied:

> We drank and we drank and we drank, and eventually Gavin's boatman said, 'Ach, hell, let's get out!' So about two in the afternoon, in an extraordinarily pissed state, we struck out. I don't think I've been in worse seas in my life. There was huge tide-rip going across between Skye and Soay and we were tossed about all over the place. My photographer became very seasick but I stuck it out and eventually we beached on Soay some time during the night – goodness knows when, I had so many drams inside me by this time I had no idea of the time. And the chap at the helm said to his mate: 'Jock, hold the wheel. If God got us across the Sound of Sleat, I'll do the rest myself.'

Meeting Maxwell came as a surprise, too. As he recalled: 'the intricacies of Maxwell's character would exercise anyone's powers. An aristocrat, a naturalist, a bisexualist, a bit of a buccaneer, he was a *very* complex character.' By comparison, Hillaby thought himself 'a very unsophisticated person', but he soon developed what he termed 'a tensile relationship' with Maxwell, and eventually an uneasy friendship came to pass, and a form of mutual respect flourished.

Soon it was time to set sail in the *Sea Leopard* on a shark-fishing voyage. Hillaby was by now well aware that things were far from going well with the business. Maxwell was running out of money and was borrowing heavily from his mother to keep the venture afloat, the crew members were agitated and he even had the impression that they had not been paid. Although he could be hard on the crew, Hillaby saw that they revered Maxwell for 'his niceness, his generosity, his humanity.' But he was astute enough to see that there were serious personality clashes between Gavin Maxwell and some of the key players, noticeably Bruce Watt and Tex Geddes. Hillaby felt that:

> Bruce was worried about Maxwell's attitude towards boats and his tendency to take unnecessary risks.
>
> Bruce Watt was a very steady, rather dour man, and he knew the limitations of the boat and was not prepared to take risks with it, but Gavin was all for derring-do. There were not infrequent arguments between them about whether the *Sea Leopard* could get into such and such a harbour or whether it had enough fuel to reach some island or other. Sometimes Bruce got so frustrated he would say to Gavin: 'All right then – *you* take the wheel.' And then there'd be all hell to pay. In some of those tiny harbours there was barely room to manoeuvre and Gavin would charge in – reverse-forward, reverse-forward – and then BANG! – he'd crash into the end of the jetty so hard he'd almost shake the end off the boat. He was possessed by the *idea,* you see, and the boat *had* to fulfill it.
>
> It was a strange, piratical sort of voyage. All the time I was on board we never got a decent shot, never caught a shark worthy of the name, and Gavin was always having violent arguments with his mercurial harpoon-gunner, Tex Geddes.

But Hillaby saw the other side of things, too; the whales and dolphins, the birds, Maxwell's eccentric friends out in the islands and the sheer adventure of it all. He also saw the extent of Maxwell's isolation and loneliness when the *Sea Leopard* arrived at one of the remote islands to be met by a group of lusty young girls. Only Watt and Maxwell would remain on board as the other crew members disappeared off into the bushes with the girls, 'Gavin in his cabin, working his way through a bottle of Scotch, waiting for his crew to come back.'

Ultimately though, for John Hillaby it was a life-changing experience of the best kind, as he was to later recall:

> For me the entire shark-boat episode had been an amazing period of almost undiluted derring-do – and an illuminating experience that influenced me enormously. It showed me what other worlds there were out there, and taught me that one has to make a decision between merely earning a living and experiencing the things life has to offer.

Anybody with a grain of adventure in their soul couldn't fail to find themselves nodding in agreement with that statement. But – sadly – that wasn't the point of the endeavour; making money was, and it simply wasn't making money, even though they had stemmed the losses. The sales revenue from the season amounted to a little over £3,000 (£125,000 in 2017) a third of which had come from sales of the salted flesh and the balance from the sale of the oil. No other markets for any other products had been established – the time wasted on trying to develop them would have been far better spent out hunting sharks. But by now, having burned their way through over £15,000 (£600,000+ in 2017), and with the recognition that an unspecified amount of further capital was needed for experimentation before they could begin to consider the venture a going concern, there was a real chance that the business was unsaleable. The figures were appalling – who would buy it? And if Gavin tried to sell it piecemeal, the individual items (apart from the boats) were almost worthless. The boats might fetch £5,000 (£200,000 in 2017) he reckoned, but even that seems highly optimistic as, by his own admission, the two main vessels were both pretty battle-scarred after the season's endeavours.

Taking stock

As both Anthony Watkins and Gavin Maxwell discovered the hard way in the 1946 season, there was still so much to learn, with many pitfalls along the way. Watkins, with his flexible approach and modular vessels that had enabled him to switch to herring fishing, was still in the game, even if his capital had been diminished by the loss of the *Gloamin'*. Maxwell's business was bankrupt in all but name. That knowledge, as he sat on the Glasgow train, off to try and find new finance to re-capitalise his failing venture, must have all but broken his romantic heart, as he reported:

> My briefcases full of accounts, receipts, prospectuses, estimated revenue sheets, trade enquiries – everything I could carry with me in justification of my faith in the possible future of my new industry. Mallaig and the shining sea slid farther and farther behind me, and Glasgow, with its drab offices and the sharp cynical brains of those who would smile superciliously as they tore my dream to pieces, filled the whole of my mind.

An apt choice of words. Dreams seldom come true in the cold light of day, and the 1946 season had shown Maxwell just how far from reality his dream was. It would take a miracle to save his business– and he knew it.

One possible avenue came from the avalanche of letters he'd received in the wake of the lurid and often hopelessly inaccurate press that his project had accrued: 'All this had attracted a growing community recently discharged from the Services and restless for adventure. They had savings, varying, usually, between five hundred and a thousand pounds, and were willing to invest it all in a project that offered release from the ennui of peacetime.'

Maxwell would have instinctively recognised that ailment, and undoubtedly sympathised with its victims, but he could not, in all conscience, take advantage of it. By his own reckoning, if he had he could have re-capitalised the company to the tune of £19,000 (£700,000 in 2017). But to do so would have exposed these new investors to the chance that they might lose everything if the company failed, and he couldn't face that responsibility. What was needed was a substantial investor or parent company that had sufficient capital to see them through the experimental stage to full-scale production.

Occasional small sums came their way around this time in the form of boat hire for the making of a film on Skye, and an intriguing connection that would change the face of shark hunting in the not-so-distant future. Maxwell was offered an advisory fee by one Charles Osborne to instruct a company on the west coast of Ireland in the ways of shark hunting. The company, headed up by a Mr Sweeney, had sea salmon fishing rights in County Mayo and was plagued by the damage caused to the salmon nets by Basking Sharks in the summer – he was looking for ways to mitigate their losses. Maxwell was happy to accept a fee of £100 pounds to tell all he knew about catching and marketing Basking Sharks, and in doing so, helped to usher in a new and more successful form of shark fishing than he could ever have imagined.

But still no wealthy backer appeared for Isle of Soay Shark Fisheries, and Maxwell was almost in despair when a sort of salvation finally appeared. His cousin had married the wealthy Duke of Hamilton, who had many personal and business connections with the Hebrides. After a short period of negotiation, the Duke agreed to buy the island and the company through the company that handled his business affairs, the Hamilton and Kinneil Estates, for £13,550 (£525,000 in 2017) in September 1946. The new business would be registered as a subsidiary of the parent company and re-named the Island of Soay Shark Fisheries Limited.

Maxwell would become the Managing Director of the new entity, with a salary of £600 per annum, a shareholding of £500 and a share of profits if targets were met. But saving the dream came at an awful cost – Maxwell, as he had predicted, 'had begun by losing all my own money in it.' His 'Island valley of Avalon', Soay, the boats, equipment and all of his original capital had gone. On the positive side, the capital his nine friends had invested was now safeguarded, and the new company was re-capitalised to the tune of £26,000 (c. £1,000,000 in 2017). He recognised that at this late stage there was far too little time to do anything other than tinker around the edges with the operation for the 1947 season.

But at least they were still in the game, although they were still saddled with the factory and all of its problems. The parent company naturally wanted to see progress on the water before investing more money in any case, and didn't want to see the factory they had just purchased abandoned at the very outset, but Maxwell knew enough now to realise the implications of that policy, and he was aghast at the prospect:

> I dreaded the repetition of those hours and days wasted by the catchers in towing
> dead sharks back to Soay from the Outer Hebrides, but I was not yet ready to
> accept the fact that it was impracticable at the present stage to market any product
> other than the liver oil.

It seems strange that after protesting his disinclination to deal with a multitude of shark products since the outset, he now states the opposite, especially after the experiences of the first season. Surely now was the time to state clearly that the policy should be to work towards a single factory vessel that had the capacity to produce liver oil only and forget all of the other products. If followed through, that policy might have seen the company go into profit after the 1947 season.

And in a further perplexing development, Maxwell made his sole appointed nominee on the board of the new company Gordon Davidson. This was, after all, the man who had advised him to pursue the flawed policy of endless diversification in the marketing of shark products that had played such a devastating role in the downfall of his original company. As Maxwell remarked of Davidson: 'In common with the rest of the world, he had, unfortunately, no experience of the materials with which he was dealing, and in his failures I used unjustly to forget that experiment by trial and error was perhaps as necessary on shore as at sea.'

But there was no time left for further trial and error – the new company had to make substantial progress fast or in no time it, too, would fail. Knowing what he knew, surely Maxwell should have picked someone more in tune with his professed belief that the 'sharks liver was the elephant's ivory'?

As it stood, Maxwell would enter the new season with the same faulty business model, trying to deal with far too many shark products, and all passing through the chaotic little plant on Soay. Of his few assets, the best he had were the excellent crews of the *Gannet* and the *Sea Leopard*, and he was about to lose the core of them. Dan MacGillivray had left to commence the start-up of a boat hiring business based in Mallaig, that he and Bruce Watt would be partners in. Watt would leave later, once they had secured the services of a new skipper. Maxwell knew that this would have serious implications for the successful running of the boats: 'Two irreparable losses: Dan with his soft island speech and calm unhurrying competence; Bruce, often dour, uncommunicative and "crabbit," but infinitely dependable.'

Neil Cameron, too, would leave, called up for his National Service. The loss of these first class men would leave big shoes to fill. The two men selected to replace MacGillivray and Watt, Harry Thomson and an individual referred to only as Jamieson, who had come as a pair from a small cargo vessel, were very much an unknown quantity. It didn't look as if the 1947 season was going to be any better than the disastrous one they had just endured. It must have seemed like a very bleak prospect indeed.

Chapter 9
The 1947 Season

In April 1947 both of the existing shark hunting operations in Scotland were as ready to go as they would ever be.

Anthony Watkins had replaced the lost *Gloamin'* with a bigger, more capable steam drifter called the *Recruit,* in which he had incorporated all of the lessons he had learned the year before. She had larger tanks and a more efficient oil-extraction plant, and the weak spot of the gutting deck had been dispensed with. In future all Basking Sharks would have their livers extracted whilst hung in slings from the derrick, in a bid to avoid accidents and injuries.

All of his fleet had undergone thorough overhauls and been converted back into shark catchers. The Kongsberg cannons were in place, and each boat was equipped with 15 of the new, improved harpoons. The core of the crews remained the same, although the *Recruit* had a new skipper and engineer. Apart from the apprentice crew aboard the *Dusky Maid,* the other two catchers had crews experienced in catching sharks, and Watkins himself would be aboard the *Dusky Maid* to show the new boys the ropes. By the Easter holiday they were ready to leave Tarbert, Loch Fyne for Mallaig.

In Mallaig, the *Sea Leopard* had undergone her annual engine overhaul, and the few improvements that money would allow had been made. The *Gannet's* gun had been re-mounted, so she was back in action.

The price being offered for the oil had continued to climb, eventually to reach a peak at £135 per ton (around £5,000 in 2017), so if ever the two operations were going to make money, this had to be the year. But the weather, as usual, had other ideas.

Watkins's fleet arrived in Mallaig in filthy weather, so bad that crossing the Minch was not an option, to find the *Sea Leopard* alongside for the same reason. Watkins, fed up at this all-too-common occurrence, headed off to the West Highland Hotel to drown his sorrows, where, in the bar, he noticed Gavin Maxwell. The two men exchanged commiserations over the weather, and Maxwell decently stumped up a cheque for £20 to settle the bet from the previous year that he 'Would catch a shark of over thirty feet' – he hadn't.

The two men took solace in whisky, to drown out the awful sound of the wind and the rain, whilst giant seas boomed against the harbour wall. This was some profession they

had chosen. It wouldn't be until early May that they were able to escape the shackles of the harbour.

Bad weather doesn't go on forever, although sometimes it feels like it, and by the time it eases fractionally you're so desperate to leave, you'll go out in almost any weather. Not only that, but Maxwell wanted to steal a march on Watkins and his fleet, and so wasted no time in casting off. But despite clear skies, the wind never dropped below near gale force (Beaufort scale 7) and the *Sea Leopard* covered over 1,000 miles searching the coasts of the Inner and Outer Hebrides without sighting a single fin. When they got back to Mallaig, word had arrived that the summer herring (associated with shark sightings) were finally being caught, and a Basking Shark had been sighted out at the Binch Buoy off Barra in the Outer Hebrides. Maxwell was first away again, followed later by Watkins and his fleet.

Maxwell was soon into sharks, catching four within an hour off the Binch Buoy, but the sharks had not yet arrived in numbers; and the rest of the week proved frustrating and costly in fuel, as they scoured the coast between Barra and Uishenish, South Uist. Nine sharks for the week was less than they might have hoped for, but the hit rate had gone up dramatically, which was satisfying after the previous year's failures.

Watkins, too, was also catching sharks, and was mightily relieved to be doing so, as the cost of replacing the *Gloamin'* had run him badly into debt. The weather conditions remained awful, but under the circumstances they had no option but to go out and look for sharks, uncomfortable and unrewarding though that might be. To their amazement they found themselves catching sharks at a fair rate. The only problem was that the sea was far too rough to risk gutting them alongside the factory ship, and eventually they ended up altering course for the shelter of Loch Bracadale on Skye to undertake that task.

The rest of the northern season was a repetition of those early days for Watkins and his crews. The weather was consistently bad, only allowing them to venture out in the lulls, and the sharks this year were generally well out into the Minch, out of sight of land where the seas were roughest. If the shoal was lost then it might take days before it could be found again. Watkins was getting good at 'reading' the shoals, but even so, it was frustrating work. But the practical man in him was happy:

> Financially the results were nothing to get excited about, but technically they were very satisfactory. My fleet was proving itself to be highly efficient. The gear, both for catching and processing, was first rate. In spite of the bad conditions, which made both gunnery and winching up very difficult, we were getting an average of nine sharks for every ten rounds fired. This was taken as a matter of course by most of the fishermen. Probably only John Paterson and Bob Ritchie, who had been with me on that first experimental voyage of the *Dusky Maid,* aeons back it seemed, really appreciated how hardly this technical know-how had been won.

By the middle of June, though, Watkins decided to return to the Firth of Clyde, to a loud cheer from his crews, as it meant that the men could at least see family and friends on an occasional basis. And the Clyde men didn't much like the northern grounds, in any case:

> The Clyde fisherman spends a large part of his life fishing the Outer Hebrides. In poetry and song it is a place of romance and beauty, but to fishermen it is a place of force-ten winds, strong tidal races and few safe anchorages. I do not think I ever met a Clyde man who was not glad when the season 'up north' was at an end.

However, they had caught 41 sharks in their short season 'up north', despite the weather. If the weather had given them a break though, with three catchers operating effectively and the factory ship capable of handling all they could catch, they could have made a killing (literally). As it stood, without the high prices for oil being on offer at the time, it might have been a much more marginal proposition. They must have fervently hoped that the more sheltered waters of the Clyde would pay off in the second half of the season.

Maxwell, of course, had no option to go elsewhere, tied umbilically as he was to the factory at Soay. But Maxwell was doing well, especially once the weather calmed down enough to help them spot sharks. Out in open water between Barra and Canna they ran into a huge shoal, far bigger than the one they had encountered in Loch Scavaig the year before:

> At one moment we counted fifty-four dorsal fins in sight at the same time, and that was, as it were, only like looking at the topmost branches of a tree that has been almost entirely submerged. We could see the fish down below us in the green water as a practically continuous mass, crossing and re-crossing, ponderous and mighty; only the topmost layer, and not all of them, was breaking the surface.

The weather conditions were perfect, so calm that Maxwell reckoned a Basking Shark could be spotted four miles distant from the bridge of the *Sea Leopard*. But the sharks were as unpredictable as ever. They had shot only three sharks when the shoal went down – not to return to the surface. The next day they found the shoal again, eight miles to the north, once more in flat calm conditions, but it had taken until early evening to do so. Two sharks were killed, but three were lost when the drums attached to their ropes burst under the pressure of being dragged down into deep water. As night fell, they were still shooting at sharks by searchlight and getting no sleep, but at last they were into a big shoal.

And so it continued, with the *Sea Leopard* towing sharks back to Soay, losing the shoal, then finding it again, before finally mist began to form, gradually getting thicker until they didn't dare let the *Gannet* out of their sight. They shot six sharks (losing one) before it closed in entirely. Maxwell confided in a letter:

> We got these five lashed alongside and started very slowly for home. It was very eerie: you know the way writhing white mist is used in films for an effect of mystery and horror – the mist, right down to the surface, was moving and twisting and re-forming, and out of it would appear again and again these great slippery fins, ahead, astern and on both sides. Heaven knows how many sharks we would have killed if we hadn't had to think about towing them home; we should certainly never have had to wait for a shot after reloading. There was enough money round about us to make us all rich for life, and we couldn't touch it.
>
> Next week – if we haven't lost the shoal. This year, next year, sometime …

Bruce Watt had finally left, but by then Maxwell was so taken with the new mate Harry Thomson that 'I did not then realise the magnitude of our loss.' Thomson certainly attacked his new role with great gusto, organising the crew into day and night watches and making Saturday a full fishing day. Maxwell thought he had found his Superman at last. The factory was working flat out, but the overtime pay necessary to keep pace with the higher catch rate steadily ate away the profit of each shark. With the catchers away doing their job, there was no transport available to carry salted flesh across to Mallaig, so it was time to test the salting tank, converted from the old icehouse.

The first load of nearly 12 tons of flesh was loaded into a three-foot depth of brine in the tank. Maxwell noticed with mild concern that the final layer of flesh was floating, which seemed strange, but dismissed his concerns as they had followed Gordon Davidson's instructions for preparing the brine to the letter. By way of a distraction that Maxwell enjoyed, Dr Leonard Harrison Matthews, Scientific Director of the Zoological Society of London and Dr Hampton Parker of the British Museum (Natural History) had arrived at Soay to conduct pioneering work on the shark's anatomy, and you get the feeling that he would have liked nothing more than to have helped them with their studies. But that would have to wait as the big shoal was still out there.

Finding it again wasn't easy, as the wind had got up once more. When they did finally find the shoal, once more right out in open water, things didn't go well, as Maxwell reported in another letter:

> We have had an exasperating week. The big shoal is still out in mid-Minch; the exasperation began on Tuesday, when we lost in the dark the float buoys of two sharks whose ropes had become entangled, and which we had left a mile away while shooting another shark. Next morning, a full gale was blowing, and we could not even look for the buoys, which were fifteen miles out in the open sea. The gale went on blowing right through Thursday, and it was too rough to let us out of Castlebay harbour until an hour or two before dusk last night, when we killed two sharks off Barra Head. We found the big shoal again at four thirty this morning, but they wouldn't remain steady at the surface, and we never got a shot. For sheer temper-trying I know nothing quite so powerful as being among a lot of sharks that won't allow themselves to be shot at.

Only three sharks had been taken at the time the letter was written, but a postscript added that they had taken four more by staying out all night and using the searchlight, so Thomson's new regime seemed to be working.

Buried at the end of the letter was a far more alarming admission, to the effect that the new salting tank was tainted, and that the 16 tons of flesh it now contained would likely have to be disposed of. Sixteen *tons* – what would they do with it?

Further gales had followed the last contact they'd had with the big shoal, and they had no idea where to start looking to find it again. They set sail for Uishenish only to find that no sharks had been seen there, so they continued further north, finally going in to the tiny island of Scalpay late that evening. The next morning they encountered a huge Basking Shark in the tide-rip off the eastern end of the island, the first outrider of a huge shoal that they would attack for the next 18 days, eventually killing nearly 50 sharks. Success at last for Maxwell one might have thought, but no, he felt that they 'could, I think, have killed three times that number if the boats had nothing else to do but kill them and bring them in.' The simple reason was that Scalpay was nearly 80 miles from Soay. And so, just as the catching team finally achieved their full destructive potential, the madness of opting for a fixed base on an island as remote as Soay was once again thrown into brutal relief.

Sharks were seemingly everywhere between the two headlands to the north and south of Scalpay. Within two days they had shot 15 sharks, and Maxwell knew that would entail four days' work for the *Sea Leopard* ferrying the sharks back to the factory, then returning to the hunt. So they decided to drag the carcasses onto the beaches at Scalpay and liver them there;

even that took a full day and involved both boats. And then the next question arose – what to do with the 10 tons of livers? They had no means of storing them.

To their amazement they learned that there was a huge quantity of barrels in a store at Scalpay harbour, left over from an abandoned plan to install a herring-curing plant there. As the barrels were no longer required, Maxwell was able to swiftly organise their purchase at a reasonable price. But having solved one problem, the next arose immediately: how to ship the barrels back to the factory. The *Gannet* was simply too small and the *Sea Leopard* lacked a hold, so the cargo was thus far too great for her to carry without a multitude of journeys that would remove her from the hunt. Transport for the barrels had to be found, and Maxwell decided to hire a 'puffer', one of the fleet of small, steam-powered cargo vessels that were in those days the backbone of small-scale transport throughout the islands. The cost would be around £300 (around £12,000 in 2017), but would easily be justified if they could maintain the current catch rate.

Then the weather broke once more, leaving Maxwell to ruminate on his predicament:

How many whose livelihood depends on the sea have thought that if they could control the wind they would ask no more? What proof of divinity more certain to touch the hearts of those early fishermen could Christ have given than the stilling of the waves? Now, when at last we were in sight of fortune, those winds that we could not still sprang up again, and for all the following week they blew and blew, until the calm, shining sea over which we had sailed so few days before seemed a memory infinitely distant.

As desperate as this ill fortune was, some of the shoal had by now infiltrated the maze of tiny channels and rocky islets that surround Scalpay, and it was just sheltered enough for the little *Gannet* to chase them. Scary stuff, being towed around by sharks in the tortuous, narrow, rock-strewn channels, but far better than being out in the grim tide race off Scalpay. Slow going, as the *Gannet* could only tow one shark at a time, but it still enabled them to kill six more sharks. These sharks then joined their fellows at the sandy abattoir nearby; the once glorious beach now ran with so much blood that the sea was crimson for hundreds of metres from the shore.

Removing the livers on the beach wasn't always plain sailing either, even though the team had developed a means of cutting a 'door' in the shark's flank, that would require only one final horizontal cut to spill the liver out. This was a delicate task:

As the horizontal cut was almost completed the door would begin to sag outward under the enormous weight of the liver and entrails that was holding it in, and one had to jump aside to avoid that ponderous slithery mass as it came rumbling out like an avalanche. Once I was not quick enough in avoiding it, and was knocked flat on my back and enveloped by it, struggling free drenched in oil and blood, with a feeling almost of horror.

Then there was the question they had of removing the harpoons from the bodies. If the harpoon had gone completely through the body, then all that was required was to undo the shackle that attached the harpoon to the wire trace. But if the harpoon was embedded in the vertebrae, or the barbs had become lodged in the mass of cartilage around it then that might take half a day of work sawing away and chopping with axes to free it.

By nightfall, when all of the men were exhausted, the barrels of liver had to be ferried across and loaded on to the *Sea Leopard*, or secured on the shore to stop the tide floating them off. But on the first night they discovered that they had broken loose and there were now barrels floating all over the harbour – some were even on their way out to sea. So that took up a further day for both vessels to round up the unruly barrels. What they would have given for a factory ship.

Finally, the puffer, the *Moonlight,* arrived from the south and the process of loading the barrels into her capacious hold began. She had the capacity for three times the number of barrels they had already prepared, so Maxwell decided to risk keeping her there in the hope of making further big catches, and to assist with the lifting of the sharks. The *Moonlight*, like all of her sisters, was equipped with a powerful steam winch and a derrick for loading and discharging her cargo in remote places, so was ideal for that task. The weather had calmed, so the two catchers were soon back in action, killing eight sharks the first day and a further 12 the next. Maxwell was within sight of success – or so he believed:

> We had killed twenty sharks in the first two days of that week, and it seemed that at last we were going to make the killing of which we had dreamed, perhaps fifty fish in the week. The price of oil had now risen to a hundred and thirty-five pounds per ton, so that if we could land that number we should take nearly two thousand five hundred pounds for the week on the livers alone. The first formal meeting of the board of directors had been called for the following weekend in Mallaig, and I felt that such a catch would ensure the factory ship for which I meant to press.
>
> When I heard the gale warning on the wireless that night I could scarcely believe that we were, after all, to be robbed of our chance. 'Iceland, Faeroes, Hebrides, Fastnet, Shannon; strong to gale south to southwest, visibility good.' It seemed that we were never to have a week of fair weather when the sharks were plentiful, and I felt all the frustration of a child who is denied, capriciously, as it appears to him, the desire of his heart.

Nothing was left but to deal with the last of the sharks on the beaches, load the barrels aboard the *Moonlight* and send her on her way to Soay. One last shark was killed in calm water in the lee of Scalpay and the *Sea Leopard* opted to tow it back to Soay across the Minch in a raging sea that set her rolling so horribly that a sea swept over her and sent water down the galley chimney, putting the stove out. The shark alongside pounded against the hull alarmingly, unnerving them all. It wouldn't be until a later date that they would be appraised of just how well-founded those fears were. But despite curtailing their slaughter of the big shoal, the revenue from their time around Scalpay came to more than £2,500 (£90,000 in 2017), the kind of figure that could not fail to impress the board of the new company – Gavin Maxwell could perhaps justifiably feel that a corner had been turned.

Calmer waters?

Anthony Watkins, now back in the more sheltered Clyde Sea, was soon catching sharks. The weather was good and his tough crews of seasoned fishermen had developed a 'nose' for Basking Sharks. They were able to spot fins in a choppy sea where others only saw waves, combined with an uncanny sense of when and where to anticipate the sharks would show themselves. The installation of VHF radio telephones across his fleet also helped enormously,

enabling the different boats to spread out over a far wider area in the search for sharks yet still keep in touch with each other, calling the fleet in if a shoal was found. With all these factors in their favour they *should* have been cleaning up …

Despite the fact that he now had a solid working fleet with good guns and a factory ship, Watkins had retained his inexhaustible interest in trying new techniques. Knowing that sharks had often been inadvertently caught in fishing nets, he decided to experiment with nets as a means of deliberate entrapment. He discovered that no net manufacturer made nets with anything like the mesh size he had in mind (28-inch mesh), and so when the weather was bad, switched his crews to net-making duty. Many versions were tried and discarded before he arrived at an effective working version: 'It was easy to handle and shoot, and no shark ever got out of it, though when you saw the water swirling and heaving with some titanic underseas battle, you wondered why.'

On the negative side, the net was expensive and after every shark caught, needed extensive, time-consuming repair. But as he was unable to identify any single site where there was a reliable run of sharks that would make netting viable, he decided that 'netting was more trouble and expense than it was worth.'

In Scotland, maybe. But it's worth noting here that the Irish firm that had sought advice from Gavin Maxwell were still keenly interested in the possibility of hunting sharks and Charles Osborne, who had come to stay with Watkins, was most definitely intrigued by the netting experiments. At their thoroughly exposed site at Achill Island, harpooning sharks was often impossible due to the weather, although they could still see lots of sharks a few feet below the surface. Netting might be an effective alternative catching method there, he felt, although he still favoured a harpoon fishery, and he pumped Watkins for as much information as possible on his range of catching methods.

One day at the very end of the season, Anthony Watkins found himself fishing right where he had harpooned his first Basking Shark 10 years before, and where he had first realised that 'you cannot catch sharks without the right equipment.' Looking around the deck of the *Recruit*, he could only reflect that that lesson had been well learned:

> It was a perfect day, so hot that most of us were stripped to the waist. All the catchers were in sight. The *Dusky Maid* and the *Perseverance* both had a fish on their line, and the *Paragon* was heading towards us with a shark on either side. The *Recruit* had one shark in the slings being gutted. Four more were awaiting their turn in a queue along the starboard side, endlessly heaving, turning and twisting on their tail-ropes. Around us the water was stained with blood; on the surface lay an oily scum as far as the eye could see, while a huge excited crowd of gulls fought over chunks of flesh and liver which littered the decks. From the burst guts of the sharks in the slings a crimson cascade of feeding poured in a steady stream into the water. From the boiling-tanks amidships arose clouds of noisome steam. The *Recruit* wallowed heavily and sluggishly in the defiled waters, for her tanks were nearly full.
>
> To me it was an idyllic scene. Even the fragrance was just right. I wished the old crew from the *Myrtle* could have been conjured up from the past. They would have appreciated it as much as I did.

Watkins loved the sight, sound and smell of money, and knew he was making it. A few days later he sat down to his weekly chore of doing the accounts. The figures looked good:

Sale of oil		£1,430	
Less:			
Wages Basic	£84		
Bonus (29 sharks)	£72		
Coal	£34		
Diesel Oil	£17		
Sundry Expenses	£35	£ 242	
Trading Profit		£1,188	

This is a staggering figure for a week, representing as it does around £45,000 at 2017 values. If there was that kind of money to be made hunting Basking Sharks, then the future looked very bright indeed for Anthony Watkins and his team.

Farther north, Gavin Maxwell now had to face the first board meeting of the new company, a 'long, interrupted and tedious' affair, during which he had to run the gamut of well-meant questions that he had already answered through trial and error years before. But he held a good hand – in the spring he had predicted a season's catch of 50 sharks, and by now they had killed 83, and he finally advanced the proposals that he felt would allow the business to develop from a position of strength. He requested a factory ship and a spotter plane, the former to allow independence from the factory on Soay and the latter to find and keep track of the shoals. Both were immediately approved in principle. One of the directors owned a Tiger Moth biplane, and offered to fly it up the following week, to test the spotting concept before the season was over.

Maxwell and his co-director took to the air on a bright and breezy day and made a circuit from Skye around the Outer Hebrides. Conditions were far from ideal, as the presence of white horses on the dark sea indicated a sea state in which Basking Sharks would more likely be submerged than 'basking' at the surface. Stopping to re-fuel at Stornoway, they then flew south down the Outer island chain in bumpy weather, before finally spotting their first shark in calmer water in the lee of a headland, then a further pair at least two metres below the surface. On the way back that evening, flying over the Inner Sound between Skye and the mainland just north of Raasay, they saw five more Basking Sharks, this time even deeper in the water. The concept clearly worked, and as a result they received the support of the board to use both a factory ship and a spotter plane in the 1948 season. And they ended the 1947 season with a record number of 166 sharks in the bag. Perhaps shark fishing was finally going to pay off for Gavin Maxwell and his enterprise?

But, back down to earth, there was a major problem that remained unresolved – the pickling tank and its noisome contents.

The golden days are over

Yes, the pickling tank. It was now filled with 16 tons of rotting Basking Shark, as the brine solution had proved to be too weak to pickle the flesh. The stench on opening the lid of the tank knocked Maxwell flat – worse, far worse than the smell of the Blackwall crypt that had sent him in search of the sanctuary of Soay all those years before. When he had regained his breath, he once more peered into the darkness to see faintly in the gloom that the surface was alive with grubs:

Those million grubs would become a million million flies; my minds eye saw the island darkened with them as with a swarm of locusts, Avalon eclipsed by the Prince of Flies that I had summoned up.

The grubs were as immortal as the evil dreams of which they seemed a part. We sprayed paraffin upon them; they flinched, as it were, but soon the fumes of the spirit had become displaced by those of the ammonia, and the grubs were as living and virulent as before.

Even quicklime failed to still the movement within the tank, and the nagging demand 'How shall we deal with the tank?' followed him wherever he went.

By now it was autumn, and the search for new vessels took on a far greater urgency. The *Sea Leopard* was found to be totally rotten; so riddled with dry rot that she was a total write-off, with the engines to be sold as working units and the hull sold as firewood. The miracle was that her elegant, military hull had withstood the battering of sharks on those long, rough passages back to Soay. Maxwell sadly collected his belongings from his cabin and moved ashore.

The plan was to replace the *Sea Leopard* with two or three ring-netters and to buy a suitable vessel for conversion to a factory ship, which would create an almost exact copy of Watkins's fleet. But time was short if they were to be ready for the following spring.

Harry Thomson came up with a potential factory ship in the shape of a small coaster called the *Silver Darling,* which he had known for some years as a working herring carrier. At £12,000 (£450,000 in 2017) he reckoned her to be a good buy, the purchase was agreed and she was bought by the parent company. A naval architect was engaged and the drawings for the conversion put in hand. But as usual with Maxwell's plans, nothing went right. The centrifugal oil-extracting machinery that Gordon Davidson had ordered was found to be unusable at sea, as it required a fixed horizontal base. Then the board of the parent company decided that the proposed conversion of the *Silver Darling* would be too costly and would reduce the value of the vessel too much, so she was put back on the market. Frustratingly for Maxwell, the board would not authorise the purchase of a replacement until she was sold. Things were slipping badly. All they had to start the new season was the old hunting gear and the battered little *Gannet*:

> We were not back where we started: we had actually lost ground, and I could not see how the small catching vessels for which we were still searching could ever show a profit alone, even at the enormous price of one hundred and thirty pounds per ton that we were offered now for the oil. The whole venture was already tottering; I seemed to detect that strange agitation which moves among the high leaves of a great tree as the saw bites in to the heart of the trunk.

Thomson proposed that the new catchers should beach the sharks, gut them and put the livers into barrels as had been done at Scalpay the year before, but Maxwell knew from bitter, exhausting experience that this wouldn't work – the catchers had to focus on catching the sharks and they needed a factory ship to deal with the carcasses. He finally accepted one obvious reality:

> At this stage I cleared my mind finally of a long-standing misconception. I recognized at last that the by-products of the shark could have no significance for us for many years; that the liver was the elephant's ivory, and that we should concentrate on this alone.

But the other directors weren't listening, and regarded this 'liver-only' approach as simply a temporary measure. The start date for the 1948 season was fast approaching, and they still had no new boats. Maxwell then proposed that two catchers be bought to accompany the *Gannet* lobster fishing through the 1948 season, with shark fishing as a partially experimental sideline. The main aim of the experimental side would be to devise a cheap, low-tech method of extracting the livers at sea, to circumvent the costly conversion that had been drawn up for the *Silver Darling*. Maxwell was finally coming around to the idea that complexity was no asset, in tune with his Damascene conversion over the by-products. The Board supported his proposal, but reiterated their fascination with the by-products.

Thomson was sent to find two suitable catchers, and Maxwell (continuing to see this from a military perspective) began to look at the potential of the Tank Landing Craft Mark IV as a platform for the floating factory. This looked promising, and there were plenty available at good prices, so the board authorised the purchase of the first one that came up at less than £4,500.

Two catchers, the *Nancy Glen* and the *Maggie MacDougall* were bought in Tarbert, Loch Fyne and overhauled at Ardrishaig. But by the time they had arrived in Mallaig, the *Maggie MacDougall* had developed serious engine trouble, necessitating time-consuming repairs. By the time they sailed in late May 1948, only the *Nancy Glen* had a gun mounted, and promptly, to all intents and purposes, both vessels disappeared out to sea. Maxwell took to spending his weekends in Mallaig in the hope they would put in there, to no avail. Nor were they seen at Soay.

It must have been increasingly clear to Maxwell that his relationship with the board was becoming strained, and by this stage he was playing no meaningful part in the day-to-day running of the business either at sea or on the ground. From this moment on it would appear that trust had entirely evaporated on both sides. Maxwell didn't trust some of the staff, but it would appear that the board did, and indeed placed increasing trust in them, which may account for the way that he was kept in the dark about their whereabouts and activities. He had a suspicion, later confirmed, that one of the board had been in negotiation with their competitors. In this he almost certainly knew it to be Anthony Watkins, who had indeed been in touch with the board of the Isle of Soay Shark Fisheries Limited with a view to merging some of the two firms' activities. In his book *The Sea my Hunting Ground,* Watkins later claimed that Maxwell had asked *him* to consider a merger in 1947, so what actually happened remains unclear. However, whoever and whatever the source of the contact, nothing came of it, although in hindsight it is easy to see why such a merger might have made good sense for both parties in some ways.

The board had clearly lost confidence in Maxwell, and he knew what was expected of him:

> My last action as a managing director was an urgent warning to my successor against relying too far upon certain of his employees. My resignation was finally accepted on July 12, and for me the adventure that had begun four years before was over.

The 1948 season closed early, with only 42 sharks caught, but there had been substantial progress in one respect – a method for removing the liver at sea had been devised. Further long-overdue improvements had been made, such as converting the muzzle-loading whaling guns to Martini firing actions that set off the main charge with a blank cartridge, thus doing away with the perennial difficulty of the antiquated hammer and cap method. A new Konsgberg whaling gun had also been purchased and – at last – the harpoons had arrived in sufficient numbers to see them well into the next season. If Maxwell thought that the board had totally lost interest in the project, this financial expenditure seems to suggest otherwise. But nonetheless, the end wasn't long in coming.

The 1949 season was barely underway when the continuing technical problems with the *Maggie MacDougall* meant she had to be declared unseaworthy, leaving only the *Gannet* and the *Nancy Glen* as catchers. Despite a successful start to the season with sharks caught on the Skye shore and off Canna, Harry Thomson somehow contrived to put the *Nancy Glen* on a rock on a falling tide in Lock Skipport whilst taking on fresh water, breaking her keel. Although they managed to refloat her eventually, she needed major repairs, and so was deemed out of action for the season at the very least. That left (once more) only the *Gannet* available as a catcher, and she couldn't work on her own. Coupled with that, the heady price of liver oil had eventually begun to fall, and was down to £80/£90 per ton. The parent company therefore finally decided that enough was enough and the decision was taken to wind up the Island of Soay Shark Fisheries Limited. The 'great tree' was finally down. It was hardly surprising – in five years of operation the two companies (Maxwell's original company and the later Limited Company) had contrived to lose something in the region of £50,000 (around £1.8 million in 2017) between them. Disastrous wouldn't even begin to describe such an outcome.

At the same time, Anthony Watkins seems to have lost some of his hitherto boundless enthusiasm for pursuing the Basking Shark. His excellent book *The Sea My Hunting Ground* ends on the highest of highs, the profit and loss account previously mentioned. That was the zenith for his operation, when oil prices were at their highest. There is no mention of the 1948 season and beyond, although we know that his group carried on hunting up until 1950. It seems that 1948 had been a difficult season for both hunters, and oil prices had begun their slide from a peak of £135 per ton to £90 per ton as whale oil become more available, and cheaper alternatives such as peanut oil came back on stream post-war. The records of the total commercial catch for Scotland show that 1949 was a very poor year (35 sharks caught), and as falling oil prices had made the hunt in Scotland uneconomical, Watkins sold his whole fleet to an Irish fishing company and sent a Scottish team over to train their crews. As a gambling man he would have known that the odds for future financial success against a combination of falling prices and declining catches were hopeless. He saw the writing on the wall, and moved to secure his capital before the inevitable fall, and moved on to new, less risky ventures.

What went wrong?

This was the first period during which what might be termed an 'industrial' approach to the hunting of Basking Sharks had been pursued in Scotland. It was largely driven by a periodic abundance of Basking Sharks throughout the area, underpinned by a period of extraordinarily high prices for the liver oil post-war. Therefore, conditions for the successful development of either company based on these parameters were as good as they ever could be. If either one, or both, of the companies involved had been really successful in killing large numbers of sharks, the damage caused to stocks might indeed have been serious; both operations certainly had the potential to do so. As it was, the numbers taken were surprisingly small, and would likely have caused little harm to stocks.

From an entrepreneurial perspective, however, both ultimately failed, although in the case of Anthony Watkins that failure was not financially catastrophic. The relative scale of the failures does warrant some scrutiny, though, reflecting as it does the personalities involved.

Watkins was certainly financially astute, and at all times looked to maximise his options.

He developed his business organically and with (largely) his own capital, only borrowing when he could see a clear way to pay it back. He experimented successfully with a range of equipment, and very sensibly opted for a factory ship that freed his fleet up from the need to return to a fixed base.

Following common practice in the fishing industry, he ensured that his fleet had a dual purpose so that they could work throughout the year, and were versatile enough that if disaster struck, his vessels could be quickly and easily converted back to take part in the profitable herring fishery. This factor alone perhaps made the biggest difference between Watkins's operation and Maxwell's. The money accrued during the war years from herring fishing provided the financial springboard for the development of Watkins's post-war fleet, and the conversion back to herring fishing for much of that fleet during the difficult 1946 season saved his business from bankruptcy. Although the financial figures are not available to confirm this, it might well be the case that overall, he made more money from herring fishing than he ever did from hunting Basking Sharks.

Watkins had his problems with equipment, notably the harpoon guns, but that was a case of *force majeure* as the Kongsberg guns that he already had experience with were simply unavailable in the aftermath of the war. If he had been able to equip his boats with new Kongsberg guns and the highly efficient harpoons the Norwegians had developed to hunt Basking Sharks immediately after the war, the outcome might have been very different. Mistakes were made, such as the faulty design of the lifting and handling gear of the *Gloamin'*, but those lessons appear to have been learned, and were not made again. He was unlucky to lose the *Gloamin'*, but lucky that the price of herring remained so high to enable him to offset his losses almost immediately. He was very well supported by his staff and appears to have appreciated that fact. And he had the clarity of thought to get out of the business with his shirt intact when it was clear that the price of oil was only going to go down from there on.

Gavin Maxwell had an altogether more romantic vision of life, a great asset in his later career as a writer, but the last thing that was needed when developing a business, especially one founded on such a tenuous basis and in such a remote place. If ever a project needed a really sound financial mind in charge, the Soay project was it. But as his friend Kathleen Raine remarked of the venture:

> Making money was not Gavin's line of country. Making money was quite
> incidental to Gavin the knight. It was the wild adventure – he was Captain Ahab
> after the white whale, after Moby-Dick. And you can't win against the white whale.

Sickly since childhood, which denied him the opportunity to see active service during World War II, everything about the Soay venture seems to have been about some form of compensatory activity to enable Maxwell to 'go into action'. The choice (and name) of the *Sea Leopard*, the language, all of the 'killing' and 'attacking', the description of the shark as a 'monster' or a 'dragon', all seemed selected to emphasise the warrior code, to enable Maxwell to conduct his own private war.

And what an unequal war it actually was, that ignored the reality that his opponent was no White Shark, bristling with teeth, intent on slaughter. The Basking Shark was a slow, docile, plankton feeding sea-cow that was almost pathetically easy to sneak up on to slam a harpoon into its back. As Mark Cocker pointed out in his perceptive book, *Loneliness and Time,* Maxwell persistently referred to the sharks as 'he' or 'him', even when he knew that the shark he had just killed was female and that the vast majority of sharks he would take would also be females. As he noted, 'Perhaps subconsciously Maxwell wished to see his opponent as male, since this suited his adversarial approach to the shark-fishing business. Killing 'women' by contrast, might in some way have diminished his achievement.'

Cocker classed Maxwell as 'The Overreacher', as he seemed to be unable to keep his plans simple and achievable, and so the whole edifice of his life constantly teetered on the verge of control. The choices he made over the factory site, so far from even the most basic services, the haphazard and sometimes eccentric choice of boats, the catastrophic decision to try to develop markets for all possible products from the shark, the faulty guns, were largely his and his alone. But throughout his account there always seems to be some reason why *it was not entirely his fault* slipped in at the last minute, that someone he had trusted had failed or let him down. Richard Frere, his friend and biographer, noted this proclivity in his sympathetic biography, *Maxwell's Ghost:*

> If Maxwell had a hangover after a spending spree, or was warned that there was insufficient money for something he particularly wanted to do, he would speak darkly of 'mismanagement of company affairs' which relieved him of complicity without actually indicting anyone else.

Not that Maxwell didn't put his personal 'all' into the business. He worked phenomenally hard at his shark-fishing venture, and gave it everything he had. He selected a first class crew of highly capable seamen who matched his work rate until they very nearly dropped. But evidently he aroused mixed feelings even amongst them. Ian Mitchell visited Tex Geddes on Soay shortly before his death to interview him for his book *Isles of the West:* 'What was he like as a hunter?' I asked. 'The truth is,' Tex said, 'that Gavin couldn't catch his own penis'.

Even if Gavin Maxwell was a failure in business terms, it could be argued that his short yet disastrous career did generate *one* major dividend. It forced him to review his life and find another outlet for his energies, which fortunately involved less financial risk and environmental destruction. He turned to writing, and his first book, *Harpoon at a Venture*, was a vivid and highly readable account of his shark-fishing days. The book became a bestseller in Britain and America, and launched his literary career in a spectacular manner. But it aroused critical attention not just from other shark hunters, but also from the legal adviser of the parent company involved in the Soay shark fishery, who pulled no punches: 'The pursuit of natural history and writing are your true metier … commercial business is not your line and therein lies one cause of the failure of the Soay venture.'

Having no involvement or interest in the personal issues behind the company failure, but viewing things simply from a legal and financial perspective, the adviser put forward the three main causes of the failure, as he saw them:

1. Failure of the sharks to appear or failure to find them with sufficient regularity or in sufficient numbers to make any such venture commercially possible in spite of their attractive livers.
2. Failure (with) boats and apparatus…
3. Failure to find really reliable captains and crews for these boats …

Which just about sums the situation up. Point 1 could equally apply to Anthony Watkins and his group, and to a lesser degree, point 2, too. Perhaps even Point 3, depending on whether the loss of the *Gloamin'* should be viewed as an accident or negligence.

Underlying the failure of both companies was flawed thinking, a lack of basic understanding of the ecology and distribution of the Basking Shark in the waters of the west coast of Scotland. It was clear to Anthony Watkins from the outset that there had not always been such a superabundance of Basking Sharks in the Firth of Clyde, and that this was a relatively recent phenomenon. Starting small as he did therefore made sense, in case the numbers encountered at that time were only a temporary opportunity. In contrast, Gavin Maxwell bet his whole inheritance on the premise that the big increase in numbers was the new reality. And, almost unbelievably, although perhaps symptomatic of their time, neither of these two highly intelligent men seemed to have given any consideration to the long-term sustainability of their activities.

On a smaller scale it was true that what little was known of the whereabouts of the sharks and their timing of arrival in the Minch was subject to great variation; and that whilst there were 'hotspot' sites, no knowledge existed to confirm that they could be relied upon except in the short-term. Such uncertainty should surely have given both men serious cause for thought before embarking on the development of such a highly capital intensive business?

Then finally there was always the weather, the final arbiter of all activities on the west coast of Scotland, ever ready to pull the rug from under the hunters just as they were getting into their stride.

Looked at from a completely objective standpoint, the chances of making a financial success of Basking Shark hunting alone were very slim in Scotland, at least on an industrial scale. The Basking Shark had been lucky – so far.

Chapter 10
The Independent Operators

Basking Shark hunting might have proved a financial impossibility for larger-scale operations (with considerable numbers of employees and substantial overheads) when the price paid for liver oil fell dramatically to £90 per ton, but that was still a highly attractive price to smaller, low-cost 'guerrilla style' operators. Disparagingly referred to by Gavin Maxwell as 'the several small freelance fishermen whose shoots grow seedily from the stump of my felled tree', these individuals consisted of men with existing experience, adventurers and successful herring fishermen. And the first and perhaps most effective of these was none other than Tex Geddes, who had left the employ of the Soay company when Gavin Maxwell resigned, with the express aim of setting up a shark fishing venture of his own.

As he described in his book *Hebridean Sharker*, he was well placed to do so, having learned the hard way what worked and what didn't during his years with the Soay company. Having largely worked as a 'harpoonier' using the little *Gannet*, he knew that a well-handled smaller boat could be an effective shark catcher vessel, but would be hampered by her lack of speed and carrying capacity to operate on her own. What he wanted was a slightly larger boat that could catch, handle and carry the liver of the shark, operated by two men to keep costs to a minimum.

Harry Thomson had been involved with a stillborn scheme to charter a boat for the Soay company from an ex-Army Captain called George Langford, and hearing that Tex was looking for a suitable boat, suggested he contact Langford. Langford was then living on the Island of Raasay and owned a 34-foot Fifie called the *Traveller* and was very keen to join Geddes in any shark-hunting venture he had in mind – Shark Fever again? Geddes had bought an old muzzle-loading whaling gun similar to his beloved 'Sugan', so they had a gun, a boat and two men, the most basic of set-ups. Now all they needed was to secure a market.

Geddes contacted Gordon Davidson of the Scottish Fish Meal Manufacturing Company who he knew from his Soay days, to see if he could be persuaded to buy the raw livers, which would free Geddes and Langford from having to buy any plant to convert the livers to oil. Davidson agreed to give it a try, on the understanding that he could not forecast what price they would get for the oil, as they didn't know what percentage of oil the livers would yield, or the price of whale oil (which would dictate the market price for the shark oil). Therefore, the two men didn't know if it would be financially viable or not. In addition, the gun did not have its firing mechanism modified as had 'Sugan' and Maxwell's own gun, and so it blew back every time it was fired, burning Geddes's face, which can't have been pleasant. Then there were always the temptations of the shore to stop them going fishing. In a candid, comedic aside, Tex Geddes admitted that: 'Worst of all, on the few occasions when we did arrive in Mallaig with a cargo of liver, it was almost impossible to get away to sea again: the local rum possessed a stronger attraction than the sharks.'

By the middle of the season, Geddes had removed the dangerous gun from the *Traveller* and he set off for Mallaig station to catch a train to London to get it fixed. A friend stopped him on the way there and told him that the Soay Company was folding and that all of the assets were to be sold off. The Company still owned 'Sugan', the harpoon gun that he dearly wanted. He knew her every idiosyncrasy and she also had the modified Martini firing mechanism that made her far more reliable and less prone to blow back.

Due to the very public failure of the Soay Company, Geddes had correctly guessed that no one else would be interested in the fishing gear, and so was able to buy both 'Sugan' and Maxwell's gun, plus all the harpoons and shafts that could be found before the auction. The remnants of the physical assets of the Soay Company were then auctioned off. Gavin Maxwell was spared the final tragedy of the auction, but was later told:

> It would have broken your heart, Major. It was a free-for all; what wasn't just about given away was taken, and who could blame the takers? Some of the stuff just lay there until it was finally disclaimed and was public property – half the garden fences in Mallaig are held up by your harpoon sticks.

Geddes eventually bought the *Traveller* from Langford, and she proved to be an ideal craft in many ways, strong and seaworthy with a good sized hold and a wheelhouse. The engine was antiquated but serviceable, and remarkably, she even had a power-driven winch on deck. Geddes strengthened the foredeck and added half a hefty car tyre under the gun mounting to absorb the recoil, and rigged a mast and derrick to enable him to fulfil his plan to lift the livers aboard. As crew he employed John McInerney, a young local man who had worked part of one season with the Soay Company. Geddes felt he 'had that certain devil-may-careness that somehow goes with sharkfishing.'

Geddes's plan to remove the shark's liver at sea depended on what the herring fishermen called a 'brailer', an oversized landing net. Once the shark was lashed alongside, the belly would be cut open and the livers floated out to be gathered in the brail net and winched aboard. Once there, the liver could be cut into manageable chunks and stored in barrels. The shark carcass would then be dumped. Having proved it was possible in reasonable weather, Geddes's tiny operation had made a major step forward that would make the little enterprise far more viable. In bad weather, they would still tow the sharks to shore and beach them, but that would be a last resort. Being able to liver at sea was the breakthrough, allowing Geddes and McInerney to maximise their catching potential. As a result, their most successful day's catch was eight sharks, almost as good as Maxwell's best ever day's catch of 12, and that had taken two boats and nine men. Geddes's 'dog and stick' operation, with its tiny overheads, couldn't fail to be profitable as long as the price paid for liver oil didn't collapse.

Geddes was catching sharks and making good progress, as Gavin Maxwell noted: 'He is now a freelance shark fisherman, with all the experience of our venture behind him, and he is doing well in a small way. If I again have capital to invest, it is to his business I shall turn as the foundation stone of a new concern.'

Tex Geddes had no need of it – he was working profitably and successfully on minimal capital with an operation that was the polar opposite of the ill-conceived, bloated Soay project. And to add another bitterly ironic twist, Tex and his wife were by then in negotiation to buy the island of Soay, and were eventually successful. This twist of the knife in Maxwell's wound must have hurt grievously.

Geddes and McInerney had so far had the Minch to themselves, Watkins having sold his fleet to Ireland prior to the 1950 season. But the plan to go hunting in the Hebrides was not unique to them, as George Langford hadn't given up on his dream of shark hunting and had been looking for potential partners, finding a ready audience in Maxwell's erstwhile Mate, Harry Thomson. Langford had a converted two-pounder harpoon gun, but no boat, but he remembered an acquaintance that had one, and made contact with him.

The acquaintance was Patrick FitzGerald O'Connor, and he and George Langford had met when both were in a military hospital at the end of the war. It would seem that they were two of a kind, as they had both forsaken the comfortable lives they had been born into to go and live a far more precarious existence, free from infringement on their personal liberties, in the islands of Scotland; O'Connor on Coll and Langford on Raasay. Langford knew that O'Connor had a boat, the *Cornaig Venture* – hence the call.

The *Cornaig Venture* was what an experienced sailor might call a 'tub'. Let O'Connor describe her:

> Thirty-six-foot-long, with a winch amidships and a wheelhouse aft, the Cornaig Venture has below decks a foc'stle, a hold and an engine-room. Afloat she looks like a small seine-net fishing boat with a rather high freeboard. But, leaning against Arinagour jetty that morning, waiting to be lifted by the incoming tide, she could be seen for what she was: a converted Admiralty harbour launch, built up and then decked in, with no beam to speak of as beam in fishing boats goes, and round-bilged, drawing a mere three foot or so of water.

Once, when waiting to leave the Crinan Canal, the *Cornaig Venture* with her harpoon gun had drawn a crowd of interested passers-by, amongst whom was a man who clearly knew boats:

> In the centre of this crowd stood an elderly naval commander. He stared at us for some time and then his face took on an apoplectic hue. At last, to no-one in particular, he burst out: 'The man's a lunatic! That boat's downright dangerous!' With which he pushed the crowd aside and stalked off.

Perhaps the *Cornaig Venture* was the least likely shark fishing boat ever to grace the Minch, but when there's nothing else on offer, you go with what you have. In any case, she would prove her worth in time.

George Langford was very persuasive, and before he could find a good reason to back out, Patrick FitzGerald O'Connor found himself out in the Minch, with Langford and Thomson aboard, chasing sharks. O'Connor had funded this first voyage with his own savings of £200, but by his own admission that was lost – the voyage hadn't gone as planned and he was financially poorer. But richer, he felt, in every other sense, from the scenery to the sea, the birds, the life, the sharks. As he explained in his book *Shark-O!*:

And there was much more.

For one thing I caught shark fever. We called it shark fever, anyway. A strangely fierce and exulting feeling of anticipation. It wasn't prompted by mere bloodthirstiness, nor yet by thought of monetary gain. No. It had its origin in a more primitive and (using the word in its pure sense) a more animal urge: the animal, masculine urge to hunt, and by hunting – perhaps dangerously – to gratify one's masculinity. Not just to stalk a deer or to cast at a trout or go flighting geese – not just to outwit – but to do all this and more against an opponent who, by virtue of his size, could possibly strike back at a small boat. And to do it in wild remote places. And to do it at sea. And meanwhile to live independent of civilization while shamelessly using its products: self-sufficient, when stocked up with stores and equipment, in charge of one's own destiny. Yes, in all those things lay the prompting of shark fever. In all those things and in the anticipation of having a surging five ton shark heaving the boat first this way and that on the end of a rope the thickness of a mans wrist, in all that lay the prompting of Harry Thomson's little involuntary tapdance of triumph up in the bows whenever one of us cried out 'Shark-O! Plenty big shark!'

O'Connor was hooked. Even though the first, experimental season hadn't worked out financially, he *had* to go again. Harry Thomson had hopes of starting his own operation and George Langford, perhaps forced by financial necessity, had taken a job 'down south'. Finance proved impossible to arrange in time for the new season, but at the last possible minute he was able to borrow £500 from a friend, supplemented with perhaps £300 of his own (£25,000 in total in 2017 money). Seven weeks to go before the season started, no gun, no crew and the *Cornaig Venture* down in the Firth of Clyde. Yes – that's Shark Fever.

The 1951 season

Money was too tight to purchase a Norwegian harpoon gun, but O'Connor was able to track down two old whaling guns. Long hours of chat in the foc's'le of the *Cornaig Venture* with Thomson and Langford the previous season had convinced him that some form of rendering plant was the way forward, with the *Cornaig Venture* as a combined catcher and factory ship. But the sight of an advert for cheap 26-foot metal lifeboats stirred an idea in his mind – why not convert one into a floating factory that could be towed behind the *Cornaig Venture*? Equipped with a simple agricultural boiler and suitably sized tanks to store the oil, that would make for a highly mobile unit that would leave the catcher vessel free to harpoon sharks. 'Cut down handling, have good equipment, use a mobile factory' was O'Connor's mantra. Simple, low cost and flexible, this approach fitted somewhere between Tex Geddes's totally basic two-man operation and Anthony Watkins's more complex approach.

With Thomson and Langford no longer available, O'Connor had to assemble a new crew. And an eclectic mix they were: Ralph, the highly educated son of a wealthy and cultured socialite friend; JB Cunningham, an eccentric engineering genius; and The Bosun, as they called him, an ex-whaler and an 'honest to goodness rogue', who had a taste for the rum. Quite a team, and their mettle would soon be put to the test in the most direct manner. Leaving the gentle waters of the Crinan Canal, they battered their way north to Coll, the *Cornaig Venture* towing the lifeboat, now named *Minnie the Moocher*, behind them.

Straight into trouble. Out by the remote Hyskeir their first shot at a shark blew the gun clean off its mountings with O'Connor still hanging on to it, and both disappeared into the Minch. Pausing only to recover their soaking Skipper, no one at first noticed that the shark was still attached. But a sudden jerk on the rope brought them back to their senses and suddenly the situation seemed far from lost, until one last jerk announced the escape of the shark. Retrieving the harpoon showed why – the barbs had snapped off, as the metal of the harpoons was too soft. No gun, no viable harpoons – a rethink was required.

The disconsolate crew headed for the nearest telephone, on Canna, to try to salvage the dream. O'Connor was able to arrange the delivery of a second gun to Mallaig, and ordered more harpoons, this time made from chrome steel. Examining the gun mount carefully, JB could see why it had failed, as the joints had only been brazed. Properly welded, and with more allowance for recoil built into the attachment at deck level should make it usable, he considered. But O'Connor knew that sourcing the metal for the harpoons would take time, so what were they to do in the meantime? Wracking his brains, he wondered whether there might be some spare harpoons in Mallaig, left over from the failed Soay Company…

And there were, five in all. The gun had arrived quickly, and here they were in the only town in 100 miles that could provide the skills and materials to put their broken project back together. Word filtered through of Harry Thomson, who had apparently suffered a similar accident with his gun, damaging his catcher boat. Mallaig was indeed the best source of news, supplies and services in the Western Highlands. And not to forget, the pubs – The Bosun was now steaming drunk and wanted to fight JB. Time to get back to Coll, thought O'Connor, and, hopefully, shark infested waters.

And what a change of fortune, as they had eight sharks alongside *Minnie the Moocher* before the weather then broke. Back in the shelter of the pretty little harbour at Arinagour on eastern Coll, the air was soon foul with the smell of rendered livers, and the gulls fought over the scraps. Back amongst friends O'Connor had enlisted local help to assist with the cutting up of the sharks: 'Hughie Handy' (who had just been made a Justice of the Peace for Argyll) joined us. "Captain, for a pound a day I'd cut anyone's liver out.'"

All seemed to be going well. From looking at the conversion rate of liver to oil that they were achieving, Patrick reckoned they would reach break-even point at 21 sharks, after which they would be in profit. With the season only just started they had every reason to feel confident they would do so. But then came a surprise – they saw a baby shark, of around two metres length. And from experience gained in his exploratory voyage, he surmised that this was not good news:

> I had known it would come, but I hadn't thought it would come so soon. Each
> shoal seemed to surface in three separate bodies. First on the scene are the obvious
> adults, smug cows in the water who know all the tricks for an idle subsistence,
> ranging from twenty-three to thirty foot in length; then come the adolescents
> ranging from twelve to twenty-three foot; and then the babies. The first two
> batches surface and do their feeding and also practice some very strange habits
> among themselves, and then they must either move on or go northwards or go
> back to the depths; the babies surface, scratch around as it were for a moment, and
> then vanish.

It was time to move on, although they took one last shark right in amongst the skerries in Arinagour harbour. It towed the *Cornaig Venture* around like a toy, nearly ripping the bottom out of her on the rocks, much to the delight of the onlookers on the pier. Time for a new crew member, too, as The Bosun had taken to the demon drink again, and had finally outstayed his welcome. The last they saw of him was a forlorn figure waving from the stern of a steamer heading out of the harbour, with, as they were later to discover, the Skipper's chequebook neatly tucked into his pocket.

It was time to face the crossing of the Sea of the Hebrides, always a potentially dangerous undertaking in a small boat, even more so in a converted harbour launch like the *Cornaig Venture*. Approaching the Outer Islands, they came across a large trawler, hove-to with no one on deck and with all of her gear stowed. This seemed strange, and they wondered if they had disturbed some smuggling operation – in the end they put it down to the bad weather. Arriving in Lochboisdale after a particularly torrid crossing, O'Connor noticed JB finally appear on deck to remark nonchalantly: "'Oh, we've arrived have we?" So beautifully matter of fact. But when he bent to the ropes Ralph and I were entertained to see a large wodge of butter and jam on the back of his left shoulder. That mess on the foc'stle floor… Ralph and I weren't the only ones who'd been upside down …'

They had arrived in the Outer Hebrides. Just in time to meet the competition.

Another boat?

North of Lochboisdale, O'Connor met the ring-net fleet, preparing for the night's herring fishing. A shout from one of the ringers brought their attention to a shark nearby, and it was 'action stations' once more. Seconds after the roar of their gun came the boom of another.

'What the devil was that?' I asked, too busy watching the coils of rope to look round and see for myself.
'It's another boat,' said Ralph presently.
Another boat? There were only two other British boats after sharks that I knew of – Harry's and an old Fifie (which had once belonged to George Langford) run by a man called Geddes – and neither of them had a gun that could make all that noise. 'Another boat?' I repeated, helping Donald to flick the coils as they rushed overboard. 'If another boat has been getting ready for sharks this year, we'd have heard all about it. It must be a whaler, poaching.'

A very good guess. And the whaler had a shark on the line.

Then O'Connor heard a strange, deliberate metallic sound across the water, the unmistakable sound of a big single cylinder diesel engine working the ship's winch, and he knew it wasn't a local vessel – no one used those engines in Britain. In a matter of moments they had lifted the still-struggling shark into the air, slit it open, removed the livers and dumped the carcass. And then their gun roared again – these were no beginners.

But where were they from? And how did they manage to catch and handle a shark so fast? The following day, the crew of the *Cornaig Venture* found out, when they found their nemesis anchored in Loch Eynort.

O'Connor eyed the 'beautiful ship', amazed yet appalled at her perfect dimensions and equipment. The crow's nest for spotting sharks, the raised wheelhouse to enable the helmsman to see the quarry right up until the moment of firing, the spring-loaded blocks for lifting the shark. He saw that 'with the blunt face of workability she combined the tall curves of grace.'

The harpoons were another eye-opener. Barbless, they were designed not to hold in the shark, but to go straight through the body and then lie flat on the other side. No shark could escape from that. The crew were astonished: 'We were talking like yokels at first sight of city delights whose offerings we could never afford.'

Then they saw that the ship was not deserted. Eight or nine faces were watching them guiltily from inside the wheelhouse. On the stern the homeport name had been crudely painted over, in a bid to conceal their origin, but it was easy to read at close range: 'Bergen'. Norwegians! That explained a lot. They were definitely poachers, and not at all welcome.

O'Connor decided to try and psyche them out, with Ralph playing the part of a bored official taking down their details. Four men eventually spilled out of the wheelhouse, and made a great play of tending to their gear. Eventually one of them spoke:

'What is this all about? We are only sheltering.'
As casually as possible I replied: 'I'm afraid you weren't sheltering yesterday. I think yesterday will cost you quite a lot of money.' I waved my hand around the ship. 'And all that gear you have will make a handsome present for some British sharkhunter.'

Round one appeared to have gone to O'Connor. But then around the headland came another vessel of the same type, then – three more! Clearly the Norwegians were making good use of their VHF radios to stay in touch. As abruptly as they approached, they swung back out to sea and were soon over the horizon.

'I wonder', said JB, 'whether that trawler we saw on our way across from Coll is part of the organization. Could it have been the parent ship – the factory boat?'
'Nothing would surprise me now.' And exasperated, I added: 'But *five* of them! And a man in London told me that Basking Sharks weren't expendible in large numbers.'

This was serious. O'Connor decided to call in the Law.

Across on the other side of the Minch, Geddes was taking a different view. Unlike Maxwell, Watkins and O'Connor he was born a fisherman, and as a result felt a solidarity with his fellow fishermen. When competition had arrived in the Minch the previous year in the form of O'Connor, Langford and Thomson, Geddes hadn't been displeased, 'For it is always a comforting thought to know that there is another boat fishing the same grounds.' And now he took the same view with the Norwegians.

Halfway through the removal of the liver from a shark they had taken off Neist Point, Geddes and McInerney looked up to see 'a lovely Norwegian fishing boat with a huge gun on her stem' making towards them. She was the *Barmoy* and her Skipper, Per Mjetanes, and his son Jacob were to become good friends with Geddes and McInerney, visiting each other when in harbour and sharing the same fishing grounds on a regular basis. 'I never considered her as a competitor and we helped each other as fishermen should,' wrote Geddes. And evidently the feeling was mutual, as later the same day the *Barmoy* approached them, knowing that the *Traveller* had no VHF radio, and told them that they had heard via their radio that there were sharks out at Eriskay.

But even Geddes must have wondered whether there should be limits to such solidarity when he crossed the Minch the following week. Approaching Lochboisdale a cloud of seagulls at the entrance of the Loch showed that serious fishing was going on. As the *Traveller* 'drew closer they could hear the WOOMPH of the Norwegian guns and the effect was reminiscent of a sea battle. Up until that time Johnny and I had no idea that there were so many Norwegians in the Minch.' Closing with the fleet they counted no fewer than 30 Norwegian shark-boats. Evidently they had really hammered the shoal, for the sea was red with blood and now the boats were chasing down the few surviving sharks. The *Traveller* was too small and slow to compete in such a race and they only managed to get one shark because it rose close inshore of them out of sight of the Norwegians. Farther south it was almost the same, although there were fewer Norwegian boats. Tex Geddes's brotherly feelings for his fellow fishermen must have been sorely tested.

Nevertheless, Geddes admired the Norwegian craft, which ranged between 50 and 80 feet long. Immensely strong, they were of double-skinned pine construction, and powered by a single cylinder semi-diesel Wichmann engine, giving them a speed of five to eight knots. Most of them were dual-purpose inshore whaling and shark-hunting vessels, the whalers distinguishable by a black band around the barrel-type crow's nest on the foremast. The wheelhouse was high, and had a second steering position on top of its roof to allow the helmsman excellent visibility even at close range. Comfortably appointed down below and equipped with the latest echo sounder and wireless equipment, they were more than a match for the Scottish ring-netters that Watkins had used.

A close inspection of the gun also impressed Geddes. Like his own 'Sugan', the gun was a muzzleloader fired by a blank cartridge, propelling a very different type of harpoon, simply a fluted iron bar with the line attached at the midpoint of the bar. The larger bore gun could fire the harpoon right through the shark, whereupon the tension would come upon it and it would lie flat against the body of the shark, unable to pull back through. Geddes thought his harpoons better, as the barbs meant that the harpoon didn't need to go right through the shark, but would grip in the flesh. But the Norwegian design meant the harpoon could very easily be cut off once the shark was ready to be dumped, unlike Geddes's harpoons that could sometimes get stuck in the cartilage. The Norwegians had even developed a method of absorbing the shock loadings generated when lifting a just harpooned, struggling shark via a system of pulleys mounted on a strong spring and a three-foot square block of rubber attached to the deck.

Geddes was less impressed with the Norwegians' skills than was O'Connor. Geddes reckoned that what they gained in being able to lift the shark out the water almost immediately to gut it was actually lost in the time it took to get the shark securely fixed in a position to do so. The shark would still be full of fight, whereas his method of tiring the shark out before trying to bring it alongside was safer and took less time in the long run. But his impression may have been based on the report he got from his friend Per on the *Barmoy*, which was one of the smaller Norwegian vessels, as the larger ones seemed to be able to handle the sharks far more effectively. Harry Thomson was told in Stornoway that when there were plenty of sharks around, each Norwegian boat was capable of dealing with three or four sharks in an hour. If that were to be multiplied by the 30 boats Geddes had encountered at Lochboisdale then the potential kill rate at five hours' fishing a day could give a total catch per day of 600 sharks …

There was no doubt that these were serious killing machines with the best equipment and highly skilled crews, and that Geddes and his fellow Scottish shark fishermen now faced a 'positive invasion of far more efficient rivals than any we had met yet.'

Geddes later met O'Connor in Scalpay and discussed the situation. Geddes was sympathetic to O'Connor's efforts to draw attention to the poaching going on under the noses of the authorities, but still had sympathy for the Norwegians. And he reckoned that O'Connor's proposed tactic of enlisting the support of the press in his campaign would backfire, as he felt that 'we had had enough publicity and it had gained us nothing.' While O'Connor fumed, Geddes did the pragmatic thing and moved northwards, where the Norwegians weren't fishing. Fishing off Stornoway, Geddes and McInerney found plenty of sharks, and by arranging to ship their barrels of liver back to Mallaig by steamer, found the fishing there to be 'money for old rope.'

The *Cornaig Venture* moved north, too, catching a few sharks in Loch Maddy before eventually finding a big shoal around Scalpay, where Maxwell had enjoyed his best days. There they met Harry Thomson, who was having yet more trouble with his own catcher boat, the gun having blown apart the bows of his boat for the third time that season. Harry was also able to tell them some further background about the Norwegians and how they had come to acquire their skills. He claimed that Norwegian boats had been fishing for Basking Sharks since 1938 off their own coasts and in the North Sea, but numbers had dropped severely, so they had decided to make the long journey to the west of Scotland. Which certainly sounded ominous for the stocks in Scotland, O'Connor felt.

A Fisheries Department cruiser, the *Minna*, finally made an appearance, although this had little effect on the activities of the Norwegians, who were able to play cat and mouse with the Authorities thanks to their VHF radios. O'Connor had *Minnie the Moocher* and her 'boiler transported up by MacBrayne's ferry, and they were able to make a start on the flotilla of corpses in a tiny bay on the south side of Scalpay they called 'The Livering Hole.' Things were going well, even though they were by now down to only three harpoons and the boiler had started acting up. Then the Norwegians caught up with them, 10 of them, appearing right in their hunting ground, no more than half a mile offshore. O'Connor and his crew were now nearing their 60th shark, but that was little consolation, as outgunned and outnumbered, they could only watch as the Norwegian boats cleaned up the rest of the shoal.

One day O'Connor decided he had had enough. A group of Norwegian vessels had taken up residence right in the north entrance to Scalpay, and he directed the *Cornaig Venture* to close with her. As they came alongside he demanded to know who spoke English. Finally, one man owned up that he did and O'Connor delivered a blunt warning to him:

> 'I give you fair warning' I shouted back, 'and you can pass it on to the rest of your friends. If I find one more Norwegian anywhere inside the three-mile limit with a shark, I shall have no hesitation in coming alongside and firing a harpoon below his waterline. Do you understand?'
> I went up to the bows and raised the loaded gun.
> 'I understand,' said the man.

O'Connor decided that as he was getting no support from the Fisheries Department, he would enlist the help of the press and contacted the *Daily Express*, who immediately agreed to send a photographer and reporter by plane. Somebody was interested at last. They decided to go out again and try for a shark to impress the men of the press, but as they cleared the Hamarsay gap they were confronted by nine Norwegian fishing boats, fishing right in the mouth of East Loch Tarbert. This was too much for O'Connor, who seized a small shoulder harpoon gun that they carried for emergency use, and with JB at the wheel, they set off for one of the smaller vessels that had a shark on the line. The last time he had fired the small gun the recoil caused by the heavy harpoon had knocked him through 180 degrees and left him flat on the deck facing the wheelhouse, so he loaded it with up with lightweight marble shot instead.

Several Norwegian vessels cut across their bows, but the *Cornaig Venture* charged on, straight for the boat attached to the shark. At a range of less than 100 yards, he let them have it, amidst a gratifying cloud of smoke. Several men took cover in the stern of the Norwegian boat, and another stood amidships waving his arms. O'Connor re-loaded, but by the time he was ready to fire again, the Norwegians had cut the shark loose and were heading for the open sea. 'JB leaned out of the wheelhouse window. "The Daily Express are going to like this," he said. 'And they did.'

The story made the front page of the Scottish *Daily Express* under huge headlines, accompanied by a series of excellent photographs of the crew, the *Cornaig Venture* and best of all, five Norwegian shark hunting vessels fishing close inshore. The paper kept the story in the air over the next few days, and it did have the effect of making the Norwegian fleet go around to the west of the Outer Islands. A victory of sorts, but other stories of their coolly outflanking the Fisheries cruiser continued to do the rounds. And one of the Norwegians even approached O'Connor to do a deal that if he would just leave them alone, they would pay him £10 for every shark they caught in the area. Evidently his tactics weren't going unnoticed. But as he knew only too well, they were not winning the war. It was all very hard to take, and cast an ominous pall over any future shark fisheries that he, and others, might choose to pursue.

Not that O'Connor and his crew were above behaviour that might harm the stocks. When the sharks finally ran out at Scalpay, the *Cornaig Venture* resumed her search for Basking Sharks, this time turning south along the shore of Harris, then west into the Sound of Harris, a shallow maze of channels, islets and rocks. On a day when the sun is shining, the Sound of Harris is an astonishing place to be, a rich mandala of colours to rival any coral reef in the world. But on a bad day… They were lucky it was a good day. Once in the Sound, they encountered a strange sight: hundreds of Basking Sharks swimming nose to tail in a long line, as far as the eye could see. Swimming very slowly, O'Connor reasoned that they couldn't be feeding, and in any case there was no sign of plankton in the clear water. Perhaps, he surmised, it might be some sort of 'love dance'?

Presently Donald brought us back to our senses.
'It's grand skipper, but which one will we take?'
For a moment my mind recoiled, as from the thought of committing a sacrilege, and I looked down at the huge body in front of our bows with shame. Then exultation took over and I heard myself replying:
'We'll run up the line, Donald. We can be choosy tonight. Let's get a monster!'

They got two, and saw a further strange thing. The snouts of both sharks, one female, one male, were rubbed raw and bleeding from abrading against the sharp denticles on the skin of the animal in front. Yet the sharks seemed oblivious to this painful process, and as the darkness fell, the line still stretched out into the distance, like religious penitents on a pilgrimage. Or lovers lost in the romance of the love dance?

By early July the 1951 season was at an end, as evidenced by the sight of two tiny fins jinking swiftly around at the surface – Basking Shark pups. They hadn't seen any sharks for four days, and the fruitless time spent searching was eating into the profits made so far. Making a mental calculation on the state of the balance sheet, O'Connor reckoned they had grossed around £2,000 (c. £60,000 in 2017), and most of that had been made in one fortnight. Less the crew's share of £200–300 each and the capital costs of around £1,000 it didn't seem bad, but then there was cousin Billy's £500 to pay back. O'Connor might have made £200–300 himself. But if there were no more sharks, then the running costs of £70 per week would soon eat that up. And he was concerned that after the slaughter of the 1951 season, with the Norwegian fleet so active, there might be no prospect for using their expertise and equipment the following year. It had all been a huge adventure but the party was over. Time to go home.

The 1952 season

Evidently Patrick FitzGerald O'Connor had had enough of shark hunting, despite the attraction of further adventures in the Minch. Harry Thomson had also decided to call it a day, perhaps unsurprisingly after the three disastrous episodes with the guns damaging the bows of his launch. Given that both ventures had to cover the costs of four men, their costs were working against them in any case. Especially when compared to Geddes. Although he only dealt in the livers and so lost some profit margin over O'Connor and Thomson who sold the oil itself (which made more money), his running costs were far lower and there were only two of them to pay. But Geddes now knew that if he came up against the Norwegians on the fishing grounds, he lacked the speed to outrun them in a chase for a shark. The answer was a simple one – install a more powerful engine in the *Traveller*.

So by the time the 1952 season was underway, the *Traveller* had undergone a major re-fit and now sported a fine 44-horsepower diesel engine, giving her the pace to outrun the Norwegian boats. Not that Geddes had changed his views on their presence in the Minch – he still liked them and viewed them as friendly competition. But now, with her new-found speed, the *Traveller* was soon well into the sharks and able to compete with the best of them.

Geddes's success hadn't gone unnoticed in Mallaig. As the shark-hunting season coincides with the low season for herring fishing, the ring-netters reckoned that they could fill their herring-gap profitably (as Watkins had) by turning to the sharks between March and July. Several of the ring-net skippers approached Geddes to try and get him to sell them his spare gun, but he was far too smart to do so, knowing that if one gun failed, without a spare his operation was dead in the water.

Jim Manson of the *Margaret Ann* managed to buy two of Anthony Watkins's Kongsberg guns, complete with harpoons, and Manson, along with Peter McLean's boat *Golden Ray*, was equipped to go shark hunting. But the harpoons proved less than perfect with the heavy ring-net boats as they put far more strain on the gear when the shark tried to tow them, causing the harpoons to pull out. Geddes, characteristically, lent them 20 of his harpoons.

This allowed them to start catching sharks straight away, and soon both boats were landing cargoes of liver at Mallaig. Jim Manson was soon to give it up, though, as he hated the damage to his lovely varnished ring-netting vessel. He also disliked the butchery involved, that had caused him on occasion to refer to Geddes and McInerney as 'those pair of bloodthirsty baskets' and the sharks as 'poor, innocent beasts', even though he had often suffered damage to his nets caused by Basking Sharks.

But his half-brother William Manson in the *Mary Manson* continued with the hunt, as did his brother-in-law William Sutherland in the *Jessie Alice*. The Manson fleet reckoned a good day's catch to be 15 sharks. After originally livering the sharks on convenient beaches, they experimented with rendering the livers at sea, using O'Connor's old boiler, but in the end reverted to simply landing the unprocessed livers as Tex Geddes did. If the catch was good, the carcasses were simply dumped, but if the catch was poor they would take the carcasses back to

Mallaig to be loaded on to railway wagons and transported down to Falkirk to be converted into fishmeal. As each carcass fetched around £60, this to some degree enabled the drop in the price of livers to be offset.

However, they had mistimed their entry into the market. During the 1952 season the price paid for the raw livers collapsed from what Geddes saw as 'a fine price' of £48, down to £25 and then a few weeks later, £15 per ton. Whale oil was abundant once more, and the price paid for competitor oils such as shark oil followed prices down. Geddes continued until the end of the season and then hung up his guns. William Manson continued to hunt throughout the season, but the other boats in their fleet transferred over to fishing for white fish. Geddes reported that Manson caught a 'tremendous number of sharks' during the following 1953 season, but as the price of livers remained resolutely depressed, William Manson also ceased hunting Basking Sharks, admitting that it had never really paid due to the high cost of equipping their boats when set against the low price paid for the livers.

The Norwegian fleet continued to follow the shoals wherever they could find them, which in 1952 included Shetland, where their success in what must have been a bumper year encouraged a local father and son team, Archibald and John Fullerton, to equip their 40-foot seiner *Brittania* with a Konsgberg gun. They caught 20 sharks in 1953, but stopped when they found that because of the collapse in liver prices, they were only covering their costs.

And so, for the time being, the Basking Shark was able to enjoy a brief respite in Scotland. Whilst the various local operators in Scotland had enjoyed varying degrees of success, none of them could compete in terms of numbers caught with the Norwegian fleet, skilled, efficient and well equipped as they were. It is reckoned that the Scottish fishermen took no more than 300 sharks in any one year, with 1947 and 1948 being the peak years. Frustratingly, it is impossible to extract the actual numbers taken by the Norwegian fleet in the early 1950s, as their catch data makes no attempt to identify the sea area that they were operating in. However, it seems likely from the eyewitness reports of both Tex Geddes and Patrick FitzGerald O'Connor that many hundreds of sharks were killed in Scottish waters each year by the sizeable Norwegian fleet, at a time when the sharks were apparently abundant. This 'hidden catch' might have made all the difference in terms of the damage down to a local stock that has reverberated down the years.

From 1954 until 1982 there would be no more local fishermen involved in the pursuit of the Basking Shark, although it seems likely that at times the Norwegian vessels may have continued the hunt, albeit opportunistically. But just as the hunt in Scotland was drawing to a close, the war of attrition on the Basking Shark was flaring up elsewhere – in Ireland, right back where the hunt had begun.

The books

No account of the hunting of the Basking Shark in Scotland from this era would be complete without some mention of the various books written by the hunters detailing their exploits. After all, not only did they leave the most comprehensive individual accounts of the pursuit of the animal through the centuries so far, but as Phillip Kunzlik rightly suggested in his useful pamphlet on the Basking Shark, they 'may well have proved more remunerative than the fisheries themselves' for the authors.

These are not books that would gladden the heart of an ardent conservationist, recounting as they do the wholesale slaughter of a now much-loved, iconic species. But these were accounts from a period of want, and although it is probable that most of the oil from the

Basking Sharks killed in the Scottish fishery ended up in margarine, in the immediate post-war era that wouldn't necessarily have been seen as a bad thing. The hunter-gatherer was still looked upon as a natural component of a wild world considered almost inexhaustible, and the stories of the derring-do of the hunters were at that time greatly enjoyed. That only one of them apparently questioned the long-term viability of their activities isn't therefore entirely surprising, when seen through that prism.

That the men who wrote these books at times found the gory business of killing sharks appalling is understandable, but the public at that time were more used to killing and dressing game, fish and livestock than today's supermarket-reared generation. The authors certainly didn't expect (or, in fact, receive) any public opprobrium as a result of telling their tales. These were first and foremost adventure stories, and were products of their times, and if with hindsight we might wish the history they report to have been different, nonetheless they should be seen in historical context.

Harpoon at a Venture

The first to put pen to paper was Gavin Maxwell. Following the disastrous failure of the Soay venture, Maxwell was to all intents and purpose destitute. Attempts to set himself up as a painter and a poet were only partially successful, and eventually he turned to the idea of writing a book about his time spent hunting sharks in the Hebrides. His friend Kathleen Raine told him that he wrote better than he painted, and having listened to him as he 'talked of Soay and the romance of his disastrous shark-fishery venture' she told him he should write the story.

Besides, as his biographer Douglas Botting suggests, Gavin Maxwell seemed to be a changed man after the bloody years of slaughtering such a harmless creature:

> As for Gavin, he seems to have undergone a sea-change. Perhaps as contrition for
> his ravages among the sharks shoals, perhaps in acknowledgement of the fact that
> his love for wild creatures was now stronger than his urge to kill them, he rarely
> hunted wild animals again except perhaps on a few occasions.

Could it be then, that the book was also driven by a need to set the record straight and enable Maxwell to begin a new relationship with the wild world he so clearly loved? Given the far more conservation-minded tone he adopted in future books like *Ring of Bright Water*, this may not be so far-fetched. And given the effect that *Ring of Bright Water* had on the development of the early conservation and animal welfare movements perhaps, ultimately, some greater good came out of his failed shark-hunting venture after all.

By then living in a remote lighthouse-keeper's cottage at Sandaig Bay at the head of the Sound of Sleat, he worked on a synopsis of what would become 'his Soay book'. Back in London, his first attempt to secure a publisher was unsuccessful, so a disappointed Maxwell returned to Sandaig to try to live off the land and sea that surrounded the cottage, and pushed the abandoned synopsis into a drawer. There it would have lain perhaps forever but for a chance find on the beach below the cottage. Scouting the shoreline for a piece of driftwood to make a breadboard, Maxwell spotted a suitable-looking piece and prised it from the sand. It was a barrel top, with the faintly legible letters I.S.S.F. – Island of Soay Shark Fisheries – embossed upon it. 'It had all the nostalgia of a *carnet de bal* tied with faded ribbon; it brought so many half-forgotten scenes so vividly to my mind that I began the book next day.'

Settling down to the grind of writing, Maxwell churned out page after page of longhand prose culled from his memory and his diaries from the *Sea Leopard*. When inspiration failed, Maxwell would visit his former shipmates, including Tex Geddes, now living at Glasnacardoch Lodge, near Mallaig, who later recalled to Douglas Botting:

> Dammit, he was almost living there. He complained that he was broke, so I told him I was broke too, but his broke and my broke were two entirely different things. Well, using my typewriter and my paper and my coffee he wrote some of *Harpoon at a Venture* at Glasnacardoch Lodge. In fact, *we* wrote *Harpoon*. He didn't know he could write and he found it a hell of a chore. But he got three chapters written and then he went off to London to try and get a contract for it.

This time he was successful, not only securing a publisher for the UK market, but also the USA. The book was finally published in 1952 and was an immediate success, with *The Times* hailing Maxwell as a 'man of action who writes like a poet.' The BBC radio programme 'The Critics' took a more analytical view of the writer behind the pages, and perhaps summed up best the appeal of Maxwell's writing:

> He has the zest, the courage, the uncalculating enthusiasm and generosity, above all the panache of the non-professional, the highly intelligent inexpert. He's a youthful version of the White Knight… a highly sympathetic and attractive character in a world of glum and narrow experts, and the White Knight's spirit is a noble and chivalrous one. It is as an adventure of the spirit that *Harpoon at a Venture* is most worth reading, most exciting.

A writer was born. Gavin Maxwell would go on to become one of the world's best known and loved nature writers from the springboard of *Harpoon at a Venture*, happily on less destructive matters than killing Basking Sharks. But not everyone was happy with his book.

Shark-O!

Patrick FitzGerald O'Connor, already a published novelist, had been busy at his own book about his shark hunting days, which was published as *Shark-O!* in 1953. In an afterpiece to the book, he took Maxwell to task for what he saw as disparaging remarks about himself, Anthony Watkins, Harry Thomson and Patrick Sweeney, and Maxwell's seeming identification of himself as the pioneer of shark hunting.

Maxwell was able to trace this back to correspondence between O'Connor and Watkins, which showed that the two men had encouraged some of Maxwell's old Soay crew to bring an injunction against his book. As a result, Maxwell sued for libel, and Tex Geddes was called as a witness for the prosecution, Geddes delighting the courtroom with his challenge to O'Connor's counsel Melford Stevenson, that he 'didn't know a thing about sharks and couldn't tell his arse from his elbow'. Letters produced in court showed that Watkins and O'Connor had indeed conspired to 'debunk this bastard somehow' and Maxwell's case was successful on the grounds that his integrity as an author had been attacked, and secured £500 damages, costs and a letter of apology in *The Times* from O'Connor and his publisher.

This shouldn't overshadow the fact that *Shark-O!* is also a great read. Another book that identifies with the adventure of shark hunting, *Shark-O!* is perhaps less polished than *Harpoon at a Venture*, but O'Connor is far more honest and open about the motives behind the hunt and his own part in it than Maxwell. It is also a very funny book in places, and O'Connor is a wry and witty commentator on human nature with an observant eye for the wild world. Whilst it is far from a book from which conservationists would draw much comfort, it is the only one of the books to express anything more than a passing doubt on the long-term sustainability of hunting Basking Sharks, especially as a result of the Norwegian invasion. Sadly, it is out of print today.

The Sea My Hunting Ground

Anthony Watkins came to the table late in the day, with his own book on his exploits, *The Sea my Hunting Ground*, not being published until 1958. As the book that covers the longest period of the hunt (including the pre-war years) it gives the best account of the sheer madness of Shark Fever and the pioneering spirit that drove them all. It is clear throughout that Watkins was an astute operator, well prepared and with a good grasp of finance, and that he also possessed that essential blessing – luck. Given his predilection for gambling and calculating the odds, it should come as no surprise that his next and final book was a textbook entitled *How to Win at Poker*. Again, it's a pity that *The Sea my Hunting Ground* is long out of print.

Hebridean Sharker

Tex Geddes published his account, *Herbridean Sharker*, in 1960, and it is the book that will resonate most with any fisherman, Geddes being a highly effective and unapologetic hunter-gatherer. Having been involved from the earliest days in the abortive Soay venture, Geddes had a clear eye for what could have been achieved with a saner approach to the whole project. His ultra-low-tech approach in the later days was undoubtedly the right one and ultimately proved him right over O'Connor and Thomson's higher cost ventures when oil prices fell. If it lacks the panache or the vivid sense of place that either *Harpoon* or *Shark-O!* possess, it makes up for it in giving a more complete picture of the men, the fishermen, who proved to have been the ablest of all hunters in the post-war era of the Scottish Basking Shark fishery.

It is always the case that writers have the chance to correct history to their own advantage, and all of the authors present themselves in the best possible light. Digs against one another, some sly, some more overt abound, and although drawn on heavily in these chapters to weave several histories into one, it is only through reading them all that perhaps the truest picture becomes apparent.

Sadly, the score settling continued for years after the hunt and the books were simple history, as Tex Geddes revealed in an interview with Ian Mitchell in the 1990s: "'I wrote my book", Tex told me later that day, "to take the mickey out of Gavin. I told him his factory was no fucking use, a waste of time.'"

Read them all, and make up your own mind.

Chapter 11
The Hunt Beyond Scotland

Of course, the publicity generated by the development of the renewed Basking Shark pre-war hunt in Scotland would not have gone unnoticed by people engaged in fisheries in Ireland. The Carradale incident and Anthony Watkins's new venture attracted considerable interest from the press, and would have certainly spurred memories in the old hunting grounds amongst people who had recently seen a traditional sun-fish hunt re-enacted on the screen in Robert Flaherty's film *Man of Aran*. But the impetus to recommence the Basking Shark hunt in Ireland didn't appear there, but farther north at another historic hunting site; Achill Island on the Galway coast.

The Irish hunters

During 1941 a 4,330-ton steamship, the *Aghia Eireini* struck the cliffs on the south side of the island, becoming a total loss. Local businessman W.J. Sweeney of Achill Sound secured the salvage rights, and took charge of the operation of to scrap the ship *in situ*. Divers employed on the project reported that Basking Sharks were impeding their work, and expressed their concerns at the presence of the fish as they had no idea how they would react to divers. In order to address those fears, attempts were made to drive the sharks off with shotgun and rifle fire and underwater explosives.

In 1944 Sweeney acquired the rights to the local salmon bag net fishery, and soon discovered that the sharks were not just around the wreck. In fact, they were widespread around the island and were regularly blundering into the salmon nets, causing costly damage. Attempts were made to stop them by setting steel anti-submarine nets in the area, but these were unsuccessful. A number of sharks were killed with rifle fire and with Sweeney sensing a business opportunity, samples of the oil were taken with a view to finding a commercial outlet for it. By 1946 a buyer for the oil had been found, and Sweeney set about developing a method for catching Basking Sharks around Achill. He was joined in the initial stages of the project by Charles Osborne, a Donegal man, who (as we have already seen) visited Gavin Maxwell in 1947 to obtain information on hunting the Basking Shark.

Charles Osborne favoured the use of the (by then) conventional means of small vessels

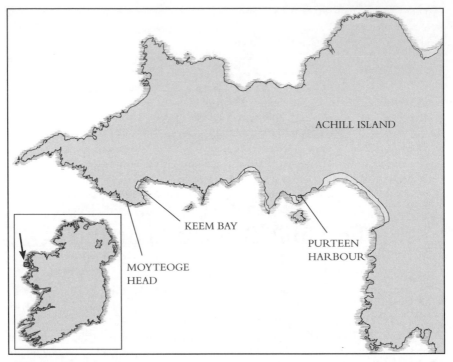

The post-war Achill Island fishery.

equipped with harpoon guns. But during a visit to Scotland in 1948 to obtain further advice on hunting techniques from Anthony Watkins, he learned of the experiments that Watkins had conducted with nets. The principle hunting ground on Achill Island was planned to be at Keem Bay, a striking place situated at the western end of the Island, where a sandy beach is flanked by cliffs and the massive promontory of Moyteoge Head. Keem Bay is well sheltered from the might of the Atlantic from the north and west but is wide open to the south and south-east. The bay shoals steadily in towards the beach, and when the weather is fine and the water is clear it is a beautiful, if elemental, place. During calm spells in early summer, Basking Sharks would regularly enter the bay, sometimes in shoals at the surface where they could easily be harpooned, hence Charles Osborne's original attraction to that method.

However, Keem Bay being such an exposed site, it was often too rough to allow harpoon-hunting activities during the season. It had also been noted that the Basking Sharks seen in Keem Bay were often only single individuals or small groups that stayed well below the surface, tantalisingly out of reach of hunters armed with harpoons. Although some sharks were taken in the early stages of the fishery with harpoons, the unique nature of the Basking Shark presence in Keem Bay suggested that nets might be a much more effective method of catching them.

Sweeney therefore set about developing a unique method of catching the sharks, using nets set from curraghs (the small, lightweight local boats) that could be launched from the beach. He had noted that Basking Sharks that strayed into his salmon nets often expired because their gill arches became entrapped, and so they eventually died from suffocation. So he had special nets constructed that would enable the sharks to be deliberately caught without suffering the extensive damage that ruined the salmon nets. These were far more heavy-duty nets, around 50 metres long and five metres deep, with a large open mesh made from 25mm diameter twine.

One of two methods of netting was usually employed. The first method relied on individuals acting as spotters from the cliffs, calling to curragh crews who would row the net out and set it around the shark in a circle like a ring net. The net would then be tightened until the shark was entangled, whereupon the crew would dispatch the shark, by spear, rifle or explosive harpoon. The second (and more typical method) involved pre-setting nets around the circumference of Keem Bay, where the sharks tended to swim. These nets were attached by a hawser or chain at one end to the shore at the base of the cliffs, and had a buoyed head-rope to enable the net to hang in the water. Three-quarters of the way to the outboard end of the net, a metal buoy was attached with lightweight line to hold the net taut, and a line led down to an anchor stone on the seabed to stop the net shifting with the tide. The remainder of the net was then doubled back at an angle of around 30 degrees in the direction of the tidal set and then buoyed and anchored, with the intention of guiding the shark into the net. When a shark entered the net, the lightweight line holding the net in place would break and the hawser would arrest the movement of the net, thus allowing the net to enfold the shark completely as it tried to swim ahead – sharks have no reverse gear. As with the salmon nets, this would incapacitate the respiratory system of the shark, which would eventually lead to death. However, the extended struggle would lead to serious damage to the net, so whenever possible the shark would be finished off by the men in the curragh by using a home-made lance to sever the spinal cord.

This second system of pre-set nets became the more popular method in later years, as it required less manpower, and allowed the use of up to three nets, left unattended overnight. If the net was undamaged, it could be re-set as soon as the shark was removed. Replacement nets were kept in store on the beach to be set in place of a badly damaged net, and at the height of the fishery in the 1950s it was reported that teams of men were kept in constant work on the beach repairing nets. Once disentangled, the carcass of the shark would be towed out to a floating raft in the bay, where the livers would be removed and taken to the tiny harbour at Purteen to be rendered down into oil via steam and allowed to settle in storage tanks before being barrelled and sent for export.

The first year of the renewed Basking Shark fishery was 1947, but this should probably be viewed as an experimental operation, as only six sharks were taken. But the fishery soon gathered momentum. By 1949 Sweeney was employing over 30 people in his operation alone and by 1950 a total of 905 Basking Sharks were killed at Achill Island. Others soon set up rival operations, and it may be that one of Sweeney's competitors bought Anthony Watkins's fleet of harpoon gun-equipped vessels in 1950. Certainly Sweeney unsuccessfully tried using a factory ship to make fishmeal in 1951, and there are reports of another stationed offshore that might well have been the *Recruit*. Given Watkins's contact with Charles Osborne, who had such an enthusiasm for the use of boat and harpoon gun fisheries for the Basking Shark, this seems entirely credible.

Charles Osborne continued to hunt Basking Sharks until 1951, when he lost his life in a tragic accident off Achill Island. He had been contacted by a film company who wanted to make a documentary on the Achill Island Basking Shark fishery, and had agreed to let them use his 7.5-metre motorboat, the *Pride of Cratlagh*, as a platform to film the harpooning of a shark. The film crew, led by Hugh Falkus, along with his 27-year-old wife of four weeks, Diana, Director Sam Lee and a cameraman Bill Brendan, were all aboard the boat with

Osborne on 12 May looking to secure footage of the hunt. During the afternoon, the tide turned, the wind began to pick up, and the sea-state worsened considerably. At some time around 3p.m. the *Pride of Cratlagh* was swept on to a notoriously dangerous rock plateau called the Daisy Rocks, and was badly damaged, taking on water heavily before sinking.

Hugh Falkus was the only survivor, and later told investigators that they had been hit by a freak wave. As the boat began to sink he lashed together a makeshift raft of petrol cans, rubber rings and wooden oars to help support them in the water. Once the boat sank he left the others clinging to the raft, and in an act of enormous bravery and endurance, swam a mile and a quarter to the shore, where he was spotted clinging to one of the moored fishing boats by fishermen. Twenty boats were reported to have launched to go to the assistance of the other members of the crew, but were too late to save them. The bodies of Diana and Sam Lee were recovered from the 10 degrees Celsius water, whilst those of Charles Osborne and Bill Brendan were never found.

The waters off the west coast of Ireland have never changed, and remained as dangerous and unpredictable as ever. Hugh Falkus went on to finish the film, *Shark Island,* and later became a well-known filmmaker and television presenter. The only possible silver lining leading from the tragic loss of his wife and friends would be further support towards netting Basking Sharks inshore at Keem Bay, adding safety to the growing evidence of the efficiency of netting.

Some small-scale groups of independent fishermen operated from the beach at Keem Bay, rendering down the livers in crude equipment on the beach, often with poor results. Eventually, all of the other operations closed, mostly for financial reasons. However, the Sweeney operation with its plant at Purteen went from strength to strength, and despite the fact that the oil price declined substantially in the early 1950s, such was the increase in the number of sharks taken that the fishery prospered. The catch climbed to 1,630 sharks in 1951 and 1,808 in 1952, and continued in the thousands until 1956 when the decline began to gather momentum. During that period the average yield of liver-oil was around 250 tons *annually*, which to put it in context, was more oil in one single year than the Scottish fishermen produced between 1946 and 1953, when the fishery eventually ceased there.

Sweeney also experimented with other markets for shark products from 1950. Foremost amongst these was the flesh, with Sweeney and an English fishmeal manufacturer going into partnership in a converted trawler equipped with machinery to convert the carcasses into meal. Due to problems with the plant on board this was abandoned, and Sweeney had to resort to men cutting the carcasses at Purteen harbour into one hundredweight (50 kilogram) pieces to be loaded onto lorries. In the five years from 1950, some 7,500 tons were sent for processing into fishmeal, initially to Lisburn in Northern Ireland, and later to Ballinasloe in the Irish Republic. Despite the volume, the price paid per ton was low (£2.50–£3.00 per ton), so by 1963 increasing labour and transport costs put an end to the marketing of the flesh once again.

Fins were sold from 1960, initially to a Spanish company. In later years a new market was identified in Asia, and the fins were air dried at the plant at Purteen for several weeks before being exported to Hong Kong. Although the price paid for the fins was not high in the early years, the sheer volume made it worthwhile, and by the time the fishery was winding down in the 1970s, fin prices were still increasing, which would have made them a far more important part of the financial value of each shark.

That the netting of Basking Sharks within the unique setting of Keem Bay was highly efficient is undeniable. It is certainly fortunate for the Basking Shark that similar opportunities

THE HUNT BEYOND SCOTLAND

didn't exist elsewhere, otherwise the drastic reduction in their numbers as a result of netting might have been replicated with catastrophic results for the overall stock.

As it was, the number of sharks taken via netting in Keem Bay plummeted from 977 in 1956 to 47 in 1966, and the fishery would be over within 10 years. In a desperate attempt to increase the catch, probably due to a spike in the value of shark oil, in 1973 Sweeney re-introduced the use of the harpoon gun with one converted vessel equipped to chase the sharks offshore, but not a single shark was caught via this method during that year. For the 1974 season, Sweeney equipped a further three 15-metre-long fishing vessels with Norwegian whaling guns to work off Galway. One of these vessels didn't catch a single fish, but the two others caught 17 sharks between them. The writing was on the wall for the Achill fishery, and in 1975 only 38 sharks were taken in total, half the number that were taken during the first full year of hunting in 1948. In just under 30 years no fewer than 12,360 had been killed, the vast majority of them in one small bay. Perhaps that alone was enough to cause the extirpation of the local stock?

Perhaps not. Those were the sharks that were reported, quite accurately by all accounts, just by the Irish fishermen. Not far offshore, however, just out of sight of land, the Norwegian Basking Shark hunting fleet was operating in the same efficient manner that it had in Scotland. Not that they were in any way apologetic about their actions, as Norwegian scientist Sigmund Myklevoll remarked cheerily in 1968, 'Some of our fishermen go as far west as west and south of Ireland, much to the irritation of the Irish who sit on shore, waiting for shark to come to them.' As Phillip Kunzlik identified, Norwegian catches increased substantially just at the time that the Achill catch declined, and whilst Ken McNally noted that there were no accurate returns of catches available to identify the proportion of the Norwegian catch originating from Ireland, nonetheless, 'Estimates put the likely total of sharks taken at several thousand annually.' This implies a far more drastic impact than the Irish figures associated with the Keem Bay fishery alone.

But the revived Irish fishery was not entirely finished. During 1973–1974 the world was undergoing a major crisis due to the OPEC oil embargo that drove up the price of commercial oils, including Basking Shark oil. By then the Irish oil was being exported to Norway, where it was achieving prices of up to £500 per ton (over £5,000 at 2017 values), and this, coupled with higher fin prices, was the driver of a short-lived resurrection of the fishery. As a result, in 1973 the Sweeney operation turned to a harpoon fishery alongside netting, equipping a single trawler with a harpoon gun to hunt along the Galway and Mayo coast. Although they failed to catch any sharks in their first year of operation in 1973, they persevered with the idea in the following two years with now three five-metre trawlers equipped with 50mm Kongsberg guns taking 17 sharks in 1974 and 36 in 1975 when W.J. Sweeney's Achill fishery finally ceased.

That Irish fishermen beyond Achill had also noticed the rise in oil prices became clear in 1974 when a new front opened up – three fishing vessels operating out of Dunmore East in the south east of Ireland mounted locally-built harpoon guns and killed 180 sharks in their local area, during a very difficult year weather wise. In 1975 another boat joined the fleet, and working between Mine Head, County Waterford and west Cork, the four boats killed an estimated 350 sharks. In a final nod to the past, two fishing vessels from Inishmore in the Aran Islands joined in and fished to the south of the islands, back in those wild waters where their forebears had pursued the sun-fish centuries before.

But by 1976 all of the Irish hunters had hung up their guns, although that was not the definitive end of the hunt, as the Norwegian fleet returned to the fray sporadically until

1986. Fishing the traditional areas off the west coast of Ireland between 1978 and 1979 and between 1982 and 1983, perhaps 400 sharks were taken by Norwegian vessels, with around 200 more being taken off south-east Ireland between 1984 and 1986. In the final year only 13 tonnes of liver were taken, which must hardly have justified the fuel cost for steaming all the way from Norway to the Celtic Sea. As at least some of the Norwegian shark-hunting vessels appeared in the Isle of Man in 1986, it may be that they considered that the west and the south-east coasts of Ireland were fished out. Whatever the cause, after nearly 250 years of exploitation, the sun-fish had a chance to bask unmolested in Ireland's waters once more.

In the Pacific

The boom in the value of Basking Shark liver oil that drove the post-war fishery in Scotland and Ireland was not a universal call to arms. On the west coast of Canada, as we have already seen, Basking Sharks were numerous in both the waters off the British Columbia mainland and the west coast of Vancouver Island, yet no substantial attempt was made to exploit this new opportunity in any serious commercial manner. Despite the fact that shark oil prices during the war were at an all-time high, the Basking Shark fisheries remained small scale, especially compared to the European shark fisheries.

As reported by Scott Wallace and Brian Gisborne, it seems that this may have been a localised factor, partly to do with the low vitamin A content of Basking Shark liver-oil. As a result the price paid for Basking Shark liver during the war years was between 3¢ and 20¢ per pound, compared to CAN 37¢ and CAN $3 per pound for dogfish liver. Dogfish liver had a far higher vitamin A content, allowing the oil extracted from the liver to be marketed as a medicinal oil as opposed to an industrial oil, and the price paid seems to have been largely dictated by the much greater demand for medicinal oil. Coupled to the usual difficulties always present in terms of the catching and handling of such an enormous fish, there was therefore less incentive to gear up to catch Basking Sharks.

Wallace and Gibson report that not everyone was put off. First Nation hunters like Huu-ay-aht whaling chief John Moses still put out to sea to test themselves against the monster fish. Moses, from Sarita on Barkley Sound, Vancouver Island would paddle his canoe up to 10 miles out to sea on calm days, to hurl a heavy spear attached to 180 metres of 30mm rope into a Basking Shark. Once attached, the shark could struggle for up to six hours, but once the battle was over, chief Moses would set off to tow it back to shore. He reckoned to obtain around CAN $80 for the two livers, which, as he remarked laconically, was 'not hay'.

By 1946 in nearby Bamfield, several small-scale operators were reported to be harpooning Basking Sharks. These included an intrepid husband and wife team, Einor and Mona Andersen, who were reported to have caught two sharks, 11 and 12 metres long, and Einor was locally acknowledged as 'the expert' on hunting the shark, having killed up to four in a day. The couple used a harpoon developed to Einor's own design, fabricated locally, fitted to a six-metre length of water pipe filled with lead that was fastened to their boat by a cable. This seems to have proved an effective system, and they reportedly used it one day to catch a 'plumb nuts' 11.5 metre-long Basking Shark that towed them for three miles before a passing pilot boat was called to their assistance. Even then it took another hour of towing plus five rifle shots to the head to kill the shark. However, similar to chief Moses, the Andersens didn't think they received adequate recompense for their efforts, only obtaining three or four cents per pound for the livers.

Barkley Sound apart, there appear to be few records of commercial fisheries for the Basking Shark elsewhere in the region. Nor do fisheries' statistical records help to elucidate

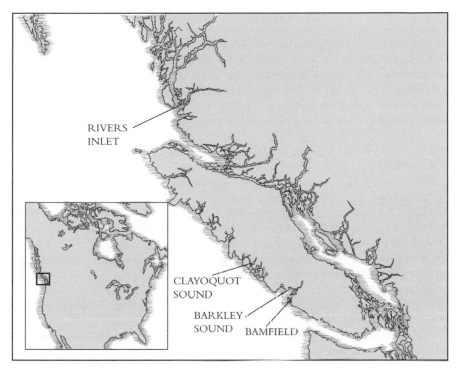

Basking Shark key sites on the coasts of British Columbia and Vancouver Island, western Canada.

how many Basking Sharks were killed at the time, as all landings were simply lumped in with those of other sporadic or low-value species under the nomenclature of 'mixed shark'. Between 1941 and 1945 over 400 tons of 'mixed shark' liver was landed, of which a proportion was Basking Shark liver.

What seems surprising is that given the apparently growing abundance of Basking Sharks in the area, a large-scale commercial fishery was not established. The market price for liver oil had risen post-war, and increasingly the Basking Shark was perceived as a serious threat to salmon fisheries, particularly in the Barkley Sound area. As had occurred in Ireland and Scotland at various times through the centuries, the Basking Shark population on the west coast of Canada was apparently increasing exponentially, to the point where they were being labelled a 'menace' and a 'curse'. Experience elsewhere would suggest that a substantial, dedicated fishery would have been a logical response to those two factors. Perhaps due to the absence of such a fishery, Barkley Sound fishermen were reported to be taking matters into their own hands and either harpooning the sharks or 'blowing them up'.

Scott Wallace and Brian Gisborne have looked at all the evidence and believe that the various commercial fisheries that existed up to 1948 may have killed many hundreds of Basking Sharks before the price of oil dropped, as whale oil and other substitute oils became abundant once more. This is hardly an extraordinary number of sharks, even compared to the failed Scottish fisheries. Perhaps the miracle is that the small-scale commercial fisheries didn't expand as might have been expected. But once the price of livers fell to a low level, it is hardly surprising that the commercial catch didn't develop further.

The making of the menace

By 1948 the salmon fishermen were clamouring for action. A Basking Shark bycaught in one of their nets and landed would be worth between $36 and $48 for the liver. But the damage to the net would vastly outweigh that, possibly up to half the value of a $350–400 net. The gillnet fishermen demanded that the government place a bounty on the Basking Shark.

At first the authorities refused to accede, claiming that the Basking Shark was simply a natural hazard, like bad weather. But as the toll of damage to nets grew with the increasing number of sharks in the vicinity of key salmon-fishing sites, that position became harder to hold, and eventually the Department of Fisheries conceded defeat and in 1949 added the Basking Shark to their list of 'Destructive Pests'. Inclusion on this list was to join other 'pests' such as seals, sea lions, mergansers, kingfishers and black bears, all species being marked down for routine control measures or outright destruction.

This didn't amount to anything serious immediately, but after a bad year for net damage in 1950 the attitude amongst the fishing community grew increasingly militant. Reports began to appear of experiments by the government to cull Basking Sharks, probably with harpoons, but these seem to have been unsuccessful, or at least on too small a scale to affect such a sizeable influx of sharks. But by 1953 the press had joined the clamour for action, reporting widespread damage to fishermen's nets in Barkley Sound. Although the nets were covered by 'shark insurance', the fishermen had had enough. So, apparently, had the authorities, and a far more aggressive course of action was to follow.

It was about to get personal. In 1955 the Department of Fisheries commissioned the fabrication of a hinged blade made from boiler plate steel, sharpened to a razor edge that was fitted to the bow of the regional fisheries protection vessel, the *Comox Post*. This simple device was designed to be lowered by a cable when the crew sighted a Basking Shark, whereupon the *Comox Post* would ram the shark with the aim of cutting it in two. Of course, the Basking Shark, being a big lazy, dim-witted beast, was no match for such tactics, and the blade soon proved to be devastatingly effective.

According to Wallace and Gisborne, the press loved it. The Basking Shark was variously described as a 'pest', a 'deep-sea pirate' and even a 'salmon-killing monster' (there were obviously not a lot of marine biologists in the press corps), and they enthusiastically supported the development of 'shark-progging'. There was no doubt whose side the press was on – the fishermen. The papers even encouraged the general public to join in the fun, with harpoons, rifles and through-ramming the enemy with boats. The tone of the coverage could be epitomised by the front-page article of the Victoria *Times* of 22 June 1955 that showed a drawing of a Basking Shark accompanied by the caption: 'This is a Basking Shark, basking and leering. But the smirk will soon be wiped off its ugly face by the fisheries department, which is cutting numerous sharks down to size.'

Indeed they were. Another press report, written by a correspondent for *The Vancouver Sun* who accompanied the crew of the *Comox Post* on a 'progging' voyage, gave a toll of 34 definite kills, five possible and 10 misses, from a 'colossal fight between the ship and the sea monsters, with the ship winning all the matches.' Which must surely come as no surprise – it was a hopelessly one-sided fight. The *West Coast Advocate* featured the headline 'Fisheries Patrol Winning war on Basking Sharks', but it would be many years before it would become clear just how completely that war was being won.

At the same time that the *Comox Post* was in action, other fisheries patrol vessels were also tasked to ram Basking Sharks whenever they were encountered. Even though these vessels were not equipped with a shark blade, the force of the impact alone would have been sufficient to deliver fatal injuries. A deckhand, Pete Fletcher, on the fisheries protection vessel *Laurier* at the time, reported that when the vessel rammed a shark, 'It was like going aground on a gravel bar, all the dishes would rattle.' An animal lover, Fletcher didn't like the work at all:

> I always loved animals, as a kid I enjoyed seeing the Basking Sharks. Back then Basking Sharks were a fact of life, as common as catching a salmon. I hated my job… killing sea lions, seals, ramming Basking Sharks. I knew it was not right. I was the wrong guy in the wrong place and that's why I eventually left.

Although no official records were kept of the number of sharks killed in this opportunistic manner, Fletcher estimated that the number was in the order of 200–300 sharks.

For 14 years the shark blade delivered its deadly greeting to Basking Sharks in Barkley Sound, amassing a tally of 413 shark kills during that period, plus others simply rammed on any occasions when the blade was not mounted. Added to these numbers should be sharks killed through entanglement in gill nets, plus those caught in trawls, pot ropes and other fishing gear. Given the frequency of the reports of entanglement, and the number of nets set in the area, these may have amounted to several hundred sharks. It seems likely that the programme of actively culling Basking Sharks only ceased once the problems the species caused to salmon fishermen had been reduced to an insignificant level, which in turn suggests a major reduction in numbers. As researchers Darling and Keogh stated in 1994, 'Basking Sharks are rarely sighted in Barkley Sound today, suggesting that the majority of the population in that area were killed.'

The public responded to the appeals from the press to join in the slaughter with gusto, encouraged along their way by press releases from the likes of the Canadian Pacific Railway to visit the region to take part in the sport of harpooning Basking Sharks. Bamfield residents joined in, ramming the sharks or harpooning them for the ride, and the local boys developed a game of ramming Basking Sharks at the surface at right angles, using them as a ski ramp to jump their motor boats off the back of the sharks. The staff at the Bamfield Cable Station reportedly resorted to shooting the sharks with rifles as target practice. One big shark, 'Old Joe', was 'riddled with so many bullets that his dorsal fin looked like Swiss cheese'. Old Joe became something of a favourite in Bamfield Inlet, with even the fishermen unwilling to harm him, but that came to an abrupt end when the new skipper of one of the cargo ferries decided to help the cause by ramming poor Old Joe. Harassment and recreational killings undoubtedly added an unknown number of kills that may have run into hundreds.

This steady war of attrition from all sides was bound to have an effect in the long term. Slowly but surely the number of sightings and reports of net entanglement dwindled. Wallace and Gisborne report that even local fishermen such as Bill McDermid were concerned:

> Oh, there were hundreds of them around, hundreds of Basking Sharks. You could look out in the harbour here (Bamfield) in August, and you could see two or three on any given day, cruising up and down the harbour. You'd go out and you'd see forty, fifty, sixty, just going up to Sarita and back. There were way too many, now there's way too few. So they eliminated them, but what was left over? I would say I've caught seven or eight I've had to destroy, like in my gillnet to get my net back. They wind themselves all up in it and drown themselves. So if you put 150 gillnetters in here and everyone gets one or two, that finishes off the few that are left. And I guess naturally, they move so god damn slow, they wouldn't find each other to mate anyhow.

Coincidental with the Canadian eradication programme, the Basking Shark was also under pressure farther to the south, along the coastline of California. The species had long been a familiar visitor to a number of coastal areas of central and southern areas of that coast, most notably at Monterey Bay, Morro, Pismo Beach and the Santa Barbara Channel. Unusually for the Basking Shark in the northern hemisphere, the peak period for sightings was between October and May – interestingly, the disappearance of the shark from these waters at that time tied in with peak numbers off the coast of British Columbia between May and July. This suggested the possibility that the eastern North Pacific stock was in fact one single stock that shifted its range northward from California to Canada following seasonal plankton abundance. Given what we know now from satellite tracking studies, such a migration (less than 2,000 kilometres) should be easily possible. If that were the case, then the establishment of active fisheries in California could undoubtedly have compounded the effects of the eradication programme in Canada.

The first recorded fishery in California for the Basking Shark was sport fishing, in the early 1920s. Between 1924 and 1938 an average of 25 sharks were landed each year, with up to 100 sharks being landed in peak years. The numbers caught, and the reported use of spotter planes to find sharks for the sports fishermen, suggest that this was no ad hoc entertainment, but a serious enterprise. By 1946 this fishery had morphed into full-scale commercial exploitation, probably to take advantage of the rise in oil prices post-war, although it is known that the flesh was sold for conversion into fishmeal as well. Three hundred Basking Sharks were killed in the first year alone, followed by an average of 200 a year until 1949 when the fisheries effort began to decline. The fishery had collapsed by 1950. Coupled to a known but unquantified level of bycatch in a range of other commercial fisheries during the period 1920–1950 up to 2,000 Basking Sharks may have been removed from the eastern North Pacific stock in California, adding to the already severe reduction they had suffered further north as a result of the Canadian eradication programme.

Across the Pacific in Japan a similar story would unfold in a fishery that would run out of control between 1967 and 1978. Based at Nakiri on the Shima Peninsula, a traditional Basking Shark hunt had existed in the area since the Edo era (1772), but as has been seen elsewhere, when oil prices rose significantly the hunt was scaled up dramatically. In response to that increase, in 1967 the local fleet began harpooning far more sharks principally for their liver oil, but also for their fins for export to Taiwan, and the flesh for human consumption or for conversion to fishmeal to be fed to livestock. The fishery peaked in 1972 when 60 sharks were sold at Nakiri market in one day, followed by a rapid decline from 1975 when 150 sharks were landed, falling then to only six sharks by 1978. The fishery ceased altogether by the early 1980s, as a result of falling liver oil prices and a lack of sharks sighted. No recovery in numbers has occurred since then, with sightings of between zero and two sharks per year now being recorded off Nakiri, in an echo across the Pacific Ocean from the west coast of North America.

The aftermath

By 1970 the war on the Basking Shark in Canada had stumbled to a halt – there was no enemy left to fight. Sightings and reports of net entanglement had dwindled, becoming a very rare occurrence throughout the waters of British Columbia and California. Given the dramatic paucity of sightings since then, the result looks more like a localised extinction than a drastic reduction in numbers. And the numbers have not recovered to this day. Off the California coast, where huge shoals of between 100 and 1,000 Basking Sharks used to be recorded, no more than three individuals have been observed at any one time between 1993 and 2009. In British Columbia where places like Bamfield once hosted shoals so large that it was at times hard to navigate through them, between 1996 and 2009 only 12 animals have been recorded. Wallace and Gisborne report a local man, Joe Garcia, commenting that maybe the war had been prosecuted too ruthlessly:

> They used to destroy the nets of the fishermen so badly, there were so many of them. I mean, it was nothing to see ten or twelve of them right in the mouth of the harbour, maybe twenty or thirty of them in the sound at one time, you know, in sight at one time. But they put a big knife on the bow of the *Comox Post* and just cut them in half. And I guess they went a little bit too far, because now there are very few left.

Given that the Basking Shark was probably never a very numerous creature even prior to the excesses of hunting, the time had come to take stock of the situation for the giant fish. By the 1980s the fisheries in Ireland and Scotland were all but over and the Norwegian catch was beginning to decline. In the North-east Atlantic at least, there appeared to be a window of opportunity to try and stem the decline and perhaps even imagine a slow recovery of the species in those places where it had once been a regular feature. In Canada and the USA, the picture was far bleaker, with an almost complete absence of sightings and even bycatch that might have hinted at some change in surfacing behaviour by way of explanation. Eventually this dire situation would be recognised at the highest level, with the shark making its appearance in the International Union for the Conservation of Nature (IUCN) Red List of Threatened Species. Whilst the global assessment for the species listed the Basking Shark as 'Vulnerable', the local assessment for the North Pacific and the North-east Atlantic regions was (and remains) at the far more drastic level of 'Endangered', a direct result of the historically high levels of exploitation in those areas.

No one was under any illusion as to the scale of the problem, yet at the time there were no conservation or protection measures in place anywhere. At best, all that had happened was that the targeted hunting and indiscriminate killing of the Basking Shark had ceased in most places, so this wasn't the most positive of foundations from which to start a comeback for the species.

However, in a growing number of minds around the globe, ideas were taking shape to turn the situation around, and as with the 'Save the Whales' campaign, try to make the Basking Shark a conservation icon that deserved protection and a chance to recover. The odds weren't good, but that couldn't be helped – the fightback had to begin, and where better than in the British Isles, where much of the damage had been done?

Chapter 12
We Set Sail

I must have been around 10 years of age when I first learned of the existence of the Basking Shark. The local newspaper in our small English seaside town in South Devon carried a report of the entanglement of a four-metre-long Basking Shark in one of the salmon nets set just offshore by local fishermen. Unable to free it in deeper water, they had towed the net into the tiny cove where we all kept our boats, and cut it free, allowing it to make its escape.

A four-metre shark! None of us kids had any idea that such things existed in *our* waters. The biggest thing most of us had encountered was a 4.5-kilogram Bass (*Dicentrarchus labrax*), and we'd thought that was a monster. The idea there was a shark out there nearly one-third bigger than my boat certainly put me off going fishing for a few days. But when I asked one of the fishermen about this shark and whether it would eat small boys, he laughed and said that it wasn't going to harm anyone, as it was only a plankton-feeding Basking Shark, and not even a very big one at that.

Not very big? It sounded big enough to me. By now thoroughly intrigued, I set about finding out all that I could about these creatures. Bit by bit I learned that the Basking Shark was a rare visitor, usually seen offshore in the spring and summer, swimming at the surface with its huge mouth wide open, straining tiny creatures from the water that passed through its gills. The picture in my little book of Britain's fishes really couldn't do it justice, but the statistics were certainly impressive. So I asked the most learned member of the town's sailing club what he knew about the Basking Shark … which turned out to be not a great deal more than the fisherman had told me. This stirred my interest even more; it was not just a shark, but a creature of mystery, too.

At about this time my parents decided to send me into exile, and so packed me off to an inner city boarding school to further my education, which was a total shock to the system. Up until that time I had been fortunate enough to live an almost feral existence in the local woods in winter and either on or in the sea in summer. Being cooped up was bad enough, but it did provide one unintended benefit in that reading became a personal passion, and I devoured books at a prodigious rate. The books all came from a narrow field though:

adventure and wild places, with anything by Gavin Maxwell and Gerald Durrell high on the list of favourites. The main consequence of which was that my school career was an unmitigated disaster, as I spent my entire time looking out of the classroom window, wishing desperately that I was out there and not stuck indoors.

From Maxwell's *Ring of Bright Water* I learned that he had hunted Basking Sharks in the Hebrides off western Scotland in the years immediately after World War II, and soon tracked down a copy of *Harpoon at a Venture*, his account of that disastrous campaign. And what an extraordinary story it was; the pursuit of a colossal wild creature in small boats crewed by wild men, all set against the backdrop of one the most elemental and beautiful places in the British Isles, if not the world. I read it again and again, each time gleaning new insights into the mysterious victim of the saga, the Basking Shark. After a while I became the proverbial 'tireless raconteur' on the subject, sending listeners to sleep or out of the door in equal numbers, as I discovered to my amazement that not everybody shared my enthusiasm. What was the matter with them?

Quite apart from the extraordinary narrative, what was of great interest to me was the wealth of information on the natural history of the Basking Shark that Maxwell had recorded. Clearly he was a keenly interested and insightful naturalist as well as a supremely gifted author, if a difficult and troubled man. He had also done the scientific world a great service (as we saw in Chapter 9) by inviting two of the most learned zoologists of the day, Dr Leonard Harrison Matthews and Dr Hampton Parker, to visit his factory on the island of Soay, where they made the first detailed, systematic examination of the anatomy and biology of the Basking Shark that had ever been undertaken. I just lapped this up.

Matthews and Parker also benefited in other practical ways from Maxwell's assistance, not least through the use of the plant and machinery at the factory to assist with the dissection of such a colossal creature, where the head alone could weigh one ton, and a careless scalpel cut could puncture the stomach and see the handler swept away by a tsunami of undigested plankton (as happened to Maxwell on one occasion). And by having access to the hunting craft, the two scientists were able to see and evaluate first-hand much of the behaviour and lore that Maxwell and his crew had reported to them.

The results of their labours were two papers presented to the Royal Society and the Zoological Society of London that were totally ground-breaking in their scope, not least in terms of reproductive biology and internal anatomy. Much of what Matthews and Parker recorded has stood the test of time, and they are still regularly referenced in scientific papers to this day, a rare occurrence in this field.

That so little research had been carried out into the life history of such a spectacular and obvious creature prior to hunting it so extensively may seem extraordinary, but is in fact not uncommon with fish. As is eloquently explained by Professor Callum Roberts in his excellent book *The Unnatural History of the Sea*, exploitation usually precedes understanding, leading to disastrous consequences for populations where no accurate assessment of recruitment and resilience has been made. A classic example in recent times has been demonstrated in the case of a commercially attractive fish called the Orange Roughy (*Hoplostethus atlanticus*), found in deep water. The Orange Roughy was very heavily fished for years, before the sudden collapse of local populations alerted scientists to study their ecology and biology to find out why. It turned out that local populations gather around seamounts, making them easy to locate and target, but the true disaster is that the species is long-lived (up to 149 years), slow growing, late maturing and only produces a small number of eggs, so that stocks could never cope with such sustained exploitation.

By now, during holiday escapes from incarceration, I had progressed to owning a small sailing dinghy, and began to make the first tentative voyages from the immediate vicinity of Boat Cove. The elixir of freedom that this delivered was strong and heady, as it enabled my friends and I to explore some of the small bays and beaches nearby that could only be reached by boat. Sailing along the coast, landing at a deserted beach and cooking up a few mackerel on an open fire may not seem much like real adventure, but it certainly did to me, generating an addiction to wild places that has lasted a lifetime.

But I still saw no sign of a Basking Shark, despite keeping a constant lookout. By incessantly interrogating the local fishermen, I discovered that they were very rare visitors, but were sometimes seen further down the coast, around Berry Head and Start Point. Members of the local sailing club who owned bigger yachts that cruised up and down the Devon coastline confirmed this general information, and some of the more adventurous amongst them sailed as far as Cornwall and the Isles of Scilly and talked of seeing sharks off Lizard Point and Land's End – mythical, alarming names that evoked visions of shipwreck and huge seas. They sounded like Cape Horn to me with my tiny boat and drastically limited experience. But slowly a pattern was emerging; the sharks seemed to be associated with headlands, where strong tides and rough seas could be expected, and where the presence of a huge sea creature was viewed as simply another hazard to navigation.

Years passed, and my horizons grew wider, until I was finally ready and able to progress to a small cruising yacht, at eight metres, big enough to cross the English Channel with a small crew of friends aboard. I can vividly recall our maiden crossing, filled with trepidation, armed only with navigational instruments that Nelson would have recognised. Just as dusk was approaching we were somewhere near the middle of the Channel between the River Exe and Alderney in calm conditions when the cry went up: 'Shark!' Seven or more Basking Sharks were dead ahead of us, right on the surface, slowly circling with their huge mouths agape, feeding quietly. We stopped the boat and gazed in wonder (and not a little fear) at the spectacle, observing for the first time the slow, sinuous, balletic movements that I would come to know so well in future years. When we finally moved on it was with a sense of awe at having spent time in the presence of something so huge and apparently prehistoric.

When Alderney came up on the horizon in the early morning we were all elated, counting ourselves as true sons of Drake, Raleigh and Cook, bold chips off the old block of those great British navigators. You'd have thought we'd discovered Australia.

All through the weekend I kept thinking back to the sharks and their mysterious ways, a slow mental replay on a loop, still trying to take it all in. Even a brief encounter on the return leg with a small pod of Long-finned Pilot Whales (*Globicephala melas*) that swam right up to us and dived under our keel at the last possible moment couldn't gain ascendancy over the image of circling sharks imprinted in my brain. I was hooked – my own Shark Fever.

The following year we made our first foray to the legendary Isles of Scilly, a cluster of tiny islands dropped like jewels in the Atlantic out beyond the western tip of Cornwall, Land's End. I'd always wanted to go there in my own boat, having been drip-fed stories of their glories by a dear old friend called Bill Robinson who made the islands his home during the long summer holidays allowed to him as a teacher. We surged our way down the south coast of Devon and Cornwall in a strong easterly breeze in big following seas before the wind started to die as we rounded Lizard Point, the southernmost tip of Cornwall.

By the time we reached Land's End we couldn't see a thing as the visibility closed in and fog descended, turning our small craft into a dripping Christmas tree. Straining our eyes to see ahead, with our little manual foghorn bleating like a goat on the way to slaughter, we inched our way out into the Atlantic, hoping against hope that the fog would lift, otherwise we would be doomed to turn back.

Blundering along in fog knowing that there are ships, fishing vessels and even lethal rocks out there in the murk is a precarious feeling, and it's hard to take your mind off your current predicament. And what better remedy for fear could there be than to find ourselves amongst sharks, ghostly fins slicing through the water all around us. This lifted our spirits immensely for a moment until we began to consider the implications of blundering through a shoal of these monsters in the fog. Just as we were on the point of turning around for fear of running aground on the rock strewn shores of the islands we were able to identify the fog signal on Peninnis Point, and an hour later were safely at anchor in the islands.

The return home was slow and uneventful, apart from nearly being run down at night by a small coaster south of Lizard Point. A late summer high had developed, and the wind slowly died away to nothing, leaving us rolling in an oily swell, forcing us to fire up our tiny, noisy diesel engine to help us home. Rounding Start Point on a drowsy summer's evening our luck was in again, and we encountered a succession of Basking Sharks. In the perfect conditions, we shut down the motor and drifted with the tide and the sharks, absorbing the wonder of the gargantuan creatures as they swam slowly by.

The calm conditions had allowed what we then called 'tide-lines' to form, long slicks of weed and debris that formed winding tracks across the sea surface. The sharks were feeding with their heads up and nose, dorsal fin and tail showing, cruising slowly along with their huge white mouths wide open, filtering out their tiny planktonic prey. We noticed that as they moved away from the surface slicks, they tended to speed up, before turning sharply back towards the tide-lines. This simple behavioural observation was a small revelation, as it suggested that the sharks were actively looking for the best patches of plankton, and were not just blundering aimlessly around, as had been suggested to me by friends. And as every single shark we saw was remaining determinedly in the vicinity of the tide-lines, these features obviously represented a favoured habitat. As darkness fell we set off again for home, to mull over an altogether momentous and fascinating voyage.

As the years slid by, marriage, children and domesticity intervened, blessings in an already lucky life. A secure job that allowed much time off in the summer meant that a second career in sailing was a possibility, which I was eager to take up, not least as an antidote to the other half of my life spent on Britain's grim road system. One day I found myself sitting in my car outside some ghastly motorway service station in the pouring rain, listening to a whale research scientist being interviewed on the radio. After a stimulating few minutes, he finished up by asking for regular seagoers to record and report to him any sightings of whales and dolphins (cetaceans) made in UK waters. Absent-mindedly I jotted down his contact details for future reference if I ever made any sightings (and if I ever got out of this car) – at that time even with a considerable number of sea miles under my belt each year, I had very rarely encountered cetaceans.

Thanks to a small inheritance, a bigger boat with real ocean capability had been purchased, and it was a very proud skipper who led his crew north to Scotland to pick up his new command, blissfully unaware that one of those rare life-changing experiences was lying in wait. Opting to sail home non-stop due to a good weather forecast, we set off down the middle of the Irish Sea in good spirits. After a couple of days, we were well on our way with a strong following wind, enjoying a great downwind sleigh-ride, when a pod of substantial, unfamiliar-looking dolphins joined us, surfing inches from our bows Remembering the request that the whale researcher had put out, I dashed below, jotted down the time and position, and grabbed my camera. Up on deck, braced against the mast, I did my best to capture the ebullient, acrobatic antics of the dolphins before they disappeared just as suddenly as they'd arrived.

Once back home, the pictures confirmed that these were not familiar animals. I'd previously seen Common Dolphins (*Delphinus delphis*) in the English Channel on a few occasions, and very rarely Bottlenose Dolphins (*Tursiops truncatus*) inshore on the Devon coast; but these were different: bigger, with a large white saddle on their backs; very distinctive creatures. Digging out the stored address, I parcelled up the best of the rather grainy photographs I'd taken and sent them off to Oxford University. Whereupon I received back a highly encouraging and informative letter from the scientist concerned, Dr Peter Evans.

These were White-beaked Dolphins (*Lagenorhynchus albirostris*), and not (at that time) often seen in such southern latitudes, being more commonly encountered in British waters in the North Sea and the west coast of Scotland. Complimenting me on getting the details down accurately, he suggested that if I should ever find myself up in Oxford, I should look him up as I might be able to help him out in the future, all of which sounded very interesting to me.

At that stage I really began to take serious notice of anything to do with Basking Sharks, and began to search for all of the available and current literature. Once again, I was amazed to discover just how little new scientific material there was on the creature, post-Matthews and Parker. There had been a few desultory attempts through the years to generate interest in restarting a British fishery, often by F.C. Stott, a marine biologist who had maintained a lifelong interest in the Basking Shark, and had collaborated with H.W. Parker at times. In an article in a 1974 edition of *Fishing News* entitled 'Why ignore the Basking Shark?' he put the case for a re-opening of the Basking Shark fishery, but happily his arguments fell on deaf ears. No doubt the memory of the unsuccessful and costly campaigns in Scottish waters still reverberated down the years.

Out and about in my new yacht, I began to take a far more practical approach to recording such encounters. Following a thoroughly stimulating meeting in Oxford, Peter Evans suggested some good books to help hone my skills, and gradually my species recognition improved to the extent that I was able to identify pretty much any cetacean that I saw, even at a distance. I had also been a member of the Marine Conservation Society (MCS) almost since their foundation, largely due to their campaign to protect the Basking Shark. But there was much more to their portfolio of activities and interests than just the Basking Shark, as a result of which I was learning far more about what was going on in the seas that I loved, both good and bad. So when they announced that they were planning to form local groups, I was all for it – I signed up straight away, joining many like-minded divers, sailors and beach enthusiasts who wanted to make a difference, and to protect our own little patch of sea.

MCS had launched a Basking Shark Recording Scheme in 1987, soliciting sightings of the species from members of the public around the British coastline. As there are always far more people walking the coastline than out on the water in boats, sightings tend to be more numerous from the shore, however limited to only as far as the eye can see. Therefore, boat-based sightings are always needed to provide balance and wider scale coverage, especially from remote sites that are seldom visited by boats, let alone walkers. So it was with this in mind that I was chatting on the phone one evening with then MCS Director Dr Bob Earll about helping to increase boat-based sightings. Ingenuously I suggested that we could run a weekend trip on my boat to record Basking Sharks. Bob roared with laughter and said, 'You can't do that – no one knows where to find them.' Feeling a little piqued by this rebuttal I said that I thought I could find them if the weather was right, and we parted on that point, Bob still chuckling at my naivety, me still puzzled by his response. By now I thought I'd got a handle on the ideal conditions to find them at one of their favourite spots, Start Point, so all we needed was some nice high-pressure weather and calm seas – how difficult could it be?

If I'd known then what I know now, I'd have been far more circumspect. But as I'd opened my big mouth, I'd have to back it up with action – and, hopefully, success.

A couple of weekends later the weather settled into a perfect pattern, so I contacted some of the local MCS team including my old friend Leon Edwards, and we set off for Start Point. The first evening out of Dartmouth wasn't ideal, with a light wind against tide chop, but we managed to find two Basking Sharks just to the east of the Point. On the second day conditions were perfect, with a glassy sea and a low swell, and we recorded five more just as before, feeding along the tide-lines. Onboard Marine Biologist Peter Wise told us these were actually 'tidal fronts', where a combination of cooler, mixed inshore water meeting warmer offshore water, together with strong tidal streams and uneven bottom topography, combined to create nutrient-rich upwellings, perfect for plant plankton to bloom and a rich food web to develop. As we were out in the main tidal stream east of the Point, and running along the southern edge of the shallow Skerries Bank, that made complete sense. These features were to become a significant role in all of our future survey work.

When I rang Bob on my return, he was surprised to say the least. But I certainly didn't feel smug about our success, as I was beginning to realise just how lucky we had been and what I could bring to the table. Finding the sharks had taken a great deal of effort, and if the weather hadn't been so good, almost certainly we would never have encountered them at all. Hanging around in such strongly tidal waters isn't recommended in any case, as even a moderate breeze against the direction of the tide will rapidly kick up a nasty sea, so wise sailors tend to avoid them if at all possible. Bob told me that he was preparing a publication on Basking Sharks around Britain, so if I'd send him our notes and observations he'd like to include them, which he did.

And then the phone rang. It was Peter Evans. Did I want to come and skipper a boat for him on a whale and dolphin survey in the Sea of the Hebrides? Did I ever. Whilst he might be surveying cetaceans, I was thinking of the opportunities to view Basking Sharks in Maxwell's historic hunting grounds. It seemed like the chance of a lifetime.

Going north at last

It was going to be a long, hard winter, tacking up and down the motorways of Britain doing my day job, dreaming only of the coming spring and the Sea of the Hebrides. Determined to be as well-prepared as possible, I ordered all of the pilot books for the areas we would be working in, vital for safe navigation in such remote and windswept areas, studded with rocks and besieged by strong tidal streams. And I got my first warning shot of what was to come.

The pilot books I was used to covered the well-charted English Channel and were very different from their counterparts for Scotland in one simple respect. When you signed up for the annual list of corrections and updates for the English Channel pilots you would receive (at most) a side of A4 paper, usually limited to simple stuff like the changing characteristic of a navigation light, or perhaps the installation of a new channel marker buoy at a local port. But the corrections for the various pilot books for the Sea of the Hebrides and the Minch amounted to pages of cautionary warnings of a far more alarming nature. 'Underwater obstruction (for which read "rock") reported in position, etc.' which instantly conjured up a vision of some poor soul 'finding' this rock with the bottom of their boat – at least they had survived to report it. Pilot books always do tend to have a somewhat discouraging air, but these pilots were downright alarming, with their endless litany of tiderips and over-falls, katabatic winds (described later in this chapter) and rotten, kelp-ridden anchorages. What had I let myself in for?

Well, not much more than I already knew, in fact, although it would be a while before I'd appreciate that. Sailing around the British Isles is never a picnic, and I'd received a basic grounding in the worst it could offer since taking up offshore sailing. The shallow shelf seas, fierce tides and endless procession of weather systems that roll in from the Atlantic were a challenge in the English Channel, let alone in the open waters to the west, where the Atlantic swell added a further complication. Passages in small sailing yachts were often little more than mad dashes between safe havens, where their crews would hole up in some welcoming hostelry, chink glasses and remark that it was 'better to be in here wishing we were out there, than out there wishing we were in here!'.

For example, to sail down west to the Isles of Scilly from our home port of Dartmouth, a succession of barriers lay in our way, all of which needed to be planned for and handled with care in order to reach a successful conclusion. The headlands that lie along the way, Start Point, the Lizard and Land's End are more than just lumps of rock to be rounded, but are true obstacles due to the tidal races lying off them, which when adverse can make progress all but pointless, hence the sailors name for them: 'tidal gates'.

The tides in the English Channel flood and ebb on a rough east–west axis twice a day, and to make best progress the small boat sailor 'works' the tide to gain a lift from the tide in the direction he wants to go. This doesn't necessarily amount to much except in the vicinity of headlands, when the tides speed up appreciably due to being diverted around landmasses, rocks and shallows. Another factor that needs to be taken into account is the lunar-influenced tidal cycle, as spring tides produce far stronger tidal streams than neap tides. The final – and deciding – factor is the wind, both its force and the direction it is coming from.

The tides off the gaunt, rock-strewn ships' graveyard of Lizard Point run hard up, especially on a big spring tide. A small boat wanting to fare west around the Lizard can therefore pick up six hours of positive tide to make great progress by being off the Point as the tide turns westbound. As long, that is, as the wind is light, or from the east (and not too much of it), then it's a magic carpet ride that will take hours off your journey. But switch the wind around and crank up the wind speed to, say, force 6 from the west, and the smooth, calm seas and powerful tide will make the next few hours some of the longest of your life. With wind against tide, the sea becomes wild and chaotic, with sudden, square, standing waves that can topple aboard your craft with savage intensity. At times it can feel like the boat is standing on its head – and you'll be glad that's the worst of it.

There's no quick way out by turning back, as with the tide now solidly against you, you're on a conveyor belt west, so you just have to batten down the hatches and work your way through it. In many cases the worst of the tidal race (as at Portland Bill) moves with the tide, so with planning you try and stay away from the worst spots at the worst moments. Other headlands have a 'race within a race'. In really serious winds these tide races are best given a wide berth of several miles at least, as some of the worst – Portland, Land's End, the Pentland Firth – have shown themselves to be capable of swallowing ships, let alone small craft. The yachtsman who through accident, inexperience, or stupidity, has found himself in the wrong place at the wrong time off one of the major headlands won't forget it quickly.

Looking at the charts for our passage to Scotland was a cautionary experience. The list of tidal gates and headlands grew longer and longer, each one with its own lurid claim to infamy. By this time, I was living in a tiny, remote cottage on the coast of north Cornwall, where, with the blink of Trevose Head lighthouse visible from the bedroom window and the sound of winter storms booming across the fields, it wasn't hard to feel a little apprehensive. But I consoled myself that I'd done that route before and survived it, and in any case, we were going to be in a big, solid boat – we'd be fine.

Peter Evans had been using an old sailing trawler, the *Marguerite Explorer*, for his summer surveys in the Hebrides, and during the winter she was based in the tiny china clay port of Charlestown on the south coast of Cornwall. The current skipper and part owner, Christopher Swann, was staying with friends up the coast from my home, so we arranged to meet up and discuss my potential involvement. As I was already working part-time as a commercial skipper, he was happy for me to take the boat north, but given that she was a completely different craft from anything I'd sailed before, I insisted that I should come as mate on the delivery trip, learn the ropes from him, and then we'd make a decision on whether I was ready to skipper the boat.

It was an instructive voyage. We left Charlestown on a fine April morning and almost immediately ran into bad weather, receiving a good drubbing rounding Lizard Point in a strong south-easterly wind, and had to put in to Newlyn to repair some minor damage to the rig. Rounding Land's End a couple of days later we made a comfortable crossing to the south-eastern corner of Ireland, where the Tuskar Rock lighthouse guards the entrance to the Irish Sea. Set too far west by the tide, we took another hiding as we battered our way around the Tuskar against a foul tide with a strong south-west wind pushing us directly into the worst of it. With the wind against the tide, conditions were memorably awful, and on a night 'as black as the inside of a cow' as the old fishermen used to say, the skipper and I found ourselves out on the end of the bowsprit trying to get the jib down. As the boat was pitching like a demented rocking horse, we were repeatedly dunked like teabags into the icy waters before being thrown skywards, hanging on for grim death. I can remember looking at the baleful wink of the lighthouse staying in the same place for what seemed like eternity, and wondering whether we would ever get around it. When I woke up several hours later, still in my full oilskins and still soaked to the skin, the now steady motion of the boat told me that we had finally made it into more sheltered waters – thank God.

This was only my second passage through the Irish Sea, and I could see why sailors seemed to take against the place. The tidal streams are for the most part fierce, many of the ports of refuge have tricky, shallow entrances, and the weather is often grim, wet and windy. It was an opinion that has never been diminished by many further passages in the ensuing years through that same piece of water. But on this occasion, it was all still fairly new to me, and in the spirit of exploration it was easy to overlook the worst of the flaws. Going off watch on a freezing cold midnight passing the lonely Maidens lighthouse north of Belfast Lough, I was filled with anticipation, dreaming of what we'd see the following morning as we finally entered Scotland's waters proper, as at that stage I'd never been further north than the Firth of Clyde.

Clambering back on deck at dawn, half asleep and stiff with the cold, I was confronted by a sight of such extraordinary beauty that I was almost knocked backwards. Off our starboard side lay the green landscape of Kintyre, ahead the pretty Isle of Gigha and away to the northeast the Paps of Jura, all lit in the bluest of skies above a skein of mist at sea level. I had never seen anything quite so striking, and that moment has never left me: my first vision of the magnificent, transcendent land and seascape that form part of the Western Isles of Scotland.

Down to business

On arrival at Oban, I took over for my first two weeks as skipper of the *Marguerite*, and with the excellent support of first mate Carlos Ogier, managed to manoeuvre my way around some of the islands without 'finding' any underwater obstructions or losing anyone over the side. Early season weather on the west coast can be either heaven or hell, but happily for me this time it was pure heaven, as the Arctic high descended over the area, giving light easterly winds, calm seas and clear skies. Of course it was bitterly cold at times, but that was a small debit against an overwhelmingly positive first experience.

At the end of the fortnight, I stepped ashore in an almost euphoric frame of mind. Having witnessed so much in such a short period – birds, whales, otters and much more – I felt a form of sensory overload. The boat had gone well, the guests seemed happy, and I was already looking forward to rejoining the boat in six weeks, this time with Peter Evans to start the survey work. If it was going to be anything like this, it was going to be easy. With hindsight I now realise how laughably little I knew, and how soon (and forcefully) I'd have that brought home to me.

Arriving in Mallaig in June was an eye-opener, with the wind blasting solidly over the pier-head, and the rain pelting down. The *Marguerite* was jammed into a corner, hemmed in amidst a gaggle of trawlers in port for the weekend, suspended in a cat's cradle of warps and festooned with tyres. How on earth was I going to get her out of there? Fortunately, my good luck hadn't completely abandoned me and after only a few minutes in the company of mate John Gyll-Murray I knew that here was a true seaman. Through deft handling of warps, and at the expense of my nerves, we soon had her out of the mass of boats and around to fill up at the fuel and water berth. The guests began to arrive in dribs and drabs, and we stowed their gear aboard and awaited the arrival of the great man himself.

Anticipation was high. As a frustrated marine biologist I had always wanted to be involved in this type of work, and the idea of 'sailing with a purpose' really appealed to me. Peter had already briefed me on the type of work we'd be doing, and had given me several of his reports and papers to read, to partially fill in the enormous gaps in my knowledge. Now the moment was finally here, and I couldn't wait to get going.

Mallaig remains the end of the line. The head of the most scenic rail route in Britain, and also the end of the road – literally; the next village to the north, Inverie, in Loch Nevis is accessible only by boat, or via a tough 20-mile hike in through the mountains. As Gavin Maxwell remarked in *Harpoon at a Venture*, 'beyond Mallaig all is gamble', and it was with that admonition in the back of my mind that we set sail on our voyage of discovery, straight into a whole new world for yours truly.

Up until that moment, sailing for me had been a well-understood, almost orderly pastime. You watched the weather, planned your route accordingly and set off toward your destination. If the weather turned bad when you arrived there, you would either stay put or make a short hop to a new safe haven, to sit it out there. None of this seemed to apply any more within this new regime. Even destinations were not necessarily solid, definable places. Instead we often had virtual destinations called waypoints entered into our GPS navigator – imaginary positions in the middle of nowhere, not off some reassuring harbour entrance.

The surveys we were carrying out were termed line transect surveys, which entailed the creation of a grid of straight lines laid across the chart that we would follow for the next fortnight, criss-crossing the Sea of the Hebrides in the south and the Minch in the north until (hopefully) we had completed them all. When 'on transect' along one of these lines, we would record our position and a range of environmental data every half-hour, including wind speed and direction, visibility, sea state, and swell height throughout the transect. This made it possible to quantify what is termed 'perception bias' where, for example, an increase in sea state or a decrease in visibility would materially affect the ability to see animals. Nowadays a transect will not be started if the sea state exceeds 3, and if conditions deteriorate so that the acceptable level is exceeded, the transect must be abandoned. An understanding of the surface and diving behaviour will also have to be taken into account to correct for availability bias when the data is being analysed.

At the start of each transect an assigned number of observers were tasked to watch for whales and dolphins from a standardised position and height of eye, with new observers being substituted throughout the duration of the transect in order to ensure clear minds and sharp eyes at all times. These observers had received basic training in spotting skills, species recognition and how to judge how far the animals were from the survey vessel. In this way, using a stable set of measurable parameters (what is called 'effort'), hard data can be derived (and compared/compiled with other data), to gain an understanding of the relative abundance of the various species in the area.

All of this was well and good for a scientist, but for a sailor, it had its challenges. The first was that the track line of a transect had to be carefully followed to avoid introducing error into the survey alignment. And that almost invariably meant using the engine rather than sailing. As the *Marguerite* was hardly the greatest sailing vessel (and went to windward about as effectively as a stone built church) this wasn't necessarily a drawback, but at the same time, rumbling along constantly under power was monotonous and the days were long. For a sailor this was no pleasure at all, and left you feeling like a glorified bus driver. The second challenge was that in the event of bad weather, you didn't just hole up somewhere and wait for it to moderate. No, out you went and moved the boat to place her in the most advantageous position for the expected weather improvement, which might be many miles (or days) away. This made entire sense when you only had a short period available for the survey, but it was a shock to the system for the crew, nonetheless. And if the weather was good, you worked from dawn until dusk without stopping and in the boreal summer that meant very long days indeed.

Not that it is was as simple as just holing up somewhere either, as true all-weather shelter was hard to come by. Almost every night found us at anchor in some remote sea loch or other, and in several of those it was very hard to get the anchor to set, either because the mud was very thin (more like Brown Windsor soup in some places) and the anchor would just pull through it, or it was just a mass of weed on the bottom that even our huge 100kg anchor struggled to penetrate. Getting the anchor to hold could be a frustrating exercise to say the least, especially with 10 pairs of tired, hungry eyes watching you, wondering what the problem was.

Shelter is a relative term in such places anyway, due to the complication of the mountains. Some of the anchorages that look ideal on the chart are far less perfect when the contours ashore are taken into account, as steep contours and valleys mean only one thing – katabatic winds. Derived from the Greek word 'katabaticos' which means 'to flow downhill', katabatic winds are formed of dense air that pours down a mountainside through gravity, often achieving wind strengths vastly in excess of the actual mean wind speed. In places like Loch Scavaig on Skye, Loch Scresort on Rum, or Loch Skipport on South Uist you might feel totally secure until the sudden arrival of vicious blasts falling down from the mountains. Gusts where the wind can rise in a few seconds from nothing at all to force 10 or more, laying the poor boat on her ear and sending her charging off like a demented bull to test the tether of her anchor and chain. There's no real rest to be had when it kicks off, as the gusts tend to be sudden and of such ferocity that they can knot the stomach of even the most sanguine skipper. Each gust might only last 30 seconds or so, but I guarantee you'll be wide awake afterwards. After a night of such fun in Loch Scresort that had seen John and I sit up all night to keep an eye out so we didn't drag the anchor (always a big risk with katabatic winds) we weren't sorry to reach the open sea again.

We set off the next morning on a long transect from the Cairns of Coll to Barra Head, the southernmost extremity of the Outer Hebrides, having already completed what I (in normal times) would have considered a full day's work. The weather forecast was awful, but Peter was determined that we give it our best shot. By the time we were halfway across, the cloud had begun to descend and a solid drizzle had set in. The wind steadily rose until our speed began to drop off as the *Marguerite* began to stick her bluff nose into the short chop of the Sea of the Hebrides.

For my long-awaited first crossing of these waters handled so well by the shark-hunters, I really didn't like this much, and I was by no means alone. Slowly but surely the number of volunteers on deck began to thin out, with many of the remaining poor souls succumbing to *mal de mer,* rendering them of little further use other than as ballast. Still, we punched our way west into the gathering gloom, with the rain now spitting menacingly in our faces. Peter

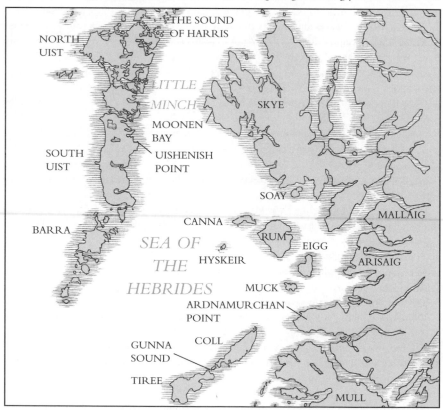

The Sea of the Hebrides and the Little Minch showing important historic and current Basking Shark sites.

insisted we keep plugging on, in the vain hope that we might make the end of the transect before the weather brought our day to a premature end.

It was not to be. Less than two miles from Barra Head the wind filled in fast and hard, and even Peter had to admit that the transect would have to be abandoned. With much relief we bore away towards Vatersay Bay and shelter, a fine supper and bed, glorious bed. Whereupon Peter dropped his bombshell. He wanted us to go on to South Uist, which would put us in a far better position to start the next section of the transects. The only haven of any significance in South Uist is Lochboisdale, a bleak place at the best of times with a notoriously tricky entrance. With the tide now against us for the next six hours, it was a long 30-odd miles away. As it was now early evening we weren't going to get in until well after midnight, which was not a welcome prospect.

Still, it was his show, and as the remaining volunteers cleared the decks, John and I tightened up our oilskins, donned our lifejackets and settled down to the slow, hard slog north. It was going to be a long night.

By the time we were able to alter course towards the entrance to the loch, conditions had got far worse; it was getting dark, the visibility had closed in, and we were already making good use of the radar to watch for trawlers following our lead and making their way to shelter. In the back of my mind was the worrying knowledge that Anthony Watkins had lost his shark catcher ship *Gloamin'* in the entrance to Lochboisdale on a bad night like this, and that was with a skipper who knew the place intimately. By the time we were into the channel proper, we could hardly see a thing through the now torrential rain, and the wind was moaning in the rigging. By now we had slowed to a snail's pace, as we scanned the radar on short range, blessing its ability to identify the coastline and the rocky outcrops ahead – at least those above the water. Plotting our position each time we got a chance, we steadily crept into safer waters at the head of the loch. We finally made our way up to the pier-head, to be confronted by rows of trawlers lying alongside one another, with not a soul in sight.

It took three attempts to get the *Marguerite* alongside, with John (who knew the boat well) taking the wheel, while I jumped aboard the nearest trawler with our warps. A few of the volunteers, revived by a fine dinner, poked their noses out like seals to watch the excitement for a moment, before wisely retreating to bed. It was now past 2 a.m., and as foul a night as could be imagined. Knowing that we were low on fresh water, John and I wanted to save ourselves a job in the morning, so set about piecing together a variety of bits of scrap hoses strewn about the pier, attached them to our hosepipe and eventually coaxed a steady trickle of water into our cavernous water tanks. We spent the rest of the night fighting an endless battle to maintain the flow as the pipes came apart or sprung leaks at every joint. By the time we were due to depart at eight in the morning we were soaked, exhausted and demoralised, and we had only succeeded in filling two out of three of the ship's tanks. But no matter, apparently – off we went.

The worst of the weather had gone through, and on a bright, breezy morning we scudded down the channel towards the open sea. It didn't look half as bad as I'd imagined in the daylight, but that was little consolation, and no sooner had we cleared the entrance than I went and slumped into my bunk after 24 hours on my feet. They could start the damn transect without me, I thought, totally whacked and not a little fed up with this whole research thing. I was going to go back to simply sailing boats for pleasure.

But as the days unfolded, I gradually began to understand and appreciate the logic of it all. The weather finally gave us a break, and in glorious conditions we sailed north past the Sound of Harris, marvelling at the exquisite clarity of the water, and the myriad shades of blue over the white-sand seabed. For several hours we sat mesmerised, watching Minke Whales (*Balaenoptera acutorostrata*) feeding in the approaches to the Sound, scattering showers of seabirds skywards as they lunged through the surface. Onwards north, through the notorious 'Stream of the Blue Men', the vicious tide race between Harris and the Shiant Isles where legend has it that the eponymous Blue Men would climb aboard small craft and demand that the cowering crew complete their song – or drown. Finally, we altered course to anchor for the night in the Shiants, spectacular basaltic outcrops that are home to an uncountable number of seabirds. That evening, sitting on deck, tantalised by the delicious aroma of roast beef wafting up from the galley, watching and listening to the coming and going of the Puffins (*Fratercula arctica*), Razorbills (*Alca torda*) and Guillemots (*Uria aalge*)as they whirred and wheeled around us before landing for the night, I had to admit that things didn't seem so bad after all.

But the weather gods hadn't finished with us yet. Another gale chased us into Lochinver on the mainland, and leaving the next day was trying, to say the least, bashing our way through a horrible head sea on our way up to the Butt of Lewis. Rounding the Butt in a mountainous swell raised a few eyebrows amongst the crew, some of whom discreetly retreated below decks. Turning back to head down the eastern shore of Lewis, at least we had some shelter from the worst of the seas, but it was clear that we were going to have to wait for better weather to commence the transects again.

Crossing to Canna, we had one of those unexpected highlights that occur when you least expect them, but are badly in need of some inspiration. The rain was bucketing down yet again, and the visibility was down to less than a few hundred metres so it hardly seemed worth looking out. But Peter, indefatigable observer that he is, refused to give up and as I saw him swing round and come almost dancing down the deck towards us, I knew that he'd seen something worthwhile. 'Orcas!' he shouted, and there, through the gloom and murk we saw one of the most spectacular wildlife sights of all, the two-metre-tall dorsal fin of a dominant male Orca (*Orcinus orca*), scything through the water straight towards us. And not just any old Orca, either, but the famous 'John Coe', identifiable by a large V-shaped chunk missing from the trailing edge of his dorsal fin, accompanied by a smaller, younger animal.

Even to an uninformed observer there is no mistaking that the Orca is a predator. There's an air of power and purpose to their movements, that made me a little nervous of their direct approach, but Peter said to hold the course steady, so I did. As they passed down on our port side I caught a glimpse of the big guy's eye, as he lifted his head up just far enough to give us the once over – a breath-taking, unforgettable moment. At such times it's impossible not to feel a connection with nature, even if it's just a one-sided affair.

As the two weeks drew on, we had accumulated an impressive tally of whales and dolphins. Minke Whale, Common, White-beaked and Risso's Dolphin (*Grampus griseus*), Harbour Porpoise (*Phocoena phocoena)* and of course, Orcas, making for a combined total of nearly 400 animals. But not a single Basking Shark, despite the knowledge that we were passing through waters where they had once been abundant. Had the hunters wiped them all out?

Finally, on the last day, as we made our way back to Mallaig, we saw our first and only Basking Shark of the voyage. I was almost ecstatic, yelled 'Shark-O!' and swung the wheel to port to turn the boat to get a better view. Whereupon the great scientist looked over the side at the shark, grinned and grumbled, 'They're just fish – I can't think why you're interested in them'. There's no pleasing some people.

Sitting on the train home, exhausted, I reflected on what I'd learned. Such as why, out of the dozens of potential skippers available, I'd got the job. The answer to that now seemed simple: no one else wanted to do it. Which was no wonder, what with the hard physical work, the endless hours on deck and the mad idea of heading out into the worst of the weather just to suit the survey schedule. It was certainly a million miles from the sailing I'd been used to. But I'd learned that Peter's resolute and dogged determination was the only way to get the job done properly, and yielded real benefits. I'd also learned an endless amount of fascinating information from him, about the survey techniques and the species we'd observed, as he was a thoroughly generous communicator. And I'd fallen totally and utterly under the spell of the Western Isles and their wild inhabitants, despite the weather and the work. On balance I thought I'd got a good deal. I was hooked.

But where were all the sharks?

Chapter 13
The Battle Begins

After two short summer seasons working in Scotland, in 1993 I decided to concentrate on using my own boat for survey work. Working for others wasn't easy and besides, I had my own businesses to attend to and a young family to raise. So I knuckled down to dedicating one week a month aboard my own boat, and set about developing a charter business based around my new home waters in Cornwall.

At that time, very few Basking Sharks were being recorded around Cornwall's coastline, yet as with so many other places, it had not always been so. By now I was an active member of Cornwall Wildlife Trust, and through them had met the wonderful Stella Turk MBE who had, with her late husband Frank, started the Cornwall Biological Records Unit (CBRU) in 1972. Stella is a remarkable and lovely person, erudite, funny and totally dedicated to her life's work, and like everyone who meets her, I was soon totally under her spell. She was kind enough to go through her records and give me a copy of all the recorded sightings since the CBRU had opened its doors. This also came with several pages of eyewitness reports of sightings culled from the local press.

My favourite amongst these was the record of a 'sportsman' in the late 1930s who had gone out in Gerrans Bay on the Roseland Peninsula in a small speedboat with a group of friends to harpoon a shark. Perhaps they had read the papers and learned about Anthony Watkins's similar exploits in the Firth of Clyde. Having successfully found their shark and achieved that aim, these worthies were taken on a Cornish version of the 'Nantucket sleigh-ride' some 16 miles out to sea before the harpoon finally pulled out of the shark. The report remarked drily that the skipper, not at all perturbed, had set out again the very next day to harpoon another shark, but with a different crew. It seemed that not everyone had enjoyed the experience.

But as observed in other areas, there were long spells when sharks went totally unmentioned, followed by shorter periods when records from around the county spoke of plentiful numbers of sharks. My friend Dr Nick Tregenza, then a local doctor, highlighted this through a novel survey method, interviewing long-term residents of west Cornwall who passed through his surgery, asking them to recall and date a number of natural history events such as Bottlenose Dolphin (*Tursiops truncatus*) and Basking Shark sightings. By comparing those reports with documented natural phenomena events such as lightning strikes and waterspouts, he was able to identify periods of abundance and even gain a broad idea of numbers of such wildlife from the past. In the case of the Basking Shark his data showed that the 1930s had seen a marked upswing in the number of sightings recorded around the west Cornwall coastline.

Following his lead, and using data from the CBRU, together with records from the Marine Conservation Society's (MCS) Basking Shark Watch project, I could see that this pattern had occurred on an irregular basis in more recent years, with another peak in the 1960s and most recently, in the late 1980s. Realising that this cyclical abundance might be important and wanting to identify the next upswing, I set up a small Basking Shark recording project within Cornwall Wildlife Trust, with the aim of ensuring that all possible records would be received and, once we had them in our hands, would be shared with our friends and colleagues at MCS. And as it was of paramount importance to secure records from out at sea as much as from the cliff top, I made a tour of yacht and dive clubs, giving talks to their members with a view to recruiting them to the cause of recording and reporting. Finally, whenever I could muster a crew and the weather would allow us, we'd set off in my 11.7-metre yacht *Forever Changes* to conduct line transect surveys in an area that was bounded by Berry Head in south Devon and the Isles of Scilly west of Land's End.

Those original surveys were so exciting and rewarding. As we were starting virtually from scratch, and there were no books detailing hunters' exploits from this area, we really had no solid idea where the key 'hotspot' sites were to be found. Sure, we had a lot of anecdotal evidence and some amazing eye-witness reports from over the years, but equally, most of the records were from land-based observers, and we had very few from vessels at

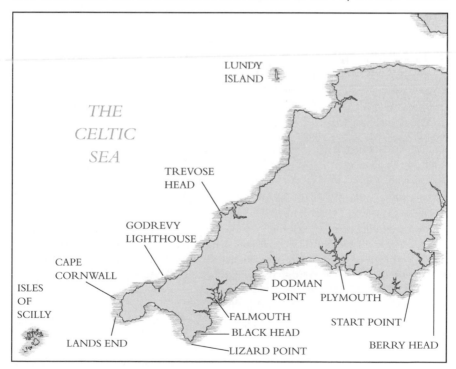

The south-west peninsula survey area.

sea. I had a good handle on the situation from my sailing contacts, but finding information from the most logical resource, the local fishing fleet, proved more awkward. Some fishermen see conservationists as meddling tree huggers and some conservationists see fishermen as damaging fish stocks and the seabed. Sometimes in the heat of the argument harsh words are spoken and non-cooperation results. This is a shame for both sides, who in fact want much the same things – abundant fish stocks and a healthy marine environment. We might disagree at times on how to secure those objectives, but ultimately we share the same aspirations.

Fortunately, I had grown up around a fishing community and had family members who were at sea with the fleet; they were kind enough to tell me what they knew. Nor was it the case that all fishermen had taken an 'oath of Omertà'. Some, such as the late Peter Pearman of Newlyn, were immensely helpful, and the late Lance Peters of Helford was another willing source of information. Ray Dennis of Newlyn was a great source of stories from the region that added more than a little colour and focus to the quest. Slowly but surely we got the project off the ground.

So through the nineties I began to gain a better understanding of the seas around my home. I was lucky enough to secure a good number of willing participants for the surveys and we made steady progress in mapping the area. Not that it was all plain sailing, not least because the waters west of Lizard Point were challenging and the weather unpredictable. But once past the tidal barrier of the Lizard into the open Atlantic the rewards were far greater, and we discovered just how productive those waters were, recording a rich and varied palette of marine life. We enjoyed the antics of Bottlenose Dolphins on occasion, occasionally encountering some of the rarely seen large offshore pods. The Common Dolphin was our regular companion once west of Land's End on the way out to the Isles of Scilly, and on a very few occasions we recorded Long-finned Pilot Whales and the shy, charming Risso's Dolphin. Graceful, with their pale pigmentation seeming almost ghost-like, these latter were, and remain, my favourite cetaceans.

And it wasn't just cetaceans that we saw. More and more frequently we encountered the Ocean Sunfish (*Mola mola*) the world's largest bony fish, that can weigh up to two tonnes and achieve a diameter of three metres. The first time you encounter one of these improbable looking fish, you can hardly believe your eyes. This is one of those creatures that looks to have been designed by a committee, with a dinner tray body and narrow, spike-like dorsal fin. Flipping and flopping from side to side, they appear to be suffering from some form of severe gyro malfunction, unable to even swim upright, but the laughter soon subsides when they take off at speed, or even, as I discovered to my amazement, leap clear of the water. In keeping with its eccentric image, the Sunfish doesn't do this the easy way, but simply lurches out of the water at any angle at all, looking like nothing more than someone playing frisbee with a rather limp pizza. 'Weird' doesn't even begin to describe the Sunfish, yet it is clearly an oceanic success story – on its own terms.

Best of all in my view were the rare sightings of the amazing Leatherback Turtle (*Dermochelys coriacea*). These magnificent ocean wanderers are right on the limit of their travels when they reach our shores, having made long Atlantic circuits from their nesting beaches in west Africa and the Caribbean (in the case of the females), in pursuit of the thinnest of diets, jellyfish. Sometimes we experience massive blooms of jellyfish, such as the enormous Barrel Jellyfish (*Rhizostoma pulmo*) that invade bays along the west coast of Britain and Ireland, and the Leatherback can then be seen on rare occasions.

It's such a privilege to see these ocean giants, endangered as they are by a myriad of threats generated by man. Leatherback Turtles regularly become entangled in fishing gear, whether it be long lines in deep water or nets and pot ropes closer to shore. Their nesting beaches in island beauty spots are often lost to hotel developments that drive them from their native nest sites. Their nests are still raided for their eggs in some parts of the world, sometimes for their supposed aphrodisiac qualities. Last and worst, they mistake the ubiquitous, discarded plastic bags for jellyfish, ingesting them and choking their gut with an indigestible morass of plastic, often with fatal consequences; yet further victims of our careless, throwaway society.

And of course we were recording Basking Sharks! We were soon able to confirm that there were indeed sites where the odds of making a sighting were higher, such as Start Point, Plymouth Sound, the Eddystone Lighthouse and most especially Lizard Point and Land's End. Numbers were still very low though, and the major question was when – or even, if – there would ever be a return to local abundance. Was this only a cyclical downturn, or a sign of a far more serious malaise? After such a long spell with little sign of increasing numbers it was very hard to convince anyone to offer us funding to help us expand our activities. Even small-scale support would have allowed us to scale up our study significantly and conduct far more extensive surveys that would have the potential to address many of the remaining questions as to numbers and distribution. But try as we might, we simply couldn't convince anyone to back us financially. Some cited the practical difficulties of studying a creature that spent the vast majority of its time under water, others felt that previous efforts had not paid off, and were wary of further investment. It was frustrating to say the least.

At this stage, it's worth remembering that in the early 1990s we knew very little about the Basking Shark (at least in terms of field studies), most of the really authoritative scientific information still being limited to what had been gleaned from the dissection and study of dead animals. As far as our understanding of population, seasonal migration and overall distribution was concerned we knew no more than the hunters of the post-World War II era. That is to say, virtually nothing, and most of what little was known was based on speculation.

There was, for example, no estimate of population on a global, regional or even local level. As far as distribution and seasonal migration were concerned, an almost total lack of concrete evidence meant that there were competing theories even in the best understood of the seas, notably the North-east Atlantic.

What *was* known from the North-east Atlantic was that Basking Sharks arrived on western coasts during the spring and early summer, across a broad geographical range from Portugal in the south to Norway and Iceland in the north. Around the British Isles the sharks tended to appear in the English Channel, off the west coast of Ireland and in Scottish waters from early May, with a general disappearance from September onwards. This movement was generally agreed to enable the Basking Shark to profit from the seasonal abundance of plankton in those inshore waters, and the possibility for widely dispersed animals to congregate and take part in mating.

It was also broadly agreed that the species was not just a shallow-water, coastal inhabitant throughout the year, owing to the presence of squalene in its liver oil. Squalene, a terpenoid hydrocarbon, is found in the livers of deep-sea sharks at high levels of concentration, where its low specific gravity helps the liver to act as a 'hepatic float' in place of the gas-filled swim bladder found in most fish. Squalene exists in the liver of the Basking Shark at levels that can approach 40%, suggesting that the species must spend at least part of its life in deep water. The liver also acts as an energy store that might be drawn upon in winter when plankton density is low. Patrick FitzGerald O'Connor discussed with two other shark hunters his own observation that the livers were lighter at the beginning of the hunting season than at the end, and all agreed that this was the case. This led some scientists of the time to speculate that the Basking Shark might migrate offshore and hibernate on the seabed during the winter, perhaps off the coast of North Africa, subsisting on the energy reserves stored in its enormous liver – a hypothesis that was soon going to come under fire.

Scientists also speculated over whether there was one single large North Atlantic stock that coalesced in deep water in winter before dispersing inshore over long distances across the summer range, or whether there were smaller, discrete, local populations that simply moved into adjacent deep water before returning over much shorter distances to the same regions in summer. This gave rise to two competing migration theories, the south to north migration and the west to east migration.

The south to north migration theory tied in well with the idea of a single stock in deep water off North Africa, and seemed to be supported by the progressive timing of arrival off European coasts, earlier in the south and later in the north, dictated by longer distances to travel. The west to east migration might still involve a single stock, but seemed more likely to be made up of local populations that migrated inshore from the deeper waters off the adjacent continental shelf.

The former theory was favoured during the earlier part of the twentieth century, and was the blueprint that Gavin Maxwell subscribed to when planning his Soay project, whilst Anthony Watkins was an early convert to the latter theory. However, by the 1960s the south to north theory had largely been discredited owing to the lack of any evidence of a return migration southwards. In an attempt to explain the collapse of the Achill Island fishery, H.W. Parker and F.C. Stott proposed an 'Inshore-Offshore' theory, suggesting a discrete local population that showed a high degree of site fidelity, spending the winter in deep water off the continental shelf before returning inshore during spring to mate and feed. The depletion of this stock in Ireland and Scotland due to overfishing was matched by higher relative abundance elsewhere such as the Norwegian coast, which tied in with the idea of very localised stocks.

In 1990 Alistair Stagg proposed a further refinement of the Inshore–Offshore theory, arguing that whilst the Basking Shark did migrate inshore to the shores of the British Isles from deep water in spring, this was on a widespread west to east axis, not a narrow base. This hypothesis suggested that the sharks actively foraged for the most abundant areas of plankton on arrival, while their later arrival farther north was dictated by the species following the plankton blooms progressively northward during the summer. This hypothesis took into account the effect on bodies of water around Britain and Ireland due to the restrictions placed by islands, headlands and tidal bottlenecks that would slow the progress of productive patches of plankton northward. For many years, variations on this hypothesis would be the nearest thing we had to agreement on the migration of the Basking Shark. It would take a later generation of scientists armed with exciting new technologies to improve upon it.

Nor did we have any idea why there had been periods of extreme localised abundance, such as had occurred in England, Scotland and Ireland from the mid-1930s until the mid-1950s. Apart from hypothesising that these 'invasions' were most likely caused by changes in plankton abundance or distribution that may in part have been due to climatic variations, there was no greater understanding. And in the light of the collapse of local fisheries that had been observed at Achill, Barkley Sound, Nakiri and elsewhere, we could only speculate whether such significant increases in abundance would ever happen again. Were the big shoals that had astonished even the hunters gone forever?

The first few years of boat-based surveys in Cornwall didn't offer much hope. Sightings remained stubbornly rare, and there was no clear pattern to be discerned from them. Out on our infrequent surveys for Basking Sharks we generally drew a blank, and so had to rely on sightings coming in from the public. The same pattern continued until the 1997 season, when there appeared to be a marked increase in numbers recorded. But it was the 1998 season that was to be the game changer. Late one afternoon in May, I was working from my home in Cornwall when the phone rang. It was one of the Cornwall Wildlife Trust supporters down on the Lizard peninsula, and he had the most amazing news. He had been out with friends fishing in a small boat earlier that day in Kennack Bay (on the eastern side of the Lizard) and had sighted a few Basking Sharks feeding at the surface. Steadily, the number of sharks increased until the boat was completely surrounded by a huge shoal of sharks. The skipper of the boat became concerned, as the sharks seemed to encircle them, and they decided to head back to Cadgwith Cove. Working their way through the shoal was unnerving, but eventually they emerged into clearer water and finally the safety of the cove. Jumping ashore, they had rung the Wildlife Trust, who gave them my number to call.

This was an extraordinary report. I'd never heard of anything like this before, and their estimation of at least 200 sharks in the shoal was phenomenal. They told me that they could still see the sharks well offshore, and if I could get myself down there as quickly as possible they would take me out in their boat to have a closer look. I couldn't wait. As I gathered my cameras and some warm clothing I glanced out of the window onto the creek outside our home and noticed the first tendrils of mist making their way up the valley. Ten minutes later we were completely enveloped by a solid wall of fog, and visibility was down to less than 50 metres. This was the worst of all possible events. The drive to Cadgwith would take two hours in good conditions, and in fog like this at least twice that time, by which stage it would be getting dark, so there seemed no point in setting off. All I could do was console myself that if we had thick fog here, it would almost certainly be the same at Cadgwith.

It wasn't. At 9.30 that evening the phone rang again and the excited voices at the other end told me all I needed to know. The fog had bypassed the Lizard altogether and the fishermen had once more been out amongst the shoal, which now stretched from Black Head right back to Cadgwith. Most of the boats had returned to shore, nervous of being out amidst so many sharks at dusk. And with good reason, as one small boat that was entering the Cove had a shark breach right behind it, far too close for comfort. All agreed that the number of sharks now exceeded 500, simply stating that beyond that they just couldn't count. And I had missed it!

With the alarm set for 3 a.m., I passed a fitful night, unable to sleep properly from anticipation. At 5 a.m. I was standing on the cliffs by Kennack Sands as the sun came up, looking out on a sea apparently devoid of all life. Nothing. Flat calm, but not a shark in sight. Out at the old Coastguard hut at Black Head the story was the same, just an expanse of smooth water, unbroken by fins – it was heartbreaking. All I could do was to retreat to Cadgwith to talk over the previous day's events with the still buzzing boat crew. I had missed the main event, but the story they told had me excited, too. The sharks might have moved on, but they had to be still out there, not far away. All we had to do was keep looking. Perhaps the upswing we had long awaited and hoped for was at last upon us.

Destructive brute or gentle giant?

By now the public perception of sharks was changing, albeit slowly. Films like the seminal 1971 documentary *Blue Water, White Death* or the amazing television documentaries of Jacques Cousteau motivated as many young people as horrified their elders, and perhaps for the first time the public fascination with sharks was turned from revulsion and fear into something more positive. At last, it seemed, people began to see sharks as extraordinary creatures that played a vital role in the marine ecosystem. Slowly but surely, the public began to take a more detailed interest in sharks, and began to understand the vast difference between the feared and ferocious White Shark and the huge and harmless Basking Shark. Although killing the former would still be justified (at that time) on the basis of protecting bathers, killing the latter began to be seen as pointless and destructive. A younger generation were seeing them more as victims than aggressors.

It was against this rapidly changing backdrop that a Basking Shark fishery re-opened in Scotland. In the late 1970s a fisherman in the Firth of Clyde, Howard McCrindle, had caught a number of Basking Sharks while mid-water trawling north of Arran, and had brought four of them ashore for processing. Oil prices continued to remain high at this time, the Norwegians were still in the market, and it must have seemed attractive enough to warrant further exploration. So in 1982 he decided to try a hand harpoon before purchasing a Kongsberg harpoon gun from one of the Irish fisheries, and devoted the summer season to hunting Basking Sharks, albeit with limited success, with only one fish taken.

But in the following year, fishing from a smaller boat, The *Franleon II,* McCrindle had what would prove to be his most successful season, and this began to attract far more attention from the press and the conservation lobby. Now equipped with the same highly effective harpoons that the Norwegian fleet used, he killed 122 sharks, following this with further good seasons in 1984 and 1985 with 92 and 40 sharks killed respectively. That success put him firmly in the epicentre of the argument about the changing use of the oceans and their inhabitants.

McCrindle set a self-imposed minimum size limit of six metres on the sharks he harpooned, to avoid killing juvenile sharks, which was commendable. But the argument against the killing of Basking Sharks had now become so heated that this apparent gesture towards conservation was dismissed as far too little, too late. Shamefully, some activists took it far too far and threatened to petrol bomb McCrindle's home. In addition, an underwater saboteur attempted to sink the *Franleon II* by drilling holes in the hull, tarnishing the reputation of conservationists and making the argument personal, which was something it should never have been allowed to become.

Not that it stopped McCrindle. In 1986 he took 38 sharks with the *Franleon II.* But from there on it was all downhill for him. Fishing from a succession of different boats, the numbers he caught fell steadily through the ensuing years until 1994, the final year that he hunted, killing only nine sharks. By that stage the financial rationale to hunt the Basking Shark had been seriously undermined in any case, as by the early 1990s the price paid for the liver-oil (exported to Norway) had fallen to £250 per ton, making it no longer a viable economic proposition to land the livers. Although McCrindle had tried to market the flesh at various times, including an attempt in 1983 to sell Basking Shark fish suppers via a fish and chip shop in Girvan, it hadn't worked out.

Shark oil was still in demand, mainly for the squalene that it contained, making up around 50% of the oil. Most of the squalene was converted into its hydrogenated liquid form (squalane), where it was used in a wide variety of cosmetics, moisturisers and health care products including haemorrhoid cream. It was also used as high-grade precision machine oil, where its low freezing point was an advantage in some specific circumstances such as high altitude aircraft oil. During this period, as well as importing, Norway was one of the main exporters of Basking Shark oil, selling 130 tons in 1997 with a value of $US 1.2 million. But the price paid for the oil had already declined dramatically after Portugal and Spain developed fisheries for the Leafscale Gulper Shark (*Centrophorus squamosus*), the Portuguese

Dogfish (*Centroscymnus coelolopolis*) and the Birdbeak Dogfish (*Deania calcea*). These small deep-water sharks were all abundant sources of squalene-rich liver-oil, supplies of which flooded the market, driving the price down to very low levels and effectively spelling the end for the Basking Shark oil market.

This left only the fins, which were now fetching colossal prices of up to £20,000 per ton, sold to Asia for the shark fin soup market. Some sets of Basking Shark fins, due to their enormous size, never made it into soup and ended up as trophies that adorned the walls of high-end oriental restaurants. Howard McCrindle was reported at that time as having sold 92 kilograms of marketable fins from a single large female, which represented a return of around £1,600 at that time.

Perhaps with one eye on the dwindling stock of Basking Sharks and the likelihood of some form of protection for the species in the near future, a Japanese fin trader, Kuniaki Takahashi, reported travelling to Norway to buy as many Basking Shark fins as possible, commencing with three tonnes of fins in 1995 at $14/kg. In 1996 he then bought the entire Norwegian fin stockpile for that year of 16 tonnes at $23/kg, and reported that his actions had aroused considerable opprobrium from his fellow traders for driving up the world price. But as we shall later see, they hadn't seen anything yet, and Mr Takahashi's smart move to corner the market soon paid off handsomely. The liver was no longer the 'elephants ivory' – now it was the fins alone.

Before he hung up his gun, Howard McCrindle took part in an excellent documentary entitled *Sharkhunter*, that allowed both sides to put their case for and against the hunting of Basking Sharks. Bob Earll spoke passionately and eloquently against the continuation of the hunt, while McCrindle put up a spirited and well-argued defence of the practice. It was a fascinating tussle without rancour, and opened up the debate across the nation. I must have watched that programme 10 times and learned something new every time.

The resumption of hunting had a number of consequences. It provided the marine conservation movement with something to mobilise against – if the Basking Shark had remained unmolested it might have been far harder to galvanise the political will necessary to provide legal protection for the animal. That this harmless, iconic, gigantic fish was now being killed for its fins alone seemed indefensible – after all, nobody was going to starve if the hunt were to cease. Yet the prices being achieved from the sale of fins meant there were still serious consequences if the shark remained unprotected, and gave the campaign for protection an impetus it had never had before. The sheer scale of the press and television coverage generated by the debate ensured that there were very few people the length and breadth of the country who were not aware of the plight of the Basking Shark.

Since the mid-1980s MCS and the World Wide Fund for Nature (WWF-UK) had been laying an unsuccessful siege to the British Government to have the Basking Shark protected by law from hunting, selecting the Wildlife and Countryside Act (1981) as the best legal mechanism to do so. There were many good reasons to seek to secure such protection, even though at that time hunting had dwindled to one individual vessel in Scotland. Conservationists were legitimately concerned about the damage caused to stocks during the previous 50 years, and the only way to start a recovery was to stop hunting. The argument was simple: as long as hunting was legal, it could just as easily flare up again (as it had many times before) when the financial value of fins or liver oil rocketed. But it was to prove a long, slow and painstaking process.

Initial support for the proposal to government to protect the Basking Shark in UK waters came from their own statutory conservation agency the Nature Conservancy Council (NCC). In fact, Bob Earll credits the original impetus to seek protection for the species to Dr

Roger Mitchell, NCC's Head of Marine Science, who persuaded him to generate support for the proposal through MCS. However, the first application in 1986, drafted by Sarah Fowler, was rejected by the Government on the grounds of a lack of scientific evidence to show that the Basking Shark was, in fact, endangered. As applications to add new species to the Wildlife and Countryside Act could only be submitted every five years in a process called the Quinquennial Review, conservationists would have to wait until 1991 to submit a further application, which didn't sit at all well with the knowledge that Basking Sharks were still being killed in British waters.

The second application, proposed by NCC's successor, the Joint Nature Conservation Committee (JNCC), met the same fate as the first. The same reason was given for the rejection, although there was widespread knowledge that in fact it had more to do with the lack of will to protect a species still being commercially fished in UK waters – by one fisherman. The move to protection was also hindered by the labyrinthine nature of fisheries politics on the international stage, in particular those that concerned European Union (EU) fisheries policy. The EU had agreed a quota of Basking Shark liver weight with Norway in 1982 of 800 tonnes (approximately 1,600–2,000 sharks), in exchange for a quota of white fish taken in Norwegian waters by EU fishing vessels. This quota was reduced to 400 tonnes in 1985 and fell to 100 tonnes in the 1990s, before being scrapped altogether in 2001, by which stage it was virtually meaningless anyway, as the Norwegian fleet were only landing the fins, and their catch data had been based on fin weight since 1992.

But conservationists had mobilised, and were now actively gathering the data that was desperately needed to further make the case to protect the Basking Shark. Largely driven by Bob Earll's passion for the Basking Shark, MCS continued to propagate research through collaboration with talented Master's students like Alistair Stagg, Helen McLachlan, Jill Strawbridge and Powell Strong and undergraduate Scott Farmery. To do so, these researchers drew on their own studies and the information held in the MCS shark sightings database developed by Sarah Fowler and Jonathan Spencer (updated by Jeff Tang), which included our own boat-based sightings data. The team, led by Bob Earll and John Turner, produced the first truly contemporary study from a public sightings scheme based on the data gathered between 1987 and 1992, published as a report to the JNCC in 1993.

There were glimmers of hope. In the Isle of Man, local marine biologist Ken Watterson had for many years conducted Basking Shark surveys around the island, and put pressure on the Manx Government to secure local protection for the animal. This was finally achieved in 1990, and as this was the first time that the species had been protected from hunting in the waters of the British Isles, it represented a significant achievement. This was followed by a similar move on the island of Guernsey – small steps, but in the right direction, that sustained hope for future efforts on a wider scale. But still the government determinedly held out, despite the fact that hunting had ceased in Scotland in 1994, and that the quota with Norway was by now merely a sideshow.

Finally, in 1998, at the third Quinquennial Review, the combined effort of all of the non-governmental organisations (NGOs) was rewarded when the late Michael Meacher, Minister for the Environment in the Labour administration, announced that the UK government accepted the argument to protect the Basking Shark, and it was to be added to Schedule 5 of the Wildlife and Countryside Act (1981). This piece of legislation protected the species from intentional killing or injury, and prohibited the sale of the shark or shark parts. Result! Although the Act only had jurisdiction out to 12 nautical miles, hunting in UK waters was effectively at an end. It was a major victory in symbolic terms, and set the scene for further conservation measures that would give the Basking Shark a chance to recover from centuries of exploitation. Now was the time for scientists and conservationists to pick up the baton and shine as much light as possible on this enigmatic and endangered creature.

The modern scientists

For many years, scientists had backed away from studying the Basking Shark, either in the laboratory or in the field. Partly this was because the anatomical studies carried out on dead sharks by Matthews and Parker at Gavin Maxwell's factory on Soay were so comprehensive, but also because the practical conveniences that the Soay factory afforded to assist those two fine scientists in handling such a huge fish were hard to come by elsewhere. Dissections were (and still are) predominantly carried out on beaches where an animal had stranded, when the fish might be less than fresh, and then only on single individuals. Matthews and Parker had, by contrast, enjoyed the luxury of dissecting a number of very fresh fish, both male and female, with the aid of many men and purpose-built machinery.

In the field, there remained the difficulty of finding a reliable supply of live Basking Sharks to study, preferably at a site inshore that offered sufficient shelter to access the animals on more than the odd occasion when the weather would allow. And there were obvious limits to what could be studied in such circumstances, given that surface activity was acknowledged to be only the 'tip of the iceberg', the Basking Shark spending the vast majority of its life down in the depths of the sea, far beyond the human eye.

Studies targeting surface behaviour, feeding strategies and regional distribution were all viable projects, but that was the limit for many years. In the absence of suitable technological solutions, the secret life of the shark looked set to remain inviolate.

Until, that was, Dr (later Professor) I.G. 'Monty' Priede of the University of Aberdeen came onto the scene to conduct an experimental programme of tagging Basking Sharks with satellite transmitters in the Firth of Clyde in 1978. His team used a small inflatable boat equipped with an outboard motor to approach a Basking Shark at the surface, and with the aid of a long pole, jab a stainless steel dart into its back behind the first dorsal fin.

The dart was attached to a wire trace that led to a buoy and a fiberglass pod containing a data collection Platform Transmitter Terminal (PTT). The internal transmitter in the PTT was programmed to send a signal to the Random Access Measurement System (RAMS) in an American NASA Nimbus satellite, allowing its position to be established. The aims of this initial experiment were to test the viability of such a device, and to establish whether it was possible to (a) track shark movements when feeding in the summer, and (b) in the long term to try and find out where the sharks went in winter.

Naturally, the attachment of the apparatus to the shark required a significant amount of boat-handling skill and strong nerves, as the shark might react badly to the prick of the dart and lash out with its tail. The helmsman had to be ready to accelerate away from the shark at precisely the moment that the pole carrying the dart was pulled back, releasing the dart, trace and the PTT, with the whole apparatus being launched over the side without entangling the boat or crew. Not that they were always successful – on one occasion, a blow from a shark's tail burst one side of the dinghy and pitched two of the crew members into the water. They were very lucky that the third crew member remained on board to rescue them, and that the other side of the dinghy remained inflated.

The first experiments were only partially successful. The fibreglass pods proved incapable of withstanding the pressure at depth and leaked, ruining the electronics within. But one of the three pods retrieved had remained attached for two months, which was encouraging. In 1979 the experiment was repeated with less success, as the pods proved too large and detached from the sharks. By 1980 the Nimbus satellite system was becoming obsolete, so no further pods were attached. Instead, the time out was used profitably to redesign the pods to accommodate more powerful lithium batteries and more sophisticated internal electronics to enable the connection to a new satellite system. With the new model, when the shark came close to a depth of one to two metres, the pod would break the sea surface and the aerial in the fin would provide an uplink to a polar orbiting US earth resources satellite when it passed overhead. The data and position location was then forwarded to the joint US-French Argos satellite system in Toulouse, to be sent to Dr Priede and his team via telex.

Despite the technical improvements contained in the upgraded PTT, the pod itself was also much smaller than the original device, providing useful benefits such as less drag and buoyancy, and so made the pod less likely to pull out its tag. This was a valuable improvement, and was vital in 1982, as due to greatly increased costs associated with the new equipment, only one pod was available. The tag was attached on 27 June north of Arran, and the first uplink was achieved five days later, 55 miles to the south, off Ailsa Craig. Over the next two weeks the position of the shark was located whenever it coincided with a satellite overpass, as it stayed at or near the surface on most days, circling Ailsa Craig. In that fortnight there was only one sunny day, during which the shark spent most of the day at the surface, apparently feeding in a body of warmer water. It appeared, on the face of it, that the concept of 'basking' behaviour linked to sunshine was not as far-fetched as some had suggested.

On the 17th day, the towing trace broke, ending the experiment prematurely. Due to the short attachment period, there was insufficient data to glean anything other than the most basic insights into shark behaviour, but overall the experiment had been a resounding success, and had proved that the concept of attaching a satellite transmitter to a Basking Shark was feasible and had the potential to yield significant amounts of data. The limiting factor at that time was the equipment itself, which was fairly unsophisticated technically, and clumsy to handle due its size. But the first major step had been taken, and it was abundantly clear that with technical improvements, this technology would lead to far greater insights in due course.

Unsurprisingly, such remarkable technology created a great deal of interest in the small sector of the scientific and conservation community interested in the Basking Shark, offering as it did the opportunity to study the species other than seasonally when it presented itself at the surface (and only when someone was looking). Successful attachment of tags to other species of shark, cetaceans and seals around the world had helped to refine the technology and had shown just how effective it might be for the Basking Shark. However, securing adequate funding for satellite tracking studies in the UK remained difficult. Not only was the equipment itself costly, largely because it was still basically custom made, but also because the satellite time required to transmit the data from tag to laboratory remained highly expensive. Although the equipment itself was steadily becoming more sophisticated, less bulky and with extended battery life, it remained prone to detachment, and there were still serious concerns over the reliability of the tags themselves, which had a high failure rate. This must partly explain why there was a gap of 14 years between Priede's pioneering study and the next satellite tagging programme, although it is also fair to say that during that ensuing period, funding for *any* Basking Shark research programme was equally hard to find.

In 1996 two scientists from the University of Durham, Dr Mark O'Connell and Dr Tim Thom, picked up the baton and spent an experimental season in the Firth of Clyde around Arran, searching for sharks to tag with a satellite transmitter.

For 1997 they developed a hybrid project, funded by Scottish Natural Heritage. Four complete sets of satellite tagging equipment were purchased, and access to the Argos satellite system in Toulouse was secured at low cost via a scientific users' group. The satellite tags were latest generation devices equipped with sensors to enable them to record sea surface temperature and depth, including any dive sequences that a shark might undertake, as well as transmitting positional information. In addition to the satellite tag, each 'array' was also equipped with a data logger to record fine-scale temperature, depth and water speed, a one-second repeater acoustic 'pinger' to enable the animal to be located and tracked at short range (less than two kilometres) via a directional hydrophone, and at the outer end a buoyant acoustic tag that included a VHF transmitter to enable tracking of the animal at the surface.

The physical attachment of each tag followed a similar methodology to that employed by Priede, using a 2.5-metre hand-held pole to drive a stainless steel dart into the back of the shark behind the first dorsal fin, then as the pole was retracted, the dart was left securely embedded. A two-metre-long wire trace attached to the dart led to the first piece of equipment in the array, the buoyant satellite tag equipped with a VHF aerial and transmitter. As only the satellite tag was intended to remain attached to the shark over a long period, the wire trace that led from the satellite tag to the other devices in the array was equipped with a 'galvanic' magnesium alloy link that would corrode through in seawater within three days. This would allow the rest of the array to float free to the surface, enabling these items to be retrieved with the aid of the VHF transmitter and the pinger.

In order to locate sharks for the project to tag, and to develop an effort-related sightings database, a team of 40 volunteer observers was recruited (largely from the Scottish Wildlife Trust) to scan a stretch of the eastern coastline of Arran, north of Brodick between Corrie and Lochranza, during daylight hours. A 10-metre-long motor boat had also been chartered to act as the main research vehicle for an 'on the water' team, cruising up and down the same stretch of coastline simultaneously, taking plankton samples at set stations twice a week.

This 10-mile long stretch of coastline was broken down into five sections, with a shore-based observer team in each section. Primarily, their aim was to conduct a visual survey to spot Basking Sharks at the surface within their sector and pass those sightings on to the boat-based team by mobile phone. Once the boat-based team had received information of a sighting from the shore-based observers, they would make their way to the position of the sighting. If they located the shark, O'Connell and Thom would attempt to tag the shark from the motorboat. In the event that this proved difficult, the tagging would transfer from the main vessel into an outboard motor-powered inflatable dinghy containing the tagging equipment, and make a careful approach to the animal and attach the tag.

The system worked well, and a total of 29 Basking Sharks were sighted over the season, five of which were approached with a view to attaching tags. The first two approaches with the motorboat proved abortive when the boat, probably due to noise from the engine, disturbed the sharks. The decision was then taken to use the inflatable boat for all future tagging attempts. The first of these attempts on 8 July was successful, and the following day another shark was approached. This time, although the tag was securely attached, the shark touched the boat, causing it to flick its tail, striking and snapping the wire trace between the dart and the array, which had a 100kg breaking strain.

Fortunately, the array was recovered, but that's where their luck would end, as subsequently sightings of Basking Sharks diminished and no further opportunities to tag animals presented themselves. Worse yet, the one tag that was successfully deployed failed to transmit any data, and after an initial acoustic contact was made, the shark moved out of range and contact was cut off. All attempts in the following days to retrieve the data-logger using VHF receiving equipment from the shore and the boat proved unsuccessful and the equipment was lost. If proof were ever needed of the difficulties likely to be encountered, this was it. Like the fishermen who hunted the animal in Scotland before them, scientists were to find that the experimental stages of any enterprise involving Basking Sharks could be both punishing and cruel.

Despite the fact that much useful data was gathered from the effort-related sightings and the plankton-sampling aspects of the project, no further attempt was made to continue the programme in the following years. It would be some considerable time before anyone would return to the costly and risky business of satellite tagging the Basking Shark.

Technology isn't everything

Whilst the goal of determining where the Basking Shark went in winter remained a fascinating if elusive Holy Grail for scientists in the mid-1990s, that didn't imply that there were not a hundred other worthwhile discoveries about the life history and biology of the creature yet to be made. With a species as poorly studied as this, the field was wide open. At which point a young scientist based at the University of Plymouth, Dr (later Professor) David Sims entered the fray.

Plymouth was an ideal place to study the Basking Shark for a number of reasons, not least due to the seasonal proximity close to Plymouth Sound of a productive thermal front (the Plymouth front) that lies approximately eight kilometres parallel to the shore. At the frontal boundary where the warm, stratified water of the western English Channel meets cooler mixed inshore water, productive upwellings of nutrients occur, encouraging plankton to proliferate along the front, with obvious attraction for many fish species including the Basking Shark. Plymouth University, with its degree course in Marine Biology, owned a yacht that could be used for on-the-water studies, and if it was not available, then there were any number of boats on offer for charter in the port. As the sharks often came into the outer sector of Plymouth Sound, the waters were relatively sheltered and could be easily and rapidly accessed by boat. It all added up to a very attractive small-scale study site.

In 1995, Sims commenced his field studies, initially concentrating on plankton and its relationship to Basking Shark occurrence. Plankton samples taken in the presence and absence of surface-feeding Basking Sharks showed that the predominant zooplankton species in all samples was a copepod called *Calanus helgolandicus*. However, there was a significant difference between the samples taken close to feeding sharks that contained 2.5 times as many *Calanus helgolandicus* individuals per cubic metre, specimens of which were 50% longer than in the samples taken where sharks were absent; this indicated that Basking Sharks actively foraged for the densest and most productive patches of plankton. Further analysis revealed that the availability of zooplankton was the key factor to affect the seasonal presence of the species, demonstrating that Basking Shark sightings declined in parallel with reductions in zooplankton abundance later in summer.

Sims followed this with a tracking study of surface-feeding sharks, showing that they stayed with the densest 'patches' of plankton for up to 27 hours, remaining with one patch as it was shifted more inshore, then offshore by the tide. And once a patch was depleted, the sharks would move on to forage for unexploited patches, tracking across the front where further patches could be found with least effort. This was ground-breaking work, showing that far from being an indiscriminate feeder that simply had to get lucky to find its planktonic prey, the Basking Shark was in fact a highly capable forager equipped to find and exploit the densest patches of plankton with remarkable efficiency.

Obviously some hitherto unrecognised sensory system enabled the Basking Shark to forage in this manner. Sims suggested that perhaps this could be explained by the heightened senses of electroreception and olfaction (smell) that are recognised to be a feature of many sharks. The Basking Shark possesses a mass of electroreceptors in pits on its snout called the 'ampullae of Lorenzini' which could be used to home in on the tiny electrical discharges given off by copepod muscle activity. And the shark also possesses nares that could enable it to use olfaction to detect dimethyl sulphide, produced when zooplankton grazes on phytoplankton. Even non-scientists began to view the Basking Shark in a new light.

In a further development from these studies, Sims demonstrated that the prey densities required for Basking Sharks' subsistence were far lower than had originally been proposed. In one of the early studies, Parker and Boeseman had suggested that feeding basking Sharks required a threshold zooplankton prey density of 1.36 grams per cubic metre of seawater (g m^3), below which they would be feeding at a net energy loss. Sims was able to show that a far more accurate threshold prey density was some 2.2 to 2.5 times *lower* than the original estimate, closer to 0.5–0.6g m^3 at which level the shark could still maintain growth rates. And as this level was below the winter zooplankton biomass estimate of 0.84g m^3 used by Parker and Boeseman, it suggested the intriguing possibility that the shark could find enough food to subsist outside the summer season.

The old shibboleth that the Basking Shark hibernated for up to four months in winter to preserve energy was based on two arguments. The first was the high threshold prey density required for the shark to subsist, that had now been shown to be based on an erroneously high figure. The second factor was the occasional capture or stranding of Basking Sharks in winter that had no gill rakers, the long, bristle-like structures used in filter feeding. The argument therefore went that they hibernated (some presumed on the seabed) in winter to save energy and regenerate their gill rakers, worn out from summer feeding. But sceptics pointed out that many of the sharks caught or stranded in winter *did* have gill rakers — maybe the animals that had no rakers were sick individuals, or the soft tissue that the gill rakers were attached to was amongst the first things to decay. It didn't necessarily mean that regular shedding of gill rakers was nonsense; it could also be that the period required to regenerate new rakers was much shorter than originally imagined. But with one of the main pillars (the high threshold prey density) that had supported the hibernation hypothesis now shown to be irrelevant, naturally the focus turned to other related arguments that looked uncertain. The hibernation theory was beginning to look decidedly shaky.

Through the use of traditional scientific survey methods, yet driven by original thinking, David Sims and his team had expanded the collective understanding of the behaviour and ecology of the Basking Shark enormously in a few short years. This in turn had a cumulative benefit in that it encouraged a new generation of students to take an interest in the Basking Shark, which would pay huge dividends in later years. For governments and scientists sceptical of the need for further protection measures, Sims's discoveries provided the kind of credible, analytical evidence to change minds. And for those of us involved in less exploratory surveys, this new knowledge provided encouragement and new directions in which to focus our attention. Knowing that there was far more to come from this innovative and dedicated team was a spur for all participants to raise the bar.

The creation of a national Biodiversity Action Plan (BAP) in 1997 set out targets and actions for recovery of the species, under the auspices of The Wildlife Trusts, WWF-UK and the recently formed Shark Trust. These exciting developments marked something of a watershed for our fledgling project. Finally, we found a niche in the funding regime, and with the support of The Wildlife Trusts/WWF-UK Joint Marine Programme we were allocated a small grant to cover one week of boat-based surveys during each of the months of May, June, July and August in the western English Channel for three years, to start in 1999. This was the breakthrough we had been waiting for, and we set about recruiting crews immediately.

This news gave us the impetus to start another project immediately, The European Basking Shark Photo-identification Project (EBSPiP). This project was created to use photographic images of shark dorsal fins to identify individual sharks, a technique known as photo-identification (photo-ID). The MCS had for several years been promoting an identification programme developed by Alistair Stagg, targeting divers who encountered Basking Sharks around Britain's shores, called STINCS (Scars, Tags, Injuries, Notches, Colour, Sex). It hadn't been a major success, largely due to the small number of reports. This was a pity, as it had demonstrated that it was possible to identify individual sharks, but the number of encounters with divers was infinitesimally small, so the sample size for statistical purposes was minuscule. During our transect surveys for surface-sighted sharks, we had seen quite a number with distinctive notches and injuries to their highly visible first dorsal fins, which suggested that by simply concentrating on the most obvious and regularly visible feature of the animals, individual sharks could be identified. As photo-ID was already successfully in use for White Sharks and Whale Sharks, we were confident that we could do the same for the Basking Shark.[1]

Widely used in the world of cetaceans (originally for Humpback Whales) photo-ID has many benefits as a research tool from a marine conservationist's perspective. It is low cost, non-invasive and accessible to anyone out on the water with a halfway decent camera and some basic skills. By identifying individuals (and noting their location) and then seeing how often they are re-sighted over time, the technique can be used as a form of mark/recapture that may yet eventually allow population estimates to be established – a first and very necessary step in understanding what conservation measures need to be put in place. With advice from prominent shark conservation expert Sarah Fowler and cetacean expert Dr Jonathan Gordon and support from the International Fund for Animal Welfare, we were able to develop a simple database, and set about soliciting photographic images of Basking Shark fins from members of the boat-owning public, including dive boat crews, survey vessels and fishing vessels – in fact anyone who was out on the water regularly and might encounter a Basking Shark. With our own catalogue of images to set the ball rolling, and the project based with our friends and colleagues Kelvin Boot and Dr Juan Romero at The National Marine Aquarium in Plymouth to give the project a public face, we waited (and hoped) for the photographs to roll in.

We didn't have to wait for long. During the 1998 season it had become clear from our surveys that shark numbers were on the increase, not least due to the shark invasion in Kennack Bay. Basking Shark sightings were being reported over a much wider area around the coastline of Cornwall. This was an exciting thought, but was tempered by the knowledge that it might prove to be a false dawn. Perhaps, too, we had simply been lucky with the weather in the weeks we were out, or had just been in the right place at the right time – it was far too early to tell. But the signs looked good, and we entered the winter of 1998 full of plans, hope and a long list of work to be done in anticipation of commencing our new survey regime in 1999.

---

[1] Initially we thought this was the first time this technique had been used for the Basking Shark. Some years later I was to discover that Canadian researcher Jim Darling had tried photo-identification out on sharks in Clayoquot Sound, British Columbia in 1992.

Chapter 14
The Early Surveys

Anyone who has ever owned a boat (or even knows someone who does) is well aware that the work involved in keeping it safe and seaworthy is costly and time-consuming. If the boat is to be maintained to the higher standards demanded for commercial use (as ours had to be) then make that 'bank-breaking' and 'endless' instead. If you have the wherewithal to have someone else do the work for you, well and good, otherwise it will all fall on you. And as we always operated on the tightest of shoestrings, that was my lot. The excitement I felt when we had first secured the funding didn't last long when confronted with the reality of the workload. It was time to take stock of what we had and what we could realistically hope to achieve.

I owned a yacht, called *Forever Changes*, an 11.7-metre-long beauty designed by Argentinian maestro German Frers Jr. A lovely, ocean-capable boat, she would provide a safe and seaworthy platform for the long passages in exposed waters that we would be facing. Able to sail in even the lightest of winds, we would be able to keep our use of fossil fuels to a minimum, in keeping with our aim to minimise our environmental impact. We couldn't have had a better boat.

Our overall goal was to establish whether hotspots for the Basking Shark still existed so that ultimately, if proven, these could be afforded some protection. The historical record suggested that there had been such sites, but more recent history showed little or no evidence of this. Whilst we knew that surface sightings were an unreliable way to arrive at a population estimate even on a local scale (as many sharks might be just out of sight below the surface), there were other, more specific benefits that our surveys could target. Sightings recorded on multiple-week, effort-based surveys provide a valuable insight into spatial and temporal abundance, especially when merged with land-based surveys that couldn't cover the more remote sites that we were able to access. Close range contact from a boat's deck would also allow detailed observation of social behaviour, more accurate measurement of individuals and permit photo-identification studies to be carried out. The long-term nature of our proposed study would also help to create a better idea of what was happening over time, essential with a creature like the Basking Shark whose comings and goings were so unpredictable. If we could achieve all of that, we would be doing very well.

I had long before realised that there was no way that I could run the on-the-water aspects of the project single-handed. During the winter there was all of the boat servicing and maintenance, funding applications to fill out and chase and the renewal of our licences to conduct our research, especially the photo-ID work which required close approach to the animals to get usable images. During the season, skippering the boat was one full-time job, overseeing the practical and logistical side of the surveys another, and cooking, cleaning and caring for the crew was a third. So I needed a first mate to take some of the load off me, and to help ensure that the volunteers played their part, too.

Finding individuals willing to act as first mate wasn't hard, but finding suitable candidates was more challenging, as I wanted people with an interest in marine conservation, preferably with a science degree background or practical experience of the type of surveys we would be undertaking, all of which narrowed the field considerably. I addition, I needed individuals with good interpersonal skills who would make up for my sometimes-short fuse when confronted with too much to do and too many decisions to make. Finding any one person who could be such a paragon of virtue seemed a major task. That the shortlist of candidates met all of the criteria, and more, was nothing short of a miracle. Like many of the hunters in whose wake I would sail, I was immensely lucky with my team.

It may seem odd that I don't mention sailing skills in the above job description, but that is partly because they all had the necessary basic skills, and the rest we could teach them along the way, either on the job or by putting them through Royal Yachting Association courses. As most of the volunteers came without any sailing experience, I needed first mates with enough ability to keep an eye on things if I was below decks and make sure that all was safe, and that the survey methodology was being adhered to religiously. The volunteers were welcome to help us sail the boat at all times, subject to supervision, which meant that either I or the first mate had to be on deck at all times – the dictum 'never leave a novice unattended' should be written across the forehead of every would-be skipper, in my view.

Beyond the sailing of the boat, the role of skipper and first mate was to train the survey crew in the basic effort-related survey techniques that had been drummed into me by Peter Evans so many years before. We used paper transect and sightings sheets rather than electronic data logging, partly because we wanted the volunteers to be actively engaged at all times, and also because it helped them to understand what we were doing in a more practical manner. And as I knew of several occasions on other survey vessels where more than a day's worth of data had been lost due to electrical failure, I preferred the reliability of paper, even if it did mean more work down the line transferring it into a database. As a result, we only ever lost one tiny fraction of usable material, when a stray gust whipped a transect sheet out of the hands of a horrified volunteer one day in the Firth of Clyde.

We carried a volunteer crew of five so that we had enough manpower to cover all eventualities, even if we hardly had room to swing a cat when we were all below decks at the same time. All aspects of the survey work were rotated every two hours, to keep eyes and senses fresh, and avoid boredom. One observer looked out to port and one to starboard at all times, one volunteer recorded environmental data and our position every half hour, and another steered the boat. That left one spare to relax or make the tea, while the mate and I kept a watchful eye to ensure that everyone was concentrating at all times and the boat was run smoothly. As Peter had taught me, diligence and attention to detail were absolute necessities to make the data gathered truly valuable.

As before, for this three-year data gathering exercise, we aimed to cover the south coast of Devon and Cornwall between the Isles of Scilly in the west and Torbay in the east, and the north coast as far as north Devon. The survey area was broken into eight sectors to allow comparison between sectors to be made, and to better identify hotspots. This made for a challenging workload, but as time progressed we found that in practice we were unable to cover the north coast in anything like an adequate manner due to logistical difficulties. The distances were just too great, and the lack of safe havens meant that we abandoned that plan very early on, which freed up more time to concentrate on the south coast.

The volunteers for the early surveys mainly came to us from the conservation groups who supported us, or from other associated marine conservation charities, and we never had any shortage of willing candidates. In order to draw attention to our activities, and to make our intentions clear, we had three-metre-high logos in adhesive sail-number material made, and had our local sailmaker glue them to our mainsail and genoa. And so we made our final preparations to start the season with a gigantic WWF panda on our genoa and an equally huge Wildlife Trusts badger on our mainsail, making us instantly recognisable even at a distance. Such visibility had its downside occasionally, as not all fishermen appreciated our activities, but at least, we felt, we were upfront about our aims. On a more positive note, we met a lot of other sailors (and indeed many fishermen) who wanted to know what we were doing, and learned a great deal from them about where they had seen Basking Sharks as a result.

Getting underway

As the end of April approached we were filled with anticipation, eagerly awaiting the first sightings of the season to Cornwall and Devon Wildlife Trusts' Seaquest project. During the 1998 season it had appeared that shark numbers were on the increase, not least due to that shark invasion in Kennack Bay. This was an exciting thought, but was tempered by the knowledge that it might prove to be a false dawn. Perhaps the good summer had helped, so that more people had been out on the cliffs, or we had just been in the right place at the right time – it was far too early to tell. But from the first survey week in 1999 we became immediately aware that we had entered a different ballgame altogether. From Lizard Point down to Land's End we had substantial numbers of sharks throughout the surveys. Apart from one survey where bad weather stopped us from heading west around the Lizard, it was the same throughout the season. Even on the north coast – there were sharks being reported everywhere.

For the first time we saw sharks breaching at close range on a regular basis – an amazing spectacle, quite unlike anything I had seen before. Not that I doubted they could do it, but with no film or photographs of this rarely recorded behaviour circulating at the time, I had no idea what to expect. Breaching was so rarely witnessed, in fact, that such enlightened 1950s scientists as Matthews and Parker had openly stated in a paper that a Basking Shark could not breach:

> It is alleged that the Basking Shark sometimes leaps clear of the water, so that [the] whole body is brought clear, and that it falls back again with a tremendous splash, an action similar to the breaching of whales. From a consideration of the usual habits of this fish, however, it appears very improbable that this statement is correct......It is probable that the stories of Basking Sharks breaching are founded upon confusion with the larger dolphins or, even more probably, with the Thresher Shark which breaches in the most spectacular way, and might well cause a mistake if seen at a distance when Basking Sharks are numerous close to the observer.

This is the kind of statement that gives academics a bad name, exuding as it does a faint whiff of disdain and arrogance. To Maxwell and his men who had assisted the scientists and reported their knowledge to them in good faith, this must have seemed like an act of extreme bad manners, and proof indeed that as some of them had long suspected, scientists were no more than 'useful idiots'. Fishermen and shark hunters alike are hunter-gatherers. As such, they rely very heavily on their powers of observation to be successful, and are quite rightly very touchy about any doubts about the accuracy of their hard-won knowledge. Having made similar errors of judgment in the very early days, I rapidly learned that their knowledge of the sea and its moods, and their understanding of the habits of its marine life and birds, is very broad indeed and should always be listened to with respect and care. As Maxwell commented in *Harpoon at a Venture*:

> To men who had watched this action as often and at such close quarters as we had, this statement seemed simply ludicrous. To us it was as if some scientist seemingly wrote that it was alleged that dogs in London sometimes lifted their legs against lamp posts, but that this was probably due to confusion with one of the rarer species of wolf, that did in fact so lift their legs in a most spectacular manner.

A mischievous-minded reader might detect a pointed element in the detail of this statement. But it would undoubtedly have been a much-savoured small victory when Matthews and Parker later unreservedly withdrew the original statement in the face of overwhelming evidence, publishing a supplementary note to the original paper accepting that Basking Sharks did, in fact, breach.

Of course I had seen many cetaceans breach over the years, even a Minke Whale, and had always marvelled at the acrobatic grace and power on display. It was almost as if they were showing off, seeing who could carve the most elegant parabolic curves to impress and outdo their neighbours. A Basking Shark breach, however, displays none of that balletic élan; it simply launches itself through the surface of the sea like an unstable missile, twisting as it goes (the Scottish term 'birling' describes it perfectly) before crashing down on its back with a massive bang and a cascade of spray. It is rare that the whole of the shark clears the water, but it is not uncommon for the same shark to breach two or three times in succession, each breach marginally less impressive than the one before.

Worst of all, unlike a cetacean, the Basking Shark shows no sign of any spatial awareness of boats at all, and so the risk of having one land on your deck like a monstrous flying fish, with potentially deadly consequences, is a very real one. Having had one breach so close to us that a deluge of water swamped the deck and doused me down below at the galley, I can attest that this is best avoided – and we never even knew there was a shark in the vicinity until it erupted from the water.

Various theories have been advanced over the years as to why Basking Sharks breach. The earliest credible theory was that they were trying to dislodge ectoparasites such as the Sea Lamprey and the copepod *Dinematura producta*, but the energy required to launch six tonnes of shark into the air is so colossal (and the parasitic burden so apparently minor) that it beggars belief that this could be the sole cause. Dr Peter Pyle has suggested that in his studies of White Sharks (*Carcharodon carcharias*) in the Farallon Islands, breaching is a form of social communication that may be linked to reproduction, and some whales are believed to use breaching to a similar end. Through photography on one occasion in Scotland, we were able to identify a large, breaching adult shark as a male, which might suggest that breaching could be a male display to attract females. However, we know that it is not just males that breach. David Sims was able to identify an adult female that breached very close to his research vessel during his Plymouth study, which he suggested may have been to advertise her receptivity to mating. And to complicate matters, even a very small shark has been seen to breach when startled in shallow water. But to anyone who has seen Basking Sharks breaching on a repeated, regular basis, the only coherent cause has to be social activity, driven by the need to find and secure a mate.

We witnessed what is believed to be courtship, too, sometimes on a grand scale with as many as a dozen sharks involved, swimming nose-to-tail and alongside, and sometimes even touching each other, seemingly lost in the moment, oblivious to our presence. One calm day we encountered a group of five lost in this activity in Gerrans Bay off the small town of Portscatho, Cornwall. Stopping the boat and just drifting, the sharks approached us, and without pause swam slowly by. The water that day was unusually clear, and it was almost by chance that I glanced deeper into the water to see literally dozens of sharks swimming below their 'friends' at the surface in perfect synchrony. It was a staggering moment, and we all lined the side-deck and peered into the depths open-mouthed with astonishment. The sharks were stacked up like a squadron of jet fighters, their beautiful python-like stripes evident even many metres below the surface. Suddenly I was taken back to a moment many years before, at school, reading Gavin Maxwell's book *Harpoon at a Venture*, and his vivid description of just such an event:

> Down there in the clear water they were packed as tight as sardines, each barely allowing swimming room to the next, layer upon layer of them, huge grey shapes like a herd of submerged elephants, the farthest down dim and indistinct in the sea's dusk… This is a shoal of fish —*fish*.

Looking down on the same sight, it suddenly dawned on me that I had entered the world that Maxwell and the other hunters had known, and that it was working the same magic upon me that it had evidently worked on them. There was no way back – I was gripped with Shark Fever, and life would never be the same again.

Not that I was the only one. The upswing in shark sightings attracted a huge amount of media attention, and much of it came my way, sometimes with interesting results. Having just returned from a week's survey I was looking forward to a few days off when a call came in from the local BBC News team that there was a large group of Basking Sharks in Harlyn Bay near Padstow, North Cornwall, and they would like to meet me there to do an interview. They gave me the phone number and address of the people who had reported the sighting and off I went. When I got there I could hardly believe my eyes. The bay was full of sharks, swimming slowly around, feeding at the surface; I counted over 30.

Harlyn Bay is seldom very calm; in fact, it is a popular beach with surfers. But the weather had been settled for days, so the sea was flat, the water was clear and the sun was shining brightly on an almost Caribbean scene – with the added bonus of sharks. The BBC crew were already there and had commandeered a boat to get out amongst the animals, some of which already had their own small posse of swimmers following them. There was no spare space on their boat, but the kind people who had made the initial report had sent around for their own boat to be brought from her mooring in the River Camel, and generously invited me to join them out on the water. An hour later, we, too, were out amongst the sharks.

It was all a little alarming. The swimmers had obviously lost all fear of the animals, knowing them to be 'toothless' Basking Sharks. I watched with amazement and not a little concern as one youngster paddled up to an eight-metre shark on an inflatable dragon (truly a meeting of monsters). Not that the shark took any notice – fortunately. Large adult sharks criss-crossed the Bay, intent on only one thing, plankton, and unusually, there was no courtship-like behaviour on display that day. Then a good-sized shark swung around in its own length and headed straight for us. Something looked odd about it, but I couldn't put my finger on it until I finally noticed that its entire dorsal fin was missing – it had to be the legendary 'Stumpy'.

I had never seen this shark before, but had been told about a very similar animal by eyewitnesses from both the south coast and off Cape Cornwall from both this and previous years. Obviously, we had been keeping a good lookout for such a distinctive animal, but in fact it wasn't that easy spotting such a shark as the dorsal fin is the most obvious feature, something that was abundantly clear with the shark before me. As the animal swam by our small boat I was easily able to see the ventral area of the body around the cloaca and confirm her to be a female. To get an accurate assessment of her length, I measured her against the boat to be just over seven metres long. We stayed with her for some time, taking still pictures and video footage for later use, before finally turning our attention to the many other sharks that were still combing Harlyn Bay. When the afternoon drew to a close the Bay slowly cleared and we, too, joined the exodus from the water. It had been an incredible experience.

Some months later, Dr David Sims and his wife Dr Emily Southall came to dinner at my house and we were comparing notes on the season just gone. I showed him my video footage, including that of 'Stumpy' from Harlyn Bay. He was amazed, and confirmed that he'd seen a very similar animal off Plymouth in 1996. Later, comparing our best still and video images, and linking notes from our respective sightings, we were able to confirm that this was the same female animal and compare her journey and increase in size over those three years; we had sufficient material to produce a useful short paper. It was a great occasion, as it showed that photo-identification of the Basking Shark was not only viable, but could have real validity.

Not all calls from out of the blue were as productive, although they still could have their entertaining side. One day the phone rang, and a commanding female voice at other end demanded to know if I was 'the Shark Man'!

'Some might call me that,' I laughed.

'Then what are you going to do about the Basking Shark on the beach below my house? The smell is appalling!'

Once we had established the whereabouts and size of the said shark, I confirmed that there was, in fact, nothing I could personally do about it. This was a big shark that weighed at least five tonnes and it was beached in an inaccessible cove. But not wanting to be appear unhelpful, I enquired:

'How did it get there?'

'The local fishermen towed it in and left it there, at the high water mark.'

'Have you spoken to the council?' I asked, 'They're usually responsible for removing stranded animals.'

'It's a private beach – they said to ask the fishermen to tow it away'.

'And what did they say?'

'They refused!' she bellowed.

Somehow I sensed that there was more than a little hidden history behind the deposit of a huge shark at this particular spot, and rather than get further involved I made my apologies, and brought the call to an end. But the following week we would be on a survey, so I made a mental note to go and find the shark while we were out.

And so it was that we arrived very late one evening off a short stretch of rocky shore indented with shingle coves on the south coast of Cornwall. Night was falling and the wind was light off the land as we stole along the shore as close as we dared, watching the depth sounder intently. I had a rough idea where the shark was supposed to be but didn't know which cove it was in, until suddenly a mephitic stench, warm and reeking, hit us like a wall. Now we knew where it was.

We anchored for the night nearby (upwind), and the following morning at first light we made our way back to the cove by dinghy with our cameras and tape measure. The cool of the night had diminished the smell somewhat, but it was still enough to make you gag at close range. A big male at just under eight metres, it was still in fairly good condition, although its belly was alarmingly distended, which didn't encourage any of us to get too close. The tell-tale marks of net entanglement were obvious; when a Basking Shark hits a gill net, it doesn't (as you might imagine) go straight through it. The net is so light and compliant that the shark will simply tow it along until it becomes aware that it is entangled, whereupon it dives to the bottom and rolls on the seabed, becoming wrapped up in the monofilament net like a mummy. It stays there struggling and trying to swim for hours before eventually expiring, leaving the pectoral fins, body and particularly the tail stock, bloody and raw from abrading on the bottom. Clearly they don't give up on life quickly or easily.

Around the area of the caudal peduncle of the shark was a becket of heavy polypropylene line that had presumably been used to tow the shark to shore. It looked like the strong easterly winds of the week before had then driven the carcass right up the beach – the shark was up at the level of the highest of spring tides. As there was no other access to the tiny beach than a gated, precipitous, pedestrian path, it looked like it was going to be there for some considerable time.

As we left in the dinghy, I glanced up to the cliff top where stood an imposing residence that overlooked the cove. There on the cliff path was a woman, gazing intently down at us through binoculars. Obviously an early riser, I thought. Even at that distance there was something in her stance that suggested outrage, so rather than provoke matters further, the Shark Man and his little brood slipped quietly away, leaving the noisome present on the beach to the wind, the gulls and the neighbours.

Sharks at last!

Our 1999 season was everything we could have hoped for. By the end of the season we had recorded 209 Basking Sharks, including our first really bumper day of 76 animals. All season we saw groups of animals engaged in what is believed to be courtship behaviour, and recorded sharks breaching on 28 occasions. This was an extraordinary advance on every front. Sometimes we seemed unable to evade the sharks. One glorious evening we came to anchor between Mullion Island and the Lizard shore, and before we'd even managed to get supper on the go, the cry of 'Shark!' went up. No more than 200 metres away, a small group of sharks was feeding around Mullion Island. Like others before us, we foolishly took our initial success as the new reality, and imagining that we were well onto an upward curve, we were filled with optimism for the 2000 season.

And found out, of course, that we were sadly mistaken. A poor summer of generally unsettled weather put paid to our hopes. Our biggest problem was that we had only one survey week a month, and if the weather for that week was poor, then our overall Sharks Per Unit of Effort (SPUE) value would suffer badly. And if we were really unlucky, our survey week would be wet and windy, while the other three weeks in the month might be perfect. If that happened every month … Wouldn't it be great if we could be out on survey every week, we dreamed? As it was, during that summer, June was the only month that really delivered, shark-wise.

However, the 2001 season was far more productive. From May onwards, Basking Sharks were present, in groups and displaying courtship behaviour. As a three-year dataset, the surveys had given us much to think about. Having been out on survey in the same waters occasionally since 1994, and consistently since 1999, a period that we could confidently say had seen a positive upswing in numbers, we had learned an incredible amount that would serve us well in the future.

We had, for example, seen Basking Sharks of every size. Although we hadn't observed a shark giving birth, we believed we had just missed such an event one flat calm day out near the Eddystone Lighthouse off Plymouth. We saw a tiny shark, up at the surface, swimming fast and erratically, and at first simply thought it was a Blue Shark (*Prionace glauca*), such as we had occasionally seen before. But then another appeared just ahead of us and we were able to see it was a minute Basking Shark, of around 2 metres in length. Both of these tiny sharks had their mouths open and appeared to be feeding just like adults. Over the next 500 metres we saw another three of these miniature Basking Sharks. As we so rarely saw these small animals, to see five of them within such a small area led us to speculate that we had just missed a birth event.

We learned over time that the smaller sharks were in general far more easily disturbed than their elders, swimming fast and diving long before a boat could approach close enough for a good photo. Most of the small sharks (less than three metres) were solitary, and it was rare that we saw adults in the vicinity of these youngsters. Due to their nervous nature, it was hard to measure them accurately, especially at a distance. But we could say with certainty that the smallest sharks were less than two metres in length, and probably very recently born.

We recognised early on that it was very difficult to measure the sharks precisely from a boat (let alone a cliff top) and that, if anything, the tendency was to over-estimate length overall, due to refraction and excitement. To try to avoid this, we established a simple method of treating our deck like a ruler, with the distance of each stanchion from the extremity of the bow being measured in advance. When we approached a Basking Shark we would send a crew member up into the bow, then carefully manoeuvre the boat alongside the animal until the tip of the snout was parallel to our bow, then a second crew member would walk down the deck until they were parallel with the tip of the tail. This distance could then be established against our marked deck and we could make a more accurate estimate of the total length.

We knew from our earlier surveys from 1999 onwards that the majority of sharks measured so far were in the four to six-metre category, with very small or very large animals being relatively rare. At the same time, we recognised that the majority of our encounters were of single animals, or small groups of up to three or four sharks. But by 1999 when we were encountering substantial shoals of sharks, we found that the size range of these groups was different, with more sharks recorded in the six- to eight-metre range, and we noticed that we saw very few smaller sharks in the shoals. When we were seeing apparent social behaviour such as courtship and breaching, we learned that the majority of the animals that we could measure would be adults (longer than 4.6–6m). The obvious conclusion from what we were seeing was that the large aggregations of sharks were predominantly adults, and their interest was not solely limited to feeding, but also the finding of potential mates.

A very small percentage of the sharks we recorded were exceptionally large. The argument over the potential maximum size that a shark can achieve still rages, with most informed observers offering a figure somewhere between 10 and 12 metres. Gavin Maxwell had lost a bet with Anthony Watkins that he would never catch a shark of over 30 feet (9.1m), but Maxwell insisted that he had seen (but had been unable to catch) sharks that comfortably exceeded that length. Reports from western Canada record that some sharks caught there were between 11 and 12 metres long, and a photograph exists that allegedly shows two sharks that had been measured at those lengths, ashore on a beach. Unfortunately, there is no object in the image that would enable the proposed lengths to be estimated against, but they are, unquestionably, very big animals.

One day we were making a transect between Round Island in the Isles of Scilly and the Runnelstone Buoy off Gwennap Head on the mainland of Cornwall when we encountered a small group of sharks engaged in courtship-like behaviour. At first I thought that they were mostly sharks in the four- to six-metre range, but a single animal amongst them that was huge. We approached the group and started taking photo-ID images and measuring the animals in the usual manner. It was then I realised that the animals I'd initially thought were all smaller, were in fact much larger, at between seven and eight metres; it was just that they looked so small in comparison with the big animal. The dorsal fin in really big sharks often tends to sag to one side, but in this animal it was almost completely folded over and dragging in the water. When we finally managed to get alongside it for measurement, we were able to confirm that it was just under 10.5m overall. In later years in Scotland we saw one or two more that we measured at 10m, so I believe I can say with some certainty that there are indeed Basking Sharks that comfortably exceed 9.1m in length – but there are very few of them.

A recent paper had documented courtship-like behaviour amongst a single group of 13 Basking Sharks on the coast of Nova Scotia that matched our own observations to date. In 2000, David Sims and his team produced a more comprehensive analysis of this critically important behaviour, describing it in far greater detail and showing that it was linked to the higher prey abundance associated with productive thermal fronts. In this study, putative courtship behaviour was only observed when groups of large animals gathered in rich zooplankton patches, suggesting that the density of sexually mature individuals was also a contributory factor to instigating this behaviour. This tied in with everything we had seen around Lizard Point and Land's End, and helped us to devise a much more targeted focus for our future work.

After three years of surveys we had established a solid amount of valuable data. Ultimately there were two elements that mattered above all: the SPUE value, and the identification of hotspots. A later study proposed that an SPUE value of $0.01h^{-1}$ would be a credible average value in open waters[1], whereas a value of $0.6h^{-1}$ would be more accurate for a frontal area, where more sharks were likely to be encountered at the surface. After 206 hours of effort-

---

[1] So, Sharks per Unit of Effort equates to the number of sharks you might see in 'one unit of effort' – typically that is one hour, but could also be one kilometre of transect, etc. So, a SPUE of $2.22h^{-1}$ for a defined area e.g. a 5km squared block, equates to seeing 2.22 sharks for each hour you spend within that block. Recognising that sharks don't come in physical packages of less than one (unless you are shark-hunting for their livers of course), if it helps, you could roughly interpret $2.22h^{-1}$ to mean that if you stayed in the block for about 10 hours, you might on average see approximately 22 sharks (i.e. very roughly, 10 x 2.22).

based surveys across the whole of the six areas on the south coast we had recorded 141 sharks on transect, allowing us to establish an average SPUE value of $0.68h^{-1}$. Although we accepted that was probably artificially inflated by the staggeringly high SPUE value established in 1999 of $2.22h^{-1}$ (which includes the 76 sharks in one day event), our overall value for the area was credible. In addition, we had established at least one potential hotspot: the area between Lizard Point and the Runnelstone buoy off the southern tip of Land's End, where we had recorded 78% of all sightings. As this was the area where we had also seen the vast majority of breaching and courtship-like behaviour, it seemed that we were on to something of real interest from a conservation perspective.

Expanding the surveys

As we reached the end of these initial surveys we were able to take stock of what we had achieved so far, look at what we had learned, what we could do better, and see where, if we could secure further funding, we might take the project from here.

The results we had achieved up to this point were impressive. We knew, too, that whilst our boat-based, effort-corrected data was badly needed in its own right, it was also of considerable value when merged with the public sightings project run by MCS. By providing an offshore dimension to the predominantly land-based sightings in their database, our data filled in a lot of gaps and made the whole far more coherent and complete. By identifying hotspots, we could better identify what threats might exist for the species within those sites, and suggest measures that might be necessary to counteract those threats. All valuable and positive benefits, but more was possible, we knew.

With only one week per month in the season, we were very vulnerable to bad weather blowing out a survey and thus skewing the data. If we could work throughout the season, that likelihood would be dramatically reduced. If we could expand the physical range of the survey, then we could employ the same tactics in other regions with, hopefully, equally positive results. But the logistics and cost of doing so would be of an order of magnitude greater. It would be a massive challenge to secure the funding and backup we would need, but wouldn't it be worth it? Dreams are cheap, however, and we were under no illusions that we faced an enormous task.

By now, WWF-UK had decided to relinquish their interest in the Basking Shark in terms of their involvement at the level of the Biodiversity Action Plan, so that they could concentrate on their global marine strategy. This was understandable, but still a great shame, as I had enjoyed working with their team immensely. The Marine Conservation Society took their place, though, which more than made up for the loss. They joined the collaboration with the Shark Trust and The Wildlife Trusts as the triumvirate responsible for developing appropriate conservation measures for the Basking Shark in UK waters. By now I was heavily involved at the administrative level through becoming chairman of both the Shark Trust and the Biodiversity Action Plan Steering Group, giving me useful insights into the world of policy, far from my usual role at sea. It also gave me a fantastic opportunity to work with some of the most inspirational people in the world of sharks.

The Wildlife Trusts wanted to expand the surveys, and the leader of the Marine Team, Joan Edwards, put her considerable energies behind identifying ways in which we might make this happen. This demanded a considerable amount of work from all involved, as funding applications had to be made and where there was any hint of interest, followed up with meetings in person. But that was no hardship, as there was a palpable sense of progress at all times, and it was exciting to be developing the project and working with such great people. Gradually, we pulled together a viable project, to be supported by the National Lottery and some of the regional Wildlife Trusts. On paper we looked good to go, but as the old saying goes, 'We'll believe it when the cheque is in the bank'.

For the first time, too, we had volunteers from Earthwatch Institute (Europe), an arm of a global organisation that recruits citizen science volunteers to assist field research programmes such as our own. For the next five years we would receive a steady flow of excellent volunteer crews through them, many from other parts of the world, which helped to spread the word on Basking Sharks far and wide.

The projected programme looked like this. For the months of May and June we would conduct weekly surveys in our original area of Devon and Cornwall. At the beginning of June, we'd leave to head north up through the Irish Sea to survey the coast of Northern Ireland for two weeks, before crossing into the Firth of Clyde for a further fortnight of surveys. In mid-August we'd arrive in the Sea of the Hebrides for four weeks of surveys, before turning for home. It was going to be an action-packed season, of that there was no doubt. But we'd be breaking new ground, and that was a really exciting prospect.

Of course, we'd miss out on some interesting sites, though. We had originally hoped to survey the coast of Wales, largely because of its historic importance as a hunting ground, but

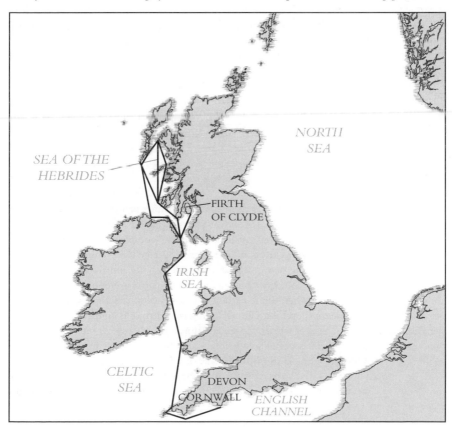

Survey routes 2002–2007.

hadn't been able to secure the necessary funding to do so. The Isle of Man was being well looked after, and didn't need us to assist, so that was no problem. But the route and the tight timing schedule in the middle period would make it difficult to cover the Ayrshire coast as well as we would have wished. But, all in all, we felt that the compromises to be made were the best that could have been wished for. We had most of the historically important sites covered.

All seemed well, except that we kept getting stalled on the funding front. Meetings were held, calls were made, and we were assured that the money would be forthcoming, but as each month passed, the deadlines slid inexorably by. But the boat had to be made ready, equipment had to be serviced and purchased, volunteers had to be recruited (and on a much larger scale than before) and my bank account was already empty. As spring came around the corner and we were getting ready to go, the money still hadn't arrived, despite all of the promises at the funders' end.

By now it wasn't only me who was getting worried. I'd been joined by my great friend David Marshall, an oceanographer with many years' experience at the Environment Agency, who had thrown in his job to embark upon a precarious career as a consultant. And his first commission was working with me as first mate on an exciting project that looked as if it might well go nowhere unless the money arrived – and soon.

And so we found ourselves sitting in our local pub in Falmouth, the night before we were due to depart on the first of the season's surveys, flat broke. My credit card was long ago maxed-out, and poor old David had to buy the beer. I'd kept a wad of cash for emergencies, such as the food for the first couple of weeks – and that was it. We couldn't call the first two surveys off, but after they were completed it would all be over, we knew, unless the money arrived. We shrugged and ordered another round – his round, yet again.

Two weeks later to the day, the cheque came through my letterbox. Talk about a sigh of relief …

The early 2002 season

We soon settled into the groove of the new regime. Recognising that the major expense was getting the boat on the water, we had piggy-backed additional marine conservation projects onto the Basking Shark programme that would not otherwise have been able to get sea-time. These included the continued sharing of our cetacean sightings and transect data with Sea Watch Foundation, plus a new one for us, a 'marine litter and jellyfish' survey for MCS as part of their Marine Turtle Project. The additional work kept the volunteers sharp and engaged, and maximised our beneficial outputs, so we were happy to take those projects on. A good thing, too, as Basking Sharks were becoming harder to come by.

We had also gathered a new supporter in the form of the Born Free Foundation, with Chief Executive Will Travers determined to see the Basking Shark achieve an Appendix II listing within the Convention on International Trade in Endangered Species (CITES). The British government had already made an Appendix III listing for the species in 2000, but that was widely recognised as just a stepping stone to a far more stringent Appendix II listing. Securing the goal of Appendix II listing would be a first for all sharks, not just the Basking Shark, and would not be easily achieved. Will Travers brought considerable drive, dedication, knowledge and experience of conservation politics at the international level to the case, assets that would prove to be major advantages when the time came for the Appendix II vote. At the local level, support from such a well-respected NGO was a big plus for us, and raised our spirits immensely. The Basking Shark was gaining some powerful friends.

We had benefited greatly in the past from a connection with Canon (UK), who had generously provided us with cameras and lenses from their demonstration models, which had greatly improved the quality of the photo-ID images we were obtaining. In the early days before digital cameras changed everything, we got through hundreds of boxes of Fuji Velvia and Kodak Ektachrome 200 a season. The resulting images gave us plenty of material for slide shows, but it was frustrating that we never got to see the results until late September each year, when I would arrive home, and have to batter the front door open to dislodge the mountain of boxes of slides that the postman had delivered throughout the summer. When Canon donated a brilliant EOS 10D digital camera and we could see the results on the spot it seemed like the greatest thing we had ever touched. We could discard all of the poor shots immediately, which was one less job to do on our return, and we were convinced that the image quality was better, too. A total revelation.

The upswing we had seen coming from 1997 onwards now appeared to be going into reverse. The extraordinary peak of 1999 was now looking like an all-too-short summer, and we were well into the steady decline of autumn; we could only hope that it wouldn't be turn into winter in the years to come. Of course, we knew well that a negative result was just as valuable as a positive one in data terms, but at the same time it didn't do much for volunteer morale; everybody wanted to see the animals – it was that simple. We had one good day off Falmouth in early May, and that was it – the next sharks we would see were off north Cornwall on our way from the Isles of Scilly up to Lundy Island.

Entertainment came in other forms, though. Anchored in the lee of Lundy Island, we felt comfortable that the weather forecast was for light winds from the west, giving us perfect shelter. We had to make our way up to Milford Haven the following day for our trek up the Irish Sea, and reckoned that the predicted weather should allow us a reasonably smooth ride around the north end of Lundy at slack water around 8 a.m. the next morning, sparing us a hammering in the tide race there. However, at 1 a.m. I was rudely awoken as the boat was beginning to pitch madly. Contrary to the forecast, the wind had swung round to the north, and the rain was already setting in. Not good news. I eased out more chain and we set an anchor watch[2] for the rest of the night. At 7 a.m. we retrieved our anchor, and motoring into a solid force 5–6 wind, we headed for the headland at the north end of the island as close to slack water as we could manage, when conditions should have been at their least worst. I'm glad we timed it right, as otherwise I think we'd have sunk the boat, as the sea conditions for the next few hours were utterly diabolical, and very few of the crew escaped seasickness.

Later, as we made our way up the coast of Wales, I began to notice a steady influx of water into the bilges of the boat. Checking the obvious potential sources of leaks, I noticed that the stern gland (where the propeller shaft goes through the hull) was leaking. Grabbing a screwdriver to tighten the jubilee clips that held the rubber bellows in place on the glass-fibre stern tube, I slithered into the engine compartment. As always on modern boats, the stern gland was buried way back behind the engine and so was almost inaccessible, making it impossible to inspect properly. I had to stretch out full length to reach it, even with arms extended. I felt the screwdriver into place on the first clip, tightened it one turn and all hell broke loose, as a powerful jet of water hit me in the face. Now we were in trouble – the stern tube was cracked, and water was gushing in. If I didn't do something fast, we were downwardly mobile.

---

[2] Where someone is on watch at all times to check the anchor is holding securely.

The best I could do was fabricate a tape tourniquet to reduce the influx, but that would only make a temporary fix and wasn't going to get us far. The nearest boat lift was more than 30 miles way at Pwllheli, but we'd just have to try and get there, pumping as we went. Fortunately, we made it, and even though it was a Bank Holiday weekend, we were out on the hard by evening. As we had to be in Northern Ireland in a few short days for a big project launch event, I spent the entire night removing the stern gland, making a plug and then glassing in a solid repair of the stern tube. The next day the engineers came and inspected it all, checking the engine alignment and my repair. The following morning, we were on our way again – shattered.

Northern Ireland 2002

Arriving for our first week of surveys in Northern Ireland was quite an event. We started the first of our surveys off the entrance to Carlingford Lough, and made our way up to Ardglass, where we swapped crews. One of the aims of our project was to help build capacity in the areas we visited, so to that end we used local crews wherever we went, training them in the survey techniques and exchanging ideas on marine conservation topics. In that way, we would leave something of lasting value long after the surveys were over. We already had a great contact in our friend Gary Burrows, and with the backing of the Ulster Wildlife Trust and the Northern Ireland Environment Agency we were set to go.

And we got off to a terrific start, sighting four Basking Sharks on the first day of our survey, one of which was right in the entrance to Strangford Lough, where the water was boiling out through the narrows. Some excellent ID shots of this big shark really gave us something worthwhile to talk about at the launch event a couple of days later, and did wonders for the team, who were all very excited to see the animals – I don't think they believed they would.

We learned the hard way that this was not going to be an easy coast to survey. The tides from St John's Point north become progressively stronger, and by the time you are north of Larne, can be really fierce. We had to work the tides to our advantage or sit looking at the same piece of scenery for hours on end. At the northern entrance to the Irish Sea, the narrow gut between the Glens of Antrim and the Mull of Kintyre known as the North Channel, the tides run like a mill race, and there are wicked overfalls off several of the headlands. Once through that maelstrom and on to the north Antrim and Londonderry coasts, things didn't get a lot easier either, as the tides were still fierce. Much of the north coast was very exposed, as you were back out into the Atlantic swell; although magnificent, we were always glad to see the headlands of the Giant's Causeway safely behind us, regardless of our direction.

Nor were there many boltholes for a deep-keeled boat like ours. Our 2.2-metre draught was too much for many of the small ports, and there weren't many all-weather anchorages along the way, so good passage planning and luck with the weather were essential elements to cover the ground safely and effectively. The good news was that the welcome we received wherever we went was astonishingly friendly, and we soon made friends in many of the tiny harbours including Ardglass, Glenarm, Ballycastle and Portrush. Our local crewmembers made many good contacts too, for future reference, which was just what we'd hoped. Difficult it may have been, but rewarding it most certainly was. We ended the first season with five Basking Sharks recorded, including one really big animal found feeding along a tidal front out beyond the remote Maidens Lighthouse off Larne, before crossing over to the Firth of Clyde to exchange crews in Largs.

The Scottish surveys 2002

Conducting our own surveys in Scotland had been such a long-awaited moment, that it felt slightly surreal to finally be there and ready to go. Not only had I read so much about Basking Sharks and the west of Scotland, but as a native Scot (albeit from the east coast) the emotional connection to these waters was strong. From Largs, looking out beyond Little Cumbrae and into the Firth of Clyde, a brief glance around the horizon on a good day would take in the distant volcanic plug of Ailsa Craig and the magnificent Isle of Arran, names redolent of the days of the hunters. I could hardly wait to get out there and get started.

Beneath all of the enthusiasm, though, lay an undercurrent of concern. I was aware that sightings in the Firth had always been variable, not just in terms of numbers but also inter-annual variation and location. Philip Kunzlik's seminal pamphlet from 1988 made it clear that large numbers of sharks only occasionally entered the Clyde, and what shoals there were might be found anywhere from Loch Fyne down to the Mull of Galloway. We knew, too, that Howard McCrindle had struggled to find sharks in his final hunting seasons, despite having a useful network of fishermen to alert him to their presence, and in more recent years, the MCS sightings database suggested that very few sharks had been sighted. This was a big area and we had a lot of ground to cover, much of which was very exposed to the prevailing winds from the west. We were going to have to work hard for any successes that might come our way.

The forecast for the week was excellent, and we headed out for our first survey in good spirits, with a keen local crew. We couldn't have asked for better conditions, and we took full advantage of them, working long days in the gorgeously endless northern light of early August. We scoured the whole of the Firth of Clyde over the next fortnight, from Upper Loch Fyne down to the Wigtownshire coast, across to Campbelltown past Ailsa Craig and up around Arran, and recorded – one Basking Shark.

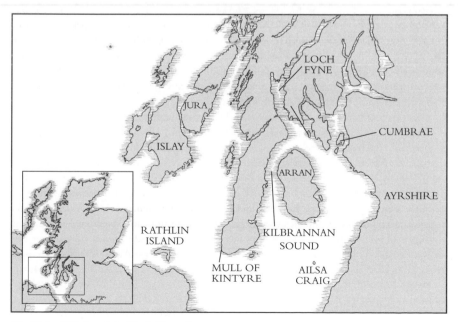

The Firth of Clyde and the North Channel.

Forewarned we may have been, but nonetheless we were dismayed by this seeming failure. One of our policies was to spend time talking to local people wherever we went ashore in the evenings, to get a feel for where and when we might encounter sharks. The stories we heard were remarkable, of buses stopping on the roadside at the head of Loch Fyne to watch the sharks, of huge shoals in Kilbrannan Sound and around Ailsa Craig. Exciting stuff, but always deep in the past, with very few contemporary reports. Many of those asked also pointed out that the herring had disappeared too, and put both drastic disappearances down to a combination of overfishing and climate change. All we could say for certain was that the sharks weren't here, at the surface, right now.

Hoping that our luck would change, we headed for the Crinan Canal and points north, enjoying the views if not the hard work of manning the manual locks. Exiting the canal, we headed immediately to Oban to take on a new crew and make our way out through the Sound of Mull to the Western Isles.

The road to the isles

I'm sure that every yachtsman has a favourite view, and mine has to be the wonderful vista that opens up as you round Ardnamurchan Point. With the low lying island of Coll away to port, and a succession of islands appearing to starboard with every mile to seaward – Muck, Eigg, Rum and eventually Skye – my heart leaps every time. Despite tens of thousands of miles at sea in many countries, I've never yet seen anything that compares to that wonderful panorama. That is, when you can see it. We were now back in the land of the wind and the rain and the fog.

But on our first day on survey, the sun shone brightly and we made our way out into open water, intoxicated by the glory all round us. It was a wonderful homecoming of sorts, as I still had some contacts in the region from many years before, some of whom were still actively engaged in cetacean surveys or marine ecotourism. If anyone might have seen any Basking Sharks around, it would be these guys and they would tell us. But the word was that Basking Sharks were now viewed as a real rarity. Other species such as Minke Whales were becoming more plentiful, and some dolphin species like the Common Dolphin were appearing more regularly, but Basking Sharks were still hardly ever seen. Not what we wanted to hear, but we had to go forward with optimism.

Six weeks of solid surveys later in the usual mixed weather conditions, and we had not recorded one shark. We had seen sociable Minke Whales galore, friendly dolphins of countless species and a myriad of wonderful seabirds against the glorious backdrop of the Highlands and Islands. But not one single Basking Shark, from Barra Head to Uishenish lighthouse, or from Iona to Neist Point. According to our contacts it was the same to the north of Skye; hardly a shark had been sighted. Dr Peter Evans was in the back of my mind, reminding me that 'Zero sightings were still scientifically valuable data', but that didn't really console me. Preparing the boat for the journey home to Falmouth in late September I had a lot to think about, not least how to find some positive side to what might appear to be an outright disaster. We had covered around 8,000 miles in our first full season, and ostensibly had little to show for it.

I had to make an urgent return home for my (non-shark but usually reliable-income) business venture, so I handed the boat over to David in Scotland and he brought her safely home. It was time to get back to earning a living over the winter. By now, most of my clients knew of my 'other' life, and viewed it with some amusement. But not all of them …

The second I got back, I jumped on a plane to join two executives from a French company to make a presentation to the UK branch of a big multinational corporation at their headquarters. This was a big deal that we had been working on for some time, and there was a lot riding on it. The meeting was to be held in their smart boardroom, with all of the key figures from the Managing Director downwards present. It was a make or break occasion.

Well, it's easy to get a haircut, shine your shoes and don a pinstripe suit, but it's far harder to undo the inevitable damage that five months of physical sea-time will inflict. The cold, damp conditions in the North Atlantic mean that any cuts and contusions hardly heal, as you are constantly knocking them open again, so that eventually they become large, ugly wounds. My hands, like a bunch of bananas at the best of times, were now a mass of raw cuts and bruises, as well as a finger broken during my last week in Scotland which was strapped to its neighbour with a bloody wrap of surgical tape. They were, it has to be said, an unholy sight.

Across the boardroom table from me sat the charming finance director, a very dapper guy who was always very smartly turned out. My presentation was next up, so I rose to my feet and began speaking, gesticulating towards the screen to help reinforce key points. All was going well until I saw the director opposite me looking at my hands with something approaching horror. From there on, I didn't know what to do with the two ghastly, flashing car wrecks on the end of my arms, as standing up I simply had nowhere to hide them. I rushed to the end of my allotted time and slumped back into my seat, hands forced back beneath the table.

As we were ushered from the building to the car park, I tried to avoid shaking hands. As we turned towards the car, a voice from behind me called my name, and I turned to face the finance director, who sheepishly enquired with genuine incredulity:

'Whatever happened to your hands? I had to ask …'

The year ended with one major piece of good news. Two sharks, the Basking Shark and the Whale Shark had been proposed for inclusion within Appendix II of CITES at the 12th Conference of Parties in Santiago, Chile. We were all immensely proud of the fact that a much-improved Basking Shark proposal had been put forward by the British government, on behalf of the European Union. Everyone knew that this was still going to be a tall order to achieve, as a two-thirds majority would be required to win the day, and many of the major fisheries nations would vote (and lean on their proxies to vote with them) against the listing of either species. This was on the grounds that accepting any marine fish on Appendix II would open the door for endangered commercial fish species like the Bluefin Tuna (*Thunnus thynnus*) to be considered for inclusion – and that would never do!

And so it proved. At the first Committee session where the two proposals were raised, both failed to achieve the required majority, although it was a very close vote. Things looked very bleak for a moment, as reversal of a vote at the Committee stage was almost unheard of, but both proposals were re-submitted during the final plenary session for a second vote. British fisheries minister Elliot Morley delayed his departure to help sway the few waverers that were needed, and conservation stalwarts like Sarah Fowler and Will Travers gave it all they had. Finally, the second vote came in – and both the Basking Shark and the Whale Shark had won through. All agreed that 'Team UK' had played a blinder.

Appendix II is the 'sustainability' part of the Convention. It doesn't prohibit international trade, as even CITES Appendix I doesn't require countries to protect listed species; it only stops commercial trade. Appendix II means that export permits can only be issued if the products were legally obtained, and if the fishery was 'not detrimental' to the survival of the species in the wild. That might not sound like much, but it was major step forward for all sharks, not just the Basking Shark and the Whale Shark. And the dissenters were right – other sharks would, and did, follow.

The 2003 season

Winter was welcome as a period of recovery, and time healed my wounds, both mental and physical. I got over the lack of Basking Sharks in Scotland, soon regained my appetite for the conservation work, and was looking forward to the new season with enthusiasm. We got off to a great start, making solid sightings during May and June in Cornwall, of good-sized groups with courtship-like behaviour and breaching recorded on several occasions. Whilst there was no longer the regularity of sightings that we had seen at the peak of the cycle, it was a solid season nonetheless. And we even had good enough weather to survey out to the Isles of Scilly on occasion, always a welcome bonus.

The Northern Ireland surveys went well, but consistent south-easterly winds made the Irish Sea coastline impossible to survey adequately, so we concentrated on the north coast, that being flat calm as more sheltered from those winds. But we still failed to find sharks up here, despite the fine weather. Fortunately, we saw lots of Harbour Porpoises around the Skerries and the Tuns Bank to keep the observers occupied and feeling that there was still life in the sea.

One of the worst aspects of this type of work is that of deadlines. Having a fixed date to leave and another to arrive at some new home port is your worst enemy. The chances of having perfect weather conditions to make the most difficult passages being approximately zero, the best you can hope for is only a moderate battering. With this in mind, we left Portrush to head towards the Mull of Kintyre in the early hours of one Friday morning, knowing we had to be in Largs on Saturday.

The forecast didn't offer much comfort. Moderate east to south-east winds in the morning, increasing by midday to Fresh from the south-east, just where we needed to go. Knowing the North Channel, there was the question of the tides, too. If we had the tide against us with the wind, it would take all day, but the sea wouldn't be too bad. If we could catch the tide down through the Channel with the wind against it would be rough, but fast. As long as the wind didn't get up too much, that is. And, wouldn't you know it, it was a big spring tide …

Pushing the boat as hard as we could, we arrived at the western end of Rathlin Island, where I planned to take a look at conditions and make the final decision as to the best route. The shortest route, being between Rathlin and Fair Head on the Antrim mainland, might seem to be the obvious choice, until you factor in a really vicious race there called Slough-na-more, that could stop us dead. It was coming up to slack water, but already the tide looked to have turned as the sea was white in places. That was it – we'd have to go round the north of the Island.

This we did, staying right under the cliffs, being sped along by a back eddy down the shore. Rathlin is steep to on the north side, so we were close enough in to 'fend off with a boathook' and spent most of our time marvelling at the spectacular seabird colonies on the cliffs – at very close range. So far, so good, but I knew the real test was going to come once we exited from the shelter of Rathlin and got into the main part of the stream that would take us through the malevolent MacDonnell Race. And we didn't want to be there when the wind got up.

We squirted out into the Race just as the tide began to turn southbound – perfect timing, at least as far as the tide was concerned. Now if only our luck will hold with the wind, I prayed.

And for a while it did, as between the rapidly increasing tide and the light south-east wind we were soon beating south at full boat speed. With the tide added, that saw us flying along at 12 knots over the ground for a moment. Still our luck held, even though by now the motion was pretty wild. Steadily the wind was rising, but there was nothing we could do but keep our fingers crossed and our course good. We were on a roller coaster ride and there was no exit.

Then suddenly we were into the worst of it: three-metre-plus waves, shaped like walls, breaking crests all around us – pure chaos, with no way of knowing where the next wave was coming from. Up until that point I think that the crew had thought I was just being dramatic, checking all of the hatches were tightly shut, putting the washboards in and making them all clip on with their safety harnesses – now they knew better. Waves collapsed on to the deck, flooding the cockpit, and then we saw flat water ahead, close under the massif of the Mull. We hadn't even noticed we were that far across. Just another quiet day at the office …

The wild weather that had plagued us in the Irish Sea continued when we were in the Firth of Clyde, but fortunately for us, the wind swung around to the west. This made it wetter, but with calmer seas for the most part, at least in the shelter of Kintyre or Arran, or up into Loch Fyne. We made the best of it, and finally found numbers of Basking Sharks in the Firth of Clyde, with a courting group near the Skate Islands – at last! And the following weeks, no sooner had we arrived in the Sea of the Hebrides than we found a group of sharks at Coll, the first of many such amazing encounters there over the next few weeks.

This was completely different from the year before. In 2002 we had scoured the whole area and failed to find one single shark. This year, everywhere we looked there were sharks, especially out at the remote lighthouse at Hyskeir. This lonely building is mounted on a solid but low-lying skerry, almost halfway between Rum and Barra. The hunters knew that the stretch of water between the Hyskeir and Garrisdale Point at the western end of Canna was productive for Basking Sharks, and it appeared that it remained so today. But being so isolated, and not on the main route to anywhere, hardly any boats passed it by, which perhaps explained why there were so few reports of sharks there. The shallow bay on the west side of the skerry is reportedly an anchorage, but it is full of rocks, and I know of at least one tourist boat that went on to a rock there as the tide fell. In short, it is best left to the wind, the birds – and the sharks. Many times we were the only boat in sight when we surveyed out there.

On a fine, clear day it is a spectacular place, with an amazing view taking in the Outer Hebrides to the west and south-west, from the Uists to the north right down to Barra. To the east lies the elemental mass that is Rum and to the north the verdant island of Canna, with its promise of shelter. And you'll be glad of that shelter if the wind gets up out there, especially if the wind is against the tide; you're right out in the middle of the main tidal stream between the Little Minch and the Sea of the Hebrides, and it is completely open to the south-west and the Atlantic. Not a great place to be in bad weather, then, as Maxwell and his fellow hunters knew only too well. As a result, it wasn't uncommon for us to have to abandon a transect to or from Hyskeir due to deteriorating conditions.

As the 2003 summer passed I began to feel that there was an upswing underway in the Hebrides. All of the positive signs I recognised from our previous experience in Cornwall were present. Basking Sharks were widespread, and we saw them wherever our transects took us[3]. This was born out by conversations we had with the ecotourism boats and our friends on the Hebridean Whale and Dolphin Trust (HWDT) yacht *Silurian* – everyone was

---

[3] Typically, we planned our transect routes to take us through areas that were both known for sharks and not known for sharks so that equal effort was spent in all areas (to try to avoid bias in our surveys).

seeing sharks. Where we encountered groups such as at Gunna Sound between Coll and Tiree, and out at Hyskeir and Canna, the average size of the animals was greater than we had ever recorded before. We recorded hardly any sub-adults, let alone juveniles, but instead many animals of eight metres or more. Many of these big, older animals had significant amounts of scarring, notches and distinctive markings (particularly on the first dorsal fin), making them prime candidates for our photo-ID catalogue. At last we felt we were making real progress.

One aspect that surprised us was that all of our contacts were adamant that the sharks had only recently arrived in the area. Earlier in the season there had been very few sightings at all, and the animals had only appeared in any numbers from late July. This was in contradiction to the records left by the hunters, who began their season at the beginning of May to coincide with the expected arrival of the sharks. By July their season was largely at an end, with very few animals taken after that time. Still, this was only one season, and we knew the sharks didn't operate to a strict timetable, so it was far too early to treat this as anything other than a possible anomaly. Seeing more sharks in the Firth of Clyde in September on our way south back to Cornwall confirmed that it was indeed a late season, at least. Exhausted though we all were, we ended the season on a high, full of anticipation for 2004.

The 2004 season

By this stage we had the logistics of the surveys down pat. With the assistance of a seasoned cadre of first mates I could keep on top of the day-to-day running of the boat, and maintain some sanity. Two things were beginning to unsettle that balance, though.

The first being that the more successful we (and others) were in making discoveries about Basking Sharks, the more people wanted to hear about it. Generating media interest in the survey work was of course part of our mission, and The Wildlife Trusts and others had excellent teams to help generate the opportunities to do so. But what had begun as a trickle of interest soon became a flood, and at times it was hard to cope with. An interview that came out at three minutes on television had probably taken up half of our precious one day off a week, and simply displaced essential work, such as servicing the engine, late into the evening. At other times, unmissable PR opportunities arose at the last minute, allowing no time for preparation, but they still had to be dealt with. Which is how I found myself sitting in a bus shelter on Plymouth seafront early one cold, damp morning, waiting to be interviewed via mobile phone by the inimitable John Humphries on the prestigious BBC *Today* programme. As the crew of a dustcart rumbled by, my only companions at that hour, I was moved to reflect on the glories that so many people imagine come with media interest. Fortunately, the interview went well …

The second was the weather. I was living in an outdoor world, for six months at a stretch, subject to the vagaries of wind and weather at all times. In our modern, technological society where most people spend the vast majority of their lives in warmth, security and comfort, mine was a completely odd, atavistic existence, of the sort that only farmers and fishermen would understand. You face changing weather conditions and adversity on a daily basis, and after a while it begins to get to you. Spells of bad weather naturally affect crew morale, and that would include whoever was first mate as much as myself. As it was our job to motivate the team, that also put extra pressure upon us both. And if we'd seen the forecast for the week and it looked diabolical, it was hard to keep up a cheery front. Make the best of it we would, but it wasn't always easy.

On the other hand, one fine, sunny day could act as a true restorative. Getting our work done in good conditions, recording sharks or other marine life provided a solid boost to our collective energies and made up for a bad spell. But slowly, inexorably, the fatigue level increased as the weeks went by.

The 2004 season in Devon and Cornwall got off to a poor start. The weather was very mixed, and there were few settled spells. By now we had become aware that after an Atlantic frontal system passed through, it would take several days at least before there was any likelihood of seeing a shark. Even if there had been groups of sharks around before the depression, the surface layer of the water would be so disturbed that whilst the sharks might still be around, they would be down in the water column with the plankton. Several days of sunny, settled weather were required to bring the sharks back up to the surface again.

So this constant unsettled weather made life very difficult, and we only recorded 21 Basking Sharks in two months of surveys. We had the underlying impression that the abundance cycle was in decline in the south-west, perhaps in favour of Scotland – time would tell.

Unfortunately, the unsettled conditions followed us north. The survey of the Northern Irish coastline was difficult due to windy weather and swell on the north coast, and as we crossed to the Firth of Clyde matters didn't improve. It was a real challenge to even start a transect on occasions, let alone finish it. One of our regular first mates, Lissa Batey, had joined me for the Clyde surveys, as I much preferred to have someone seasoned along, and Lissa, petite, fearless and a favourite with the volunteers, was an excellent choice. Compared with Devon and Cornwall, the north was no place to train beginners.

Towards the end of our first week in the Clyde Sea we had battled our way down to Campbeltown, having sighted two sharks along the way. The forecast for the following day was grim, but not the worst, with winds up to force 6 from the south-south-west with spells of heavy rain – typical Scottish summer weather. What we found as we arose at 6 a.m. to leave for Largs was somewhat different though and as we nosed our way out of the Loch we really began to feel the wind. It was a steady 25 knots from the south, and was regularly gusting to well over 30 knots, with much stronger squalls. The sky was as black as your hat, and the rain horizontal and cold. With the wind in the south, Kilbrannan Sound was wide open and already there was a good sea running. I had originally planned to go around the south of Arran, but with the wind as it was, that would be really uncomfortable in the exposed waters of the Outer Firth. So instead we took off up Kilbrannan Sound like an arrow from a bow.

Conditions were safe enough to deliver us and the crew back to Largs for the end of their week, and although conditions meant we weren't able to survey, I wasn't particularly concerned, as *Forever Changes* was superb in such conditions; everyone was safely clipped on, and the steering duties would be handled by Lissa and I alone. We were trucking along well, surfing occasionally, and had a cheering moment as we overtook a Minke Whale that surfaced momentarily alongside us off Carradale. The volunteers seemed to be comfortable and in good spirits, so all was well. The wind got up a notch as we neared the head of the Sound, so I changed the staysail for the storm jib, and we charged on at undiminished speed.

As we altered course to round the Cock of Arran, I could see the wind shadow in the lee of the island as anticipated. As we drew nearer, we slowed right down, as the wind dropped light in the lee of the land. All that we had to watch for were the regular katabatic gusts that were blasting down off the mountains above us like Aeolian grenades, sending wild catspaws of spray flying ahead of them – we didn't want to get hit by one of those.

Steadily we made our way around to the east of the island and were nearly in the clear, when I saw a williwaw heading straight for us. I had instructed one of the volunteers, an experienced sailor, to dump the mainsheet in just such an event, and, as the wind bore down on us I told him to let the sheet go, now! He froze. I sighed. The wind hit us with astonishing ferocity and poor old *Forever Changes* simply fell over on to her side and lay there like a toy duck in a bath, held down by the finger of a capricious child. It seemed like an age before the squall passed, and just as suddenly we popped upright as if nothing had happened. All the crew pronounced themselves alright, if a little surprised ('Does that happen often?'), so on we went, battling our way upwind into the short, steep seas of the Firth of Clyde.

When we arrived at Largs the wind had shifted and had increased yet again. As we roared through the entrance to the marina we set up for our berth, which would now entail a wild downwind approach. With all warps and fenders ready to go, we careered into the berth with the engine roaring, hard astern. Lissa was on to the pontoon in a split second and secured the spring – the rest was easy. I joined her, and was just adjusting the stern line when I noticed one of the crew climbing unsteadily over the guard-wires. I thought nothing of it, but when I turned around I found the guy on his knees, kissing the pontoon! As he arose I saw his face, which was ashen. 'Thank God for that,' he said, 'I've had a fabulous week, but I'm never, ever going out on a boat again!'

Which was not a choice we had. The following morning, mindful of picking up the next set of willing volunteers, we went straight back out into it, bound for Arisaig.

There, we entered a whole new world, at least as far as Basking Sharks were concerned. Lissa had performed her first mate stint admirably (as always) and was off to relax on a train back to Devon; new first mate Louise Johnson (who had volunteered with us the previous year) replaced her, on what was to be the first of many weeks sailing together. On our first circuit of Canna and the Hyskeir we had a couple of shark sightings in the morning, and then by evening as we travelled south, we ran into our first big shoal of the season, in fact the biggest shoal we had ever encountered to date. In little more than an hour we would record several breaches and many small groups involved in courtship, all within a close radius of the Hyskeir Lighthouse.

It was a spectacular sight. All around us were sharks, all well up at the surface in flat calm conditions. There were the long surface slicks characteristic of tidal fronts stretching off in every direction, and there was the familiar dank smell of plankton that we'd encountered before when around big shoals of sharks. The sky was just beginning to cloud over and the sea was quite dark as a result, making navigation through the shoal a nerve-wracking affair. With the light fading there was no time to waste trying to manoeuvre alongside each animal. With Louise on the helm, and me on the camera, we moved systematically within and between the small groups as swiftly and safely as we could, taking images for photo-ID purposes as we went. When a big shark breached very close to us we were able to confirm it was a male, and guessing that it might breach a second time, I focused in on the spot where it had landed. Twenty seconds later up it came again, and I had captured one of the first decent images of a Basking Shark breaching ever recorded – my lucky moment.

After an hour of feverish yet somehow methodical data gathering, we stopped the boat on the edge of the shoal. The sea was now so dark that it was impossible for the lookout in the bow to see a shark below the surface any more, although we knew they were there. Out to the west, beyond the Hyskeir there were more sharks, but even with the long Hebridean twilight, time was not on our side. I shut the engine down, and we settled back to simply watch and absorb this extraordinary moment. By now there was not even a breath of wind, and the sea was as calm as a millpond. The only sound we could hear was the occasional swish of a tail as a shark swam by us, oblivious to our presence. Looking out around the horizon through binoculars, all I could see were tiny black triangles criss-crossing the surface of the water. In every direction there were sharks visible. Quietly I passed round the binoculars to the others, for them to take in this astonishing spectacle. We were all completely overawed.

There was simply no way to accurately estimate the number of Basking Sharks that were present that evening. In the shoal we were with, we had documented 49 animals, but that was only one shoal – we could see at least three more, one of which was certainly far bigger than the one we were with. But our policy was to only count the animals that we could get to and confirm as 'new' animals to avoid double counting, so we could only guess at the real total. There were certainly many hundreds of animals, but with night approaching we were denied the chance to confirm that accurately. And I knew that we'd probably never know, as the forecast was for the weather to break overnight and by daybreak the conditions out here would be wild, driving the shoal from the surface. Like the hunters before us, the weather would always have the final say, and just as before, would serve to remind us of that at our moment of triumph.

What an occasion. It was like seeing a herd of elephants in the African bush for the first time. Nobody wanted to leave, but there was nothing for it but to start the engine and slowly extricate ourselves from the shoal. Once clear, we turned north for Canna Harbour, lost in quiet thought over the evening's glories. It was nearly dark when we finally dropped anchor, too late to cook supper. In the early hours of morning the quiet pattering of rain on deck woke me momentarily, as if to remind me of the challenges we still faced. When I got up at 8 a.m. to put the kettle on, *Forever Changes* was already swinging to the gusts, even in the shelter of the harbour. Yes, I could well understand the frustration the hunters must have felt.

It would be another 10 days before the sea conditions settled long enough for us to find sharks at the surface again, once more on an evening out between the Hyskeir and Canna. Further smaller shoals followed around Coll at the end of August, and again at Hyskeir in September. As before, these groups were made up predominantly of large adults, with only a few sub-adults and very small animals. Finally, we had to turn south to start the long journey home, but even then we still encountered sharks around the west coast of Mull and the Treshnish Islands. These were all smaller animals, with only the occasional adult, and we noticed that they were all spread out singly or in couples and there was none of the social behaviour that we had seen regularly with the big shoals. The hunters believed that once the small animals appeared, the shoal had moved on; it certainly seemed to make sense to us now that we had seen it on several occasions. But this was September, so that was to be expected, but yet again – another very late season.

Was this the new regime? And what could be causing such a change in seasonal abundance? It seemed we still had many more questions than answers … but luckily there were a lot of other people also on the case.

Chapter 15
Seeking Answers About
a Sea Monster

This was certainly becoming a period of discovery and changing perceptions for the Basking Shark. Many of the mysteries that had long challenged scientists due to the species' prolonged periods of absence from the surface were becoming answerable at last, through technological advances that would have astonished the earlier hunters. Public interest was at an all time high in favour of enhanced protection for the animal, stemming primarily from Britain but becoming accepted more widely. The wheels of global conservation politics were slowly grinding towards measures that would help to safeguard the future of the Basking Shark.

While we were conducting our simple, methodical line transect surveys, others were exploiting the remarkable potential of next-generation electronic tags and satellite communications to develop a greater understanding of the behaviour of the Basking Shark when at depth. Many years had passed since Professor Monty Priede's pioneering attachment of a first satellite tag on a Basking Shark, and the subsequent advances in tag technology had been phenomenal.

At the beginning of the new millennium, the state of the art tags were Pop-up Satellite Archival Tags (PSAT, or PAT tags) compatible with the Argos satellite network. In comparison with the bulky, heavy tags used in both studies in the Firth of Clyde, these tags were a fraction of the size and weight, being 175mm long and weighing only 76 grams. This drastic reduction in scale meant less drag when attached, minimising discomfort to the animal and reducing the chance of premature detachment. It also simplified the method of attachment, as there was only the one unit, linked to the tether via a monofilament and wire trace. In the first of these new studies, attachment to the shark was made by firing a stainless steel dart into the first dorsal fin via a rubber-powered crossbow.

These tiny marvels of electronics and engineering combined a data logger that could measure pressure to record depth to 1,000 metres, temperature between −40° and 60°C, and use daytime light levels to identify local noon and measure day length to provide an estimate

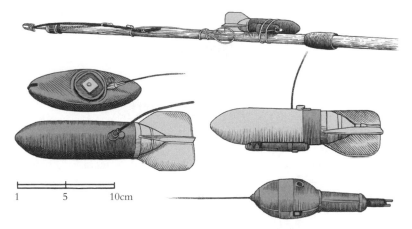

Examples of the latest generation of satellite tags used for tracking Basking Sharks.

of position. Improvements in battery technology allowed the tags to be attached for long periods. A tag could be set to release from a shark at a pre-programmed date, whereupon it would float to the surface. Argos satellites could then geolocate it and receive summaries of its transmitted data. Better yet, if the tag could be recovered, all of the raw data could be accessed, giving a more complete picture of the shark's position, depth and temperature.

During 2001 David Sims attached five of these tags to Basking Sharks in UK waters: three in the English Channel off Plymouth and two in the Firth of Clyde.

One tag from each area was set for short-term release later that summer, while the others were set to release in the longer term during winter. The geolocations of one of the sharks tagged off Plymouth were compared with data obtained from daily SeaWiFS[1] satellite images to compare the sharks' position with concentrations of plankton productivity. The results were remarkable.

They showed beyond doubt that Basking Sharks did not hibernate during winter, but instead made extensive horizontal journeys of up to 3,400km, and vertical movements to depths of as much as 750m, to exploit productive continental shelf and shelf-edge habitat in summer, autumn and winter. They would make foraging journeys of up to 460km to localised, seasonally productive fronts at the shelf-break, and all of the sharks remained around shelf waters, suggesting that they were feeding at all times.

The dive patterns of the sharks showed that they made regular vertical movements in both shallow and deep waters, that seemed consistent with foraging behaviour for productive patches of prey, perhaps using chemical clues identified via smell or taste. And the sheer depth that the animals were prepared to dive to (considerably in excess of those made by baleen whales) suggested that the Basking Shark might have an advantage over those competitors for exploiting deepwater assemblages of zooplankton. As we've already seen, Basking Sharks show a preference for energy-rich copepods such as *Calanus finmarchicus* and *C. helgolandicus* that overwinter on the continental shelf and shelf-edge in deep water between 100m and 2,200m. This suggests that the Basking Shark may have evolved to occupy a useful ecological niche.

---

[1] SeaWiFS: Sea-Viewing Wide Field-of-View Sensor is a satellite-borne sensor designed to collect global ocean biological data.

The movements of the tagged animals from south-west England to Scottish waters and vice versa seemed to contradict the belief that local populations exist, (unless we re-think what we understand as 'local'). However, one shark tagged off Plymouth did return to the English Channel in winter, after foraging along the shelf edge in the Bay of Biscay, suggesting that some sharks at least might exhibit fidelity to certain sites.

One worrying aspect of the study from a conservation perspective was how little time the animals spent in protected waters. As highlighted previously, the protection afforded by the Wildlife and Countryside Act only extended as far as the limits of UK territorial waters; that is, up to 12 nautical miles from the shore. As David Sims suggested, this could have serious implications for the Basking Shark in the event of active resumption of directed hunting outside UK waters. A later study quantified this, showing that the tagged animals only spent 22% of their time within the UK protection zone, and suggested that the Basking Shark should be added to the Convention of Migratory Species of Wild Animals (CMS, *aka* the Bonn Convention) that would offer protection for the species throughout EU waters. Fortunately, this loophole was closed in November 2005, when the Basking Shark was listed within the CMS, and as a result, in 2007 fisheries for the Basking Shark within EU waters were banned under the Common Fisheries Policy. The Basking Shark would eventually be covered by the OSPAR, Bonn, Bern Conventions in Europe and the Barcelona Conventions in the Mediterranean Sea.

Why is the Basking Shark so vulnerable?

As the second-largest fish in the world, it is perhaps unsurprising that the Basking Shark has few natural predators. The only non-human predators we have solid evidence of are the Sperm Whale (*Physeter catodon*) and the Orca.

According to records, a juvenile Basking Shark of 2.5m length was recovered intact and in good condition from the stomach of a Sperm Whale captured in a whale hunt in the Azores in 1956. Over the years I have heard credible eyewitness reports, from both the Land's End peninsula in Cornwall and Ardnamurchan Point in Scotland, of Orca attacks on Basking Sharks at the sea surface, and other reports of similar attacks are recorded from California and New Zealand. Great White Sharks have been reported scavenging on the carcasses of dead Basking Sharks, but no known reports exist of attacks on living Basking Sharks, although that shouldn't be ruled out, especially in the case of juvenile Basking Sharks. Although we have no idea of the potential scale of any such predation, presumably it is by no means a new phenomenon, and so can be assumed to have little impact on the overall mortality rate.

The only other predator on the Basking Shark that we know of is man. With its habit of 'basking' at the surface, especially during sunny, calm weather – optimal conditions for harpoon fisheries – the species is uniquely vulnerable to human predation. As we have seen from the North-east Atlantic and North Pacific, targeted harpoon fisheries for the species appear to be capable of causing drastic long-term reductions in apparent numbers on a regional basis during a very short time frame. This seems almost unbelievable on the face of it, especially in a world where mankind considers that 'there are always more fish in the sea'. Yet all of the evidence suggests that these population declines are indeed the case – the question, therefore, is why?

Basking Shark life strategies
– reproduction and vulnerabilities

There are basically two evolutionary strategies that have developed over the millennia, termed r-selection and K-selection. The former, r-selection, is used by species that tend to be small, live shorter lives, suffer elevated levels of mortality and therefore produce high numbers of offspring to compensate. Fish species such as herring and salmon are good examples of species that use r-selection in our seas. K-selection is used by species that live in a stable environment, are large, slow to mature, have few natural predators, live long lives and produce few offspring. The Basking Shark is a classic, perhaps extreme, K-selection species.

Best estimates suggest that the male Basking Shark may reach sexual maturity at between 4.6 and 6.1m length and between 12 and 16 years of age, with the female reaching sexual maturity at 8.1–9.8 metres and 16–20 years. The male is equipped with a pair of claspers of around one metre in length, and copulation is via the insertion of the male clasper via the female's cloaca into the vagina, where it is held in place by a cartilaginous claw on the end of the clasper. The extended clasper forms a groove, through which the male passes many clear packages of sperm called spermatophores amidst copious amounts of seminal fluid, to instigate fertilisation.

The length of the gestation period is unknown, but has been suggested to be between one and 3.5 years, which, given that the lifespan of the Basking Shark has been estimated at 50 years, should allow for a female to produce many litters of pups in her lifetime. However, it has also been suggested that the interval between litters may be two to four years – if this is correct (and taken in conjunction with even an *average* gestation period of two years), then the number of litters produced in an average lifetime may be far less than originally thought. Recruitment rates may therefore be very low. Marine biologist Dr Daniel Pauly at the University of British Columbia looked at hunting data from north-west Europe and analysed mortality rates for various sizes of Basking Shark. He concluded that there had been 'an exploitation rate that no fish – especially not a long-lived, low-fecundity fish such as the Basking Shark – can withstand for long'. In other words, the shark's rate of reproduction simply could not keep pace with the rate of commercial exploitation, and therefore could not withstand targeted fisheries for any length of time.

Curious factors concerning the way those populations segregate regionally and seasonally by size and sex also may have increased their vulnerability to exploitation. Groups of sharks may be made up of both males and females, but the balance of sharks killed in the Scottish fishery was badly skewed towards females, with Watkins quoting a female to male ratio of 18:1 and Maxwell reporting a ratio of 7:1. This was not unique to Scotland as in the Japanese fishery 65–70% of the sharks killed were females. By contrast, evidence from bycatch of Basking Sharks caught in sub-surface gill nets off the Newfoundland coast showed a 2:1 bias towards male sharks, and similar evidence of segregation by sex was also found in New Zealand from trawl fisheries bycatch.

In the Scottish fishery it was reported that nearly all of the females examined showed signs of recent copulation, such as injuries around the cloaca. This not only suggests that copulation takes place in inshore waters during spring and early summer when plankton productivity is high, but also that the brunt of the fisheries effort fell upon the section of the stock that the species could least afford to lose – the reproductive females.

There may have been a simple pragmatic reason for this. The hunters always sought to shoot the biggest fish, as they had larger livers and so were more profitable. When targeting fish engaged in what we now believe to have been courtship, the latter-day hunter William Manson of Mallaig reported that 'when they came across a pair at the surface they would try to pass the male in the rear to take the female' as they believed the female had 'a much richer liver.'

It is intriguing that the hunters didn't catch any pregnant females, apart from a single record of a live birth incident in Norway in August 1936, when two fishermen, Hans Goksoyr and Jonas Sordal, were out hunting and caught a female Basking Shark. Whilst towing the live shark into Teigboden, she gave birth to six pups, five alive and free-swimming, with one stillborn. Sordal[2] caught one of the pups and estimated the length at around 1.5–2.0 metres – he also noted that there was no sign of a yolk-sac or an umbilical cord (Basking Sharks are believed to be ovoviviparous, where each pup will break out of the egg and subsist on the yolk-sac before birth). So why so few records of pregnant females? Given that the gestation period may be two years or even more, it would seem logical that at least *some* pregnant females would have been killed while at the surface to take seasonal advantage of abundant plankton to support the developing pups. A scientist friend of Anthony Watkins made the cynical observation that all of the 'Females were cut up by commercial fishermen who would not recognize an embryo if it was handed to them on a plate' – but that seems unlikely. The pregnant females are out there some place – but where?

Shark movements and vulnerability

In an earlier paper, David Sims and Professor Philip Reid had challenged the widespread view that highly localised stocks existed, using the collapse of the Achill Island fishery as an example. For many years this had been held up as a clear example of a localised stock wiped out by overfishing. Using data from the Continuous Plankton Recorder (CPR[3]) survey of zooplankton abundance from the sea areas adjacent to Achill Island, they showed that the catastrophic decline at Achill may not have been entirely caused by overexploitation of a local stock, but also by a regional decline in zooplankton over the same period as the fishery.

The data showed a positive correlation between the number of Basking Sharks caught and the decline in copepod abundance. It was also pointed out that at the same time that the Achill fishery was at its peak (1949–1958), the Norwegian fishery catch remained low, but then when the Achill fishery began to decline from 1958, the Norwegian catch increased and then remained constant until around 1980. This was presented as possible evidence that Basking Shark distribution shifted northwards during that period.

There is no question that this argument has merit. However, many questions remain as to whether it represents the whole picture. In its earliest days, the Achill fishery had active harpoon and set-net fisheries. Within a short period, the harpoon fisheries were abandoned, and some consolidation of the companies engaged in the fishery took place, basically leaving only W.J. Sweeney's enterprise and a few independent, small-scale operators active. At the

---

[2] Sordal described the snout of the pup as 'narrow in front with a little bend downwards', as has subsequently been described in juvenile Basking Sharks up to 2.6m in length. (Later in life, the shape of the rostrum changes, gradually becoming less pointed as the animal approaches sexual maturity, before becoming rounded and bottle- like in adults.)

[3] The CPR Survey monitors the near-surface plankton of the North Atlantic and North Sea on a monthly basis, using Continuous Plankton Recorders on a network of shipping routes that cover the area.

height of the fishery up to 13 crews were actively engaged, which declined after the peak catch levels of the fishery to between one and three crews. The question, therefore, is whether the fishery declined because there were fewer sharks to catch, so crews dropped out, or whether there were fewer crews willing to work in the fishery, so catch rates declined.

There doesn't appear to be any absolutely clear-cut answer to the above question. It certainly sounds as if the Norwegian fleet continued to frequent the waters offshore from Achill after the decline in catches at Keem Bay had commenced. Sharks caught in this way would not show up in the Achill catch data, but within the Norwegian data. During a visit to Achill in the early 1980s I met several of the men who had been active during the last years of the fishery there. They told me that there had been a steady decline in recruitment to the crews (especially of younger men) from the 1950s onwards, which may have contributed in part to the reduced catch per unit effort. McNally suggests that changes had to be made to the pay award structure after the peak catch era, as a result of the decline in the number of fish taken and the infrequency of their appearance. That seems, in turn, to suggest that the decline in sharks came first, making it harder to recruit and retain crews.

Perhaps both factors, overfishing and a change in plankton distribution, may have played their part. But it's true that an apparent increase in the abundance of zooplankton identified from CPR data in the same waters long after the cessation of the Achill fishery has not been matched by any recovery in the number of Basking Sharks recorded around the Island. And as the evidence from places like Barkley Sound and Clayoquot Sound in western Canada suggests, there is ample reason to be concerned that any recovery may be a very long time coming … if it ever comes at all.

Are the sharks smarter than we thought?

The advances in science continued unabated, many of which had direct implications for surveys such as ours. For the first time a link was established between an abiotic factor, Sea Surface Temperature (SST) and the number of Basking Sharks recorded at the surface. Dr Peter Cotton at the University of Plymouth compared data on shark sightings held in the MCS Basking Shark Sightings database (that included our own data) and SST data from the British Atmospheric Data Centre and North Atlantic Oscillation (NAO) index data from the National Centre for Atmospheric Research in the USA.

Comparison of the MCS sightings database with SST data showed a close relationship, with peak seasons coinciding with higher mean SST levels. Further analysis showed a positive relationship between the SST and shark sightings, but only a very weak correlation to zooplankton density. The study also suggested that the relationship of the NAO to SST was an important factor, as short-term variations in SST were largely linked to the NAO. A positive NAO index causes warmer land and sea temperatures in Europe, whereas a negative index causes cooler conditions. A positive NAO was known to have a positive effect on the abundance on *Calanus helgolandicus*, albeit with a one-year time lag. Taken as a whole, this implied that climate variability plays a major role in the number of Basking Sharks sighted in coastal waters during a season, and should be considered when analysing sightings data.

This suggested that the number of Basking Sharks observed at the surface was driven by broad-scale responses to variations in temperature (an abiotic factor), whilst previous studies had shown Basking Shark surface sightings at the finer scale were largely driven by zooplankton abundance (a biotic factor). The implication is that Basking Sharks use scale-related foraging behaviour to access the best possible prey aggregations. This hypothesis was later supported by a study that compared the prey encounter success rates of real Basking Shark movements with modelled shark movements across a prey-field derived from CPR data. It was discovered that the movements of the real sharks resulted in consistently higher prey encounter rates than 90% of the simulated random movements.

Tracking their prey

Another factor that would have to be taken into account in future was the discovery that the shark varied its diving pattern according to habitat. Now that it was clear that the Basking Shark could actively orientate itself horizontally to the best aggregations of plankton, was that the end of the story? David Sims's earlier studies had shown how the sharks could forage at finer scales at the surface, but could they also do so in deep water? The dive profiles established from the satellite tagging programme showed how the animals made regular dives to sometimes extreme depths, presumably foraging vertically to locate patches of exploitable plankton, but was that behaviour random or did it also show design and purpose?

Zooplankton makes a daily migration called Diel Vertical Migration (DVM). In deep, well-stratified waters, the normal DVM pattern is of dusk ascent and dawn descent, the kind of pattern that was well known to fishermen targeting planktivorous fish like herring, who set their nets on the shoals by night as they came to the surface. But in shallow, mixed waters around tidal and headland fronts, zooplankton reverses that pattern, exhibiting reverse DVM[4], where zooplankton follows a dusk descent, dawn ascent pattern.

Examination of the dive patterns followed by tagged sharks showed a similar picture, with Basking Sharks in deeper water out at the continental shelf conducting conducted normal DVM patterns, whereas sharks in shallow waters inshore around thermal fronts exhibited reverse DVM patterns, in both cases presumably following plankton. In another study, sharks varied their DVM behaviour as they crossed boundaries between well-stratified water and mixed water[5].

These studies offer an insight into why and where the species 'basks'. If plankton move towards the surface by day in frontal waters, followed by predatory Basking Sharks, then it is no surprise that the shark numbers observed at the surface will be higher. Sims found that the effects of these opposing types of DVM on the likelihood of encountering Basking Sharks at the surface across the two different habitat types (within and outside of frontal areas) could result in a difference of at least two orders of magnitude. This would have serious implications for any attempt to establish an abundance estimate, unless adequate bias reduction factors were factored into the equation.

And of course, that knowledge added vital background as to why key areas that we had already established showed high levels of surface sightings. Where tidal fronts were a regular feature it would be logical to expect that we would see more sharks there. In terms

---

[4] This reverse DVM of the sharks' favoured prey has been put down to complex predator–prey interactions involving predatory chaetognaths (arrow worms), as zooplankton such as copepods are believed to be able to vary their migration pattern according to the position of predators like chaetognaths in the water column.

[5] David Sims subsequently suggested that marine predators like the Basking Shark may use a form of statistical search pattern when diving, similar to what is called a Lévy flight, when foraging for plankton patches that are unevenly distributed. Such search patterns have been shown to maximise the likelihood of successful prey encounters.

of confirming areas where the sharks would be at their most vulnerable to hunting and disturbance, this new information also had considerable value.

Dr Emily Southall led a joint project to examine how all of the various survey methodologies compared and overlapped each other. Unsurprisingly, the public and boat-based surveys showed an inshore bias, with sightings clustered around areas that were recognised as tidal front regions. Whilst satellite tagging provided data from the tidal front regions, it also identified two distinct hotspot areas offshore in deeper water in the Celtic Sea and the Western Approaches to the English Channel. Tagged animals spent considerable amounts of time in these two areas, whereas they hardly figured at all in the public sightings database or the boat-based surveys. This suggested that the animals were exhibiting habitat-specific foraging patterns and so were spending far more time at depth than at the surface. She concluded that future satellite tagging studies should be greatly expanded to enhance all of the existing information on Basking Shark movements and habitat usage, with a view to calculating bias reduction data for surface sightings that could help to establish accurate distribution and abundance data for the species. But given the huge cost implication of such an undertaking, it was hard to be optimistic that it would occur anytime soon.

Funds to develop new tagging programmes were being found elsewhere, however. On the eastern seaboard of the USA, Dr Greg Skomal had successfully attached PSAT tags to three Basking Sharks, with highly significant results. One shark was tagged during the first year, which migrated 800km southwards through the summer season down the US coast as far as North Carolina. During the second season, two sharks were tagged off Nantucket Island, the first of which travelled 1,600km southwards as far as Jacksonville, Florida, while the second shark covered 2,500km in the same direction to the waters between Haiti and Jamaica. Like the Basking Sharks tagged by David Sims, these animals remained in shelf waters, presumably feeding, and none of them moved out into deep ocean waters. It seemed that on both sides of the Atlantic, the animals remained part of a regional–local population, with no interaction between the two sides of the ocean.

Slowly but surely the mysteries of the Basking Shark were being unravelled.

Chapter 16
The Best of Years: 2005–2006

It was against this background that we returned to the fray in 2005, with our role in the research ecosystem put into perspective. We lacked the resources to become involved in satellite tagging, and indeed believed that others were better placed than we were to cope with the technical challenges. But we knew that we could continue to make our contribution towards the identification of hotspots on a local scale with our boat-based, effort-corrected surveys, especially in areas where public sightings were deficient due to their sheer remoteness. We could examine what threats existed for the species at those sites, and suggest strategies to mitigate against them. If we continued to encounter large aggregations of sharks, we could greatly expand the sample size in the photo-identification database now held at the Shark Trust, essential if we were to develop photo-identification as a method to eventually establish a population estimate. With improved camera equipment courtesy of our long-term supporters the Swiss Shark Foundation, we were ready to get to work. And of course, we would continue to devote time to raising awareness of the plight of the Basking Shark at home and abroad. We still had much work to do.

We had been encountering steadily diminishing returns in the south-west of England for a number of years now, which was a concern. Not in terms of the individual surveys themselves, as the lack of sightings didn't necessarily invalidate them; as we always reminded ourselves, a negative result is as valid as a positive result. But the reduced level of sightings did have an impact on the photo-ID study, as we were simply not getting adequate quantities of images of individuals to build the number of animals in the database. It was hard on the volunteers, who knew that they might not get to see a Basking Shark. They were prepared for that, but we knew and understood that it was a disappointment for the week's crew every time we drew a blank. Fortunately, we had very few weeks when we didn't see some wonderful marine life, Common Dolphins or Bottlenose Dolphins being big favourites – but they still weren't the 'big fish' that all the volunteers had signed up for and wanted to see.

As we had now completed the three-year survey cycle of Northern Ireland and the Firth of Clyde, we decided to alter the structure of the overall survey to allow us more time in the Western Isles of Scotland. For the next three years we would finish the surveys in south-west England at the end of June, allow two weeks for the delivery up to Arisaig, then conduct two months of surveys in the Sea of the Hebrides and the Minch. By now all of our volunteers were being drawn from The Wildlife Trusts and Earthwatch Institute (Europe), and we had no shortage of volunteers lined up to fill each week's crew. In fact, we were lucky that we never had a week when we didn't have a full crew, and always had a waiting list stretching into the following season.

When we had first started the surveys, we had enjoyed the luxury of choosing our participants, whether from local conservation organisations, or students we knew would benefit from field-time. When the method of funding the project changed to involve the use of paying volunteers, there was some discussion between myself and the roster of mates on whether that would result in a change in the relationship between the volunteers and ourselves. Or, indeed, whether the volunteers would be as effective as the handpicked teams. We needn't have worried on either count. The new generation of volunteers was just as motivated and effective as before, and many of them brought additional skills to the project.

This gave us the confidence to experiment with different volunteer groups. With support from Earthwatch Institute (Europe) we welcomed teams from corporate groups, the teaching profession and a drug and alcohol rehabilitation unit. If ever we had any doubts that such a diverse range of participants would work, and could be trusted to perform at the required level, they were soon put to rest.

Watching the effect that the marine world and its inhabitants had upon all participants was a great pleasure for the mates and myself. Spending time in a region like the Western Isles is to undergo sensory overload, through the mountains, the islands, the magical light and the ever changing sea. And that's before you introduce the wildlife that brings the scenic backdrop to colourful, vibrant life. It wasn't uncommon to find some captain of industry sitting next to a volunteer who had spent several years as a homeless person, both lost in companionable silence, struck dumb by the glory of it all. Watching their reaction to the various marine species was entertaining and rewarding.

When confronted with dolphins the team would be exultant, cheering, and laughing. Whales generated a more awe-inspired response, and tears on more than one occasion. But the Basking Shark was different, being welcomed with a quiet form of wonder bordering on astonishment that such apparently prehistoric creatures could still exist, and in our own waters. As Philip Hoare describes them in his wonderful book *Leviathan,* 'The sharks gape blindly as they feed on unseen plankton, their bodies browny yellow and mottled, almost reptilian, a mark of their own ancientness.' Basking Sharks truly engender a unique reaction in onlookers.

Perhaps it has something to do with the pre-programmed mental images people have of sharks from countless wildlife TV documentaries that have portrayed them as mindless killing machines, or perhaps an atavistic fear of sharks as old as humankind, but there always seemed to be a slight reserve towards a Basking Shark. Not to those of us who had seen them many times – to us it was like being around old friends.

But whatever the species we encountered, it was always a pleasure for us to make such introductions possible, and to be able to send the volunteers home with some life-enhancing memories, as well as new skills.

The 2005 season

We started 2005 as we always did, full of enthusiasm. We had a responsibility to the volunteers to start every week as if it would be the best ever in any case, despite any misgivings we might have over the weather forecast. As it turned out, we would need all the optimism we could muster, as it was a really tough season in Devon and Cornwall, and although we worked long hours, the rewards were very slim, with only seven sharks recorded. The weather didn't give us many breaks either, and when the time came to make the long hike north the wind swung round and blew furiously from – the north. We sat in Falmouth with our delivery crew aboard for nearly a week waiting for a lull, climbing the walls in frustration. The moment the wind dropped we were off, crashing our way around Land's End into the last of the wind, spray flying everywhere, yet by first light the next morning the wind had dropped to virtually nothing and we started the engine. And it stayed running until we glided out through the North Channel and picked up a light westerly breeze that ghosted us up to Gigha. Just 36 hours later we were helping the first volunteer crew of the Scottish leg to pack their kit aboard.

The first few weeks were also testing. Conditions weren't bad, but there was no sign of any Basking Sharks, and the local whale-watching boats were unanimous that it had been a very poor year for sharks so far. Now, after many years in the field, I had developed a real feel for the sea and the life unseen beneath the waves. When there's food present in the form of plankton, the birds let you know it for one thing, with Gannets (*Sula bassana*) plunging from the sky and auks like Guillemots, Razorbills and Puffins gathering and diving, chasing the fry that feed on zooplankton. Localised groups of these birds, sometimes called 'hurries' are always a good sign that positive things are happening below the surface.

Gannets diving in also gave us some indication of what they were feeding on, using the old fisherman's yardstick that if they were making vertical dives from great heights then they were chasing herring that would be lower in the water column by day – whereas shallow, angled, dives would mean mackerel, higher in the water column, giving us an indication of where the zooplankton might be. If we saw hurries of birds at a distance, and could see Gannets plummeting into the water, we'd be sure to focus our attention on the sea beneath them. Maybe a Minke Whale would be lunge feeding amongst them, or Common Dolphins ripping through the shoal, or maybe even a Basking Shark or two, serenely filtering out plankton, untroubled by all of the fuss.

Having seen so often how the 'mood of the ocean' could shift from empty expanses to bursting with life in a very short time, we watched for such tell-tale signs constantly. So when the Gannets began diving towards the end of the month we knew a change was coming. The next thing we expected to see would be the occasional breach. Often the breaching animal would be some distance away, and so positive identification as a Basking Shark was not always possible, but the chaotic style of the breach and the lack of any further sighting almost certainly ruled out any cetacean species. The more of these 'random breaches' we saw, the more certain we would be that we were going to see a shoal – and if the weather held, very soon.

And we didn't have long to wait. On 5 August we saw six Basking Sharks at Coll, nearly all large adults – a good sign. Three days later we were into a big shoal, and had the best day of recorded shark sightings we would *ever* have, with 93 animals, the vast majority of which were also large, mature sharks.

Most people imagine that a big shoal implies one single solid mass of sharks, but that is seldom the case. In my experience, allied to information gleaned from the hunters' books, each shoal is made up of smaller groups within a much larger whole. These smaller groups come and go at the surface, seemingly tracking the patches of plankton they are focused upon, which makes it very hard to accurately count animals and avoid double counting. In such circumstances we would take great care to work our way methodically through a shoal, and try to keep tabs on where the groups we had already counted were in relation to our position. Sometimes the shoal is spread over a wide area, which is a great advantage when it comes to taking images for photo-ID purposes, as it allows you to concentrate on the individual animal and not worry about running into others.

It may seem logical to imagine that Basking Sharks in a shoal will react negatively to the approach of a boat, but that's not always the case. Having watched shark reaction for years, and been involved in one of the only studies to evaluate Basking Shark reaction to boats, I'm convinced that solitary animals are far more easily disturbed than animals in a group. This appears to be at least partly due to plankton density. Solitary sharks tend to spend less time on the surface, swim faster and react to the approach of a boat at far greater distances.

When plankton patches are dense, the sharks seem to be fixated solely on the available food source. They are almost oblivious to the presence of boats, and are actually very hard to disturb. On such occasions sharks will sometimes swim right up behind a boat, and follow in its wake. It is then that your boat handling skills have to be at their best, as other sharks are likely to be in the vicinity, and vulnerable to any sudden changes of speed or direction you might make. Whenever possible we would wind up our genoa and simply sail the boat under mainsail only, even in very light winds, as the boat handled well like that and the lack of noise made for a calmer atmosphere on board. The sharks benefited, too, as there were no risks of injury from the propeller while it wasn't turning. It can be nerve wracking, though, for the helmsman, especially after a few hours of vigilant manoeuvring.

Our extraordinary day at Coll was like that, on 8 August 2005. The best of the shoals were (as usual) around the Cairns of Coll at the north end of the island and Gunna Sound, the narrow, tidal channel that separates Coll and Tiree, where small fronts are an almost constant feature in calm conditions. But during the course of the day we circumnavigated the island, and encountered sharks all the way around the island, giving us a very worthy recorded total of 94 sharks. We recorded our first shark at 10.48, and the last at 18.02, by which time we had all had enough, and needed a breather. The stress of handling the boat safely around the animals had left me frazzled, and first mate Louise and the crew were shattered too, having filled up numerous memory cards with good quality photo-ID shots, and recorded all of the data in a surprisingly understandable manner ... But what a day! Sharks breaching close by us, much courtship activity, and some big, battered, scarred sharks with very easily identifiable markings – it was truly a day to remember.

And there was no let up. Three days later we encountered a smaller group up at the Hyskeir, followed by a well-spread-out shoal south of Barra in the Outer Hebrides. The weather was perfect, so we enjoyed a slow sail down to Berneray past the magnificent islands of Sandray, Pabbay and Mingulay, recording sharks all the way. Rounding Barra Head, we entered the Atlantic proper, but on that day there was hardly any swell and so we were able to continue our journey around the sheer, bird-whitened rock face and up past the spectacular cliffs on the western side of Mingulay. Even on such a calm day, the sound of the swell booming in the caves and gullies along the shore lent an air of menace to the journey. How many poor souls had made their last sight of land through the breakers out here?

Remembering that the Norwegians had on occasion hunted sharks in the waters to the west of the Outer Hebrides made me wonder whether we shouldn't try to conduct surveys out here; but knowing that today's conditions were far from typical, I soon put that thought safely to bed. I was more than happy to be tucked up in one of our favourite anchorages in Vatersay Bay that evening, where the sea was so still that it felt as if we were aground. Up at first light to continue to take advantage of the perfect conditions, a cow and two calves meandered slowly along the water's edge – a timeless scene in a glorious place. It was hard to leave.

As we made our way back across the Sea of the Hebrides the weather was already changing. As often happens in these waters, even in summer, a really strong gale blew in from the west, confining us to port for several frustrating days. It would be nearly three weeks before we would see another shark, in the early days of September, when we once again met a big shoal.

As usual we had seen some distant breaches in the preceding days, so we were hopeful that – weather permitting – we'd soon be back into a shoal. And just as before, the breaches proved to be a good indicator, as out at the Hyskeir late one afternoon we recorded a shoal of 33 animals. Once again these were predominantly big adults, and many displayed the telltale signs of ardent courtship: the white, raw rostrums and the leading edges of their fins white from rubbing against the abrasive skin of other sharks in the 'Love Dance' as Patrick FitzGerald O'Connor called it. Maybe he was right, and perhaps at least some of these markings may occur during the sexual act itself, as occasionally we saw large Basking Sharks with extensive white patches on their pectoral fins. In many shark species the male shark bites one of the female's pectoral fins to maintain his position while copulating and whilst they might be very small (only five to six millimetres in length) and almost vestigial, Basking Sharks do have teeth. And as we noticed on inspection of a big, stranded male shark, those teeth can be worn. Maybe those tiny teeth still have a functional use, and the wear comes from pectoral biting?

It was the last gasp of the season. As local people will tell you, summer comes late and leaves early in such northerly waters, and the best of the weather is over by late July. We had been unbelievably lucky, and we knew it. It was a wrench to draw the season to a close knowing that there were sharks still out there.

But we did have something to look forward to. We had successfully argued the case for spending the entire summer season in Scotland in 2006, which would enable us to assess whether the change in the seasonal timing of arrival that we had observed so far was real or imagined. We hoped it would also maximise our opportunity to take photo-ID images. Scottish Natural Heritage, the Scottish Government's nature conservation agency, had expressed an interest in the data we had gathered so far, and we felt we were making real progress in building a case for identifying hotspots. Another advantage of beginning the season in Scotland was that we were spared the long slog home in late September that could often prove to be a gruelling voyage into shorter days, colder conditions and fierce autumnal gales. Instead we could leave *Forever Changes* in Scotland, load up the van with all of our gear and head home by road. It would mean a spell in the frozen north during the winter working on the boat, but we would be ready to start fresh in Scotland the following April. We couldn't wait for the new season to start.

The 2006 season

We launched *Forever Changes* one bitterly cold April morning and after a couple of days on a mooring preparing the boat for sea, we set off into the Firth of Lorne. The west of Scotland was sitting under a large area of high pressure and the skies shone with a clear, crystalline intensity. The backdrop of white-capped mountains stood out in stark relief in the low spring sunlight, and not a breath of wind disturbed the glassy surface of the water as we motored slowly south. Rounding the lighthouse on the island of Lismore to head down the Sound of Mull, familiar sights and places served only to heighten my anticipation for the coming months. Picking up a mooring for the night at Tobermory, we were glad of our paraffin heater as a cold clammy mist descended on the harbour – summer felt a very long way away.

As there were still a couple of weeks before the season began in earnest, Louise and I had decided to make an exploratory voyage around the islands while the settled weather held. It seemed too good an opportunity to miss. So, loaded down with a week's shopping, we slipped our mooring and headed for the open sea and the delicious adventure of a season in the Hebrides.

The sea was still as calm as a millpond as we rounded Ardmore Point on Mull. Gulls dotted the surface, visible at hundreds of metres. An occasional Harbour Porpoise broke the glassy stillness from time to time, showing a stubby little dorsal fin before diving to avoid our boat, helping us to get our spotting eyes calibrated. Out past Caliach Point the vista opened up dramatically. Away to the south lay the Treshnish Isles, beyond which we could see the strange sight of an island in the sky, a classic *Fata Morgana*, lending a mysterious air to the morning. Up ahead we could just make out the low-lying Cairns of Coll, our first sight of the season of one of my favourite islands in all of the Hebrides.

Many of the islands in the Hebrides are mountainous and wild, giving them an almost universal appeal. Coll and its immediate neighbour Tiree are exceptions to the rule, being almost flat, the highest point across both islands being Ben Hynish on Tiree, at only 141 metres. To some they lack the grandeur of their fellow islands, but they have many appealing features of their own, not least a host of magnificent, deserted beaches and in season, glorious swards of colourful flower-rich machair. But they are wild places indeed from a weather perspective, with some of the highest average wind speeds in the British Isles, and are beset by strong tides that can create horrific seas in places. Construction of the Skerryvore lighthouse that guards the ships' graveyard to the south-west of Tiree took six years, due to the conditions that prevail there. These are waters to be treated with caution.

Yet on a calm day it's hard to equate the lonely beauty and peaceful waters on the eastern sides of the islands with such savage weather. The main harbour on Coll, Loch Eatharna is little more than a large cleft in the rocks, but is open to the south and is often affected by swell. Nevertheless, it provides some of the best shelter in both islands. In April, the village of Arinagour at the head of the Loch was as quiet as a crypt, with none of the buzz of summer, but that wasn't why we were here. To our amazement, the reason we were here was right beside us. We had only just anchored when I noticed a sleek, dark silhouette just below the surface of the water swimming alongside *Forever Changes* – our first Basking Shark of the season. Then another appeared, the distinctive flash of sunlight glittering off a wet dorsal fin over by the ferry pier, then a third shark – if it was going to be this easy, we laughed, maybe we should just run the surveys in the anchorage here? It seemed like a good omen, and at least we knew that there were some sharks around this early in the season.

It didn't last. The weather broke, and one murky morning we flew up the coast past Ardnamurchan Point with a strong wind behind us and swung through the narrow, tortuous entrance to the beautiful village of Arisaig, our home for the summer, to await the arrival of the first of the volunteers.

The first few weeks were the usual mix of rain and wind, grabbing the opportunity to complete transects whenever we had the chance, dodging bad weather in between times. Basking Sharks were few and far between, hardly surprising given the changeable weather, but there were no evident signs of other marine life either. We enjoyed one good day down at Coll, with yet another shark in the anchorage in Loch Eatharna, but then despite better weather, we endured a sightings drought until late May when we had two sharks off Caliach Point on Mull. If we had been hunting Basking Sharks, we would at this stage have been getting very worried about our finances.

Finally weather conditions became more settled and as expected, we began to see signs of life, with Gannets diving and occasional Minke Whale and Harbour Porpoise sightings. By the end of the first week we had recorded a small shoal of Basking Sharks off Canna followed by a good day at Coll with a shoal of 19 animals around Crossapol Bay and Gunna Sound at the southern end of the island, another of our usual sure-fire sites. Despite excellent weather allowing us to cover lots of ground, up to the north and out to the Outer Hebrides, the sharks seemed to be contented to stay around Coll; we had further good days there, with 37 animals on one occasion and 10 on another. But despite several visits, Hyskeir remained quiet and we drew a blank down around Barra and the southern chain of the Outer Islands. We re-sighted several animals amongst the shoals recorded around Coll on a number of occasions, suggesting maybe that the animals we were encountering were part of a large group that was remaining in the area. This perhaps explained the lack of animals farther north.

Finally, there were signs that the shoal was moving on. The last good day we had around Coll in July, we found a group of 15 sharks in and around Gunna Sound with, for the first time, a number of small animals in their midst. We observed several juveniles in amongst the rocks off the small island of Gunna, far too close inshore for us to reach them. It was almost like watching hungry children scooping up the last of the ice cream from a bowl, the young stragglers polishing off the last of a huge patch of zooplankton while the adults had moved on. But to where?

We found out the next day. Departing early in the morning, we had a final couple of sharks on our way to the Cairns of Coll, and then altered course to the north-west towards the Hyskeir. Conditions were ideal and we were optimistic that we might find the big shoal again, either around the lighthouse or up around Canna. We picked up three sharks a mile or so short of the Hyskeir and then ran into the shoal, big sharks well up at the surface over a wide area, courting and breaching constantly. Once again, many of the sharks displayed the characteristic white abrasion on the nose and upper jaw of animals that have recently been engaged in courtship-like behaviour – it seemed we had arrived just as the party was getting into full swing.

During the course of the afternoon we once again moved steadily through the animals, photographing fins and measuring individuals wherever possible, but the shoal was very dense, and finding a way through it was tricky at times. One group of 33 animals was packed like giant sardines in a very small area. They were totally fixated on exhausting the dense plankton, turning on a whim and often in their own length, seemingly avoiding their fellow sharks and us by inches. Fortunately, I had my good friend Mick Roberts with me as mate, an excellent photographer, so I could just leave him to the photo-ID work, allowing me to concentrate on getting the boat safely through the shoal. Courtship and breaching were occurring constantly, with the various outliers erupting from the water right out on the extremity of the shoal –something that we noticed on many occasions. It seemed almost as if the breaching animals on the periphery of the shoal performed some sort of corralling function, helping to keep the shoal intact.

By early evening we had moved on further north towards Canna and our anchorage for the night, but still we had sharks in sight almost constantly. As we reached Canna we found a group of nine animals off Garrisdale Point, amongst which were a number of small animals; a rare sight around a shoal. What was intriguing was that we had seen small sharks only the day before around Coll, and now we were seeing them again the other side of a big shoal off Canna. What could it mean? Was the big shoal still on the move, or would it stay around this new location. We couldn't be sure that the animals we had seen that day were part of the same shoal we had been seeing around Coll, but it did seem likely. But above all, we hoped that this new sighting of small animals didn't signify the end of the shoal. That would be too bad. Nobody was complaining, though, as we had recorded no less than 80 sharks that afternoon, in one of the densest shoals we had ever encountered.

But alas, the small sharks proved to be a bad omen. Over the next few days we saw no more sharks. It was an astonishingly abrupt ending to what had been a spectacularly successful month. Coll had been extraordinarily consistent until finally the action cooled there and the sharks seemed to move north to Hyskeir. We had recorded more courtship and breaching than ever before, and these two sites were increasingly standing out across the whole of the survey area as the obvious hotspots. We could only hope that the season wasn't over. Our own previous experience suggested that there should be more to come, but the hunters post-World War II would have expected the season to be drawing to a close at around this time.

During the last two seasons of 2005/2006 we had been experimenting with extended 10-day voyages. A frustrating element of the standard surveys was that we were restricted in terms of exploring the extremities of the study area, especially if we had the usual mixed bag of weather. The six day survey platform also meant that we were often restricted in the anchorages we could use, ruling out the most sheltered ones as time spent getting in and out of them would cost us valuable survey time. Anchoring in more exposed and often less comfortable bays and roadsteads allowed us to get back into action almost immediately the next day, but meant that I had to sleep with one eye open and the steady accrual of fatigue took an inevitable toll. The 10 day schedule, however, allowed us far more scope to undertake the long legs to the Outer Hebrides, and gave us a little more early morning and evening downtime to relax in more sheltered bays.

What we hadn't bargained for was the extra strain the 10-day voyages would put on the team, crew and volunteers alike. During one such survey I had the flu, a parting gift from one of the previous week's volunteers, and as a result felt like death warmed up for the duration, virtually unable to speak and coughing like a cannon. At the end of the survey I almost wished I had died. But even without the flu, by the time we were nearing the end of a 10-day survey everyone was completely exhausted. On one occasion Louise and I were the only ones awake as we motored the last few miles to Arisaig, with the entire crew of five volunteers, whose ages ranged from 16 to 65, snoring peacefully on deck – which made for a great photo.

Even the standard six-day surveys could knock people out. One week late in the season we had left Arisaig a little later than usual, meaning that our planned anchorage was too distant to reach in daylight. As a result, the only accessible alternative was Soay Harbour, where Gavin Maxwell's failed factory lay in ruins. Normally this was an anchorage we avoided due to the restrictions placed on entering and leaving caused by the tidal bar in the entrance, but on this occasion the tide was perfect for a late afternoon entry and early morning departure. We slipped in over the bar and moved around until we found a small pool where we would not touch the bottom at low tide, anchored up and then dinghied ashore to explore the remains of Maxwell's dream. Battling through clouds of midges, we surveyed the rusting factory wreckage strewn around the old buildings, and marvelled at the madness that must have seized the man to try to develop his venture on this remote island.

During the night the wind dropped to nothing, and in the still, sheltered waters, *Forever Changes* lay as unmoving as a house. The only sound was the occasional 'kraaack' of a Heron (*Ardea cinerea*) in the trees, and I found it hard to sleep thinking of the huge significance of this strange, lonely place in the annals of the Basking Shark in Scotland. At 4.30am, David

Marshall and I were up on deck to catch the high water over the bar, moving around as quietly as possible to avoid waking the volunteers. That was easier said than done, though, as we had to start the engine to generate the power to lift 100kg of anchor and chain with our anchor windlass in order to leave. Once we had the anchor aboard, we then had to feel our way down to the narrow entrance in the inky darkness. Up in the trees on the western side of the anchorage an island resident had thoughtfully provided a transit made up of two pieces of timber with reflective tape attached to them. Once these two posts were in line, we had to keep them in line constantly while we turned onto a course that would keep us in the safest water out over the bar. The only problem with this being that we had to light up the transit with our searchlight every 20 seconds or so, with David covering his eyes every time, to avoid ruining his night-vision.

It was a great relief to cross the bar and get back into deep water. As it was only 5 a.m. and there was no sign of any of the volunteers, we quietly agreed to head up the short Sound of Soay to anchor for a few hours in Loch Scavaig. It was the most peaceful of mornings as we slowly motored up the Sound. High up in the Cuillins we spotted a tiny beam of light from some bold soul on their way to climb the Ridge – the only evidence of human life in sight. As the dawn light slowly lit up the magnificent mountains, we slipped into the Loch and as quietly as we could manage, dropped the anchor and returned to our bunks.

At 8 a.m. I was back on deck to find one of the volunteers who had been asleep in the fore-cabin sitting on deck, looking up at the mountains that surrounded us, seemingly puzzled.

'I don't remember it looking anything like this,' he said. 'Am I dreaming?'

'No, just exhausted,' I chuckled.

As the others appeared on deck the reaction was much the same – they had all slept soundly through the shift from Soay to Loch Scavaig, windlass, anchor, engine and all, yet at home they'd probably have woken up at the sound of the kettle going on. Wonderful stuff, sea air.

August saw us cover much ground, including Coll and Hyskeir, with inconclusive results. Yes, there were a few sharks still around, but they were few and far between – and small, including several young born that year. It wasn't looking good. Then a temporary reprieve as we had a good day on our way out from the Hyskeir to the Outer Hebrides, including a group of eight sizeable animals. From then until the end of the month we recorded sharks sporadically, but mostly small and sub-adult animals, mainly around Canna and the Hyskeir. The shoal had indeed moved on.

In early September the pattern was the same, this time down at Coll, where we consistently encountered small groups of juvenile animals, with very few adults. It certainly felt like the season was coming to a close at last. But one last high spot remained when we recorded 10 Basking Sharks around Garrisdale Point, Canna with the last of the big adults, in what proved to be our final shark sighting of the season. Perhaps the big shoal had finally dispersed, and was now returning south or heading out into deeper water for the winter.

It had been the most extraordinarily productive season, spending the whole summer in Scotland. In terms of the number of animals photo-identified, shoals encountered, courtship and breaching recorded and the sheer consistency of the hotspot sites, it had exceeded anything and everything we had seen before. As we made our way slowly back to Arisaig on the final day of the season I was the happiest man alive. We had accomplished all of our goals and more. A photograph taken at the time shows our gallant skipper with the biggest smile on his face, jubilant that the season had gone so well and glad that the long, exhausting gig was about to be over – for a few months, anyway. As we slipped through the rocks in the Rhu Arisaig I turned to Louise and said, 'We'll never do better than this. Maybe we should just quit now, while we're ahead ...'

A throwaway remark, spoken in jest at a heady moment, but one that would come back to haunt me in the not too distant future.

Chapter 17
The End of the Line

Having said goodbye to the last of the volunteers and held our usual end of season party it was time to put *Forever Changes* to bed for the winter. So on a bleak September morning we battled our way out of Arisaig. Sailing south around Ardnamurchan Point on a wild day with a huge swell running in from the west and without a single boat in sight, we felt a little exposed. It wasn't hard to be thankful that we weren't going farther than the Firth of Lorne to lift her out. Winter was upon us.

Once we had her out of the water the usual chores began: stripping off the sails to be sent for valeting, servicing everything mechanical and cleaning the boat from truck to keel. Time, too, to get our regular survey carried out, as demanded by the regulations relating to small commercial vessels. Not that we had any concerns, as *Forever Changes* had always passed all of these structural examinations with flying colours, and she did so once again. Apart from the demands of a new regulation to do with our gas installation we had no unexpected work to do, so we loaded up our old van with our gear and headed for home.

Having the boat at the other end of the country was in many ways a blessed relief. No winter days in the damp, dank shell of a boat in a dripping wood in Cornwall, instead just an early return to Scotland in the spring to bring the boat back down south to start the new season. It felt like I was going on holiday.

Despite our hope that we could carry out another full season in Scotland to capitalise on our successful work so far, we had been obliged to return to the south-west for the early part of the 2007 season. There was nothing for it but to make the best of it; however, it did help to focus my mind on the future. I had by now decided that this would be my last full season, partly because the workload had become too punishing, and also because I needed to get back to work with my other business interests and rebuild my finances. I also felt that we had gathered a satisfactory dataset to demonstrate that the hotspot areas existed and therefore back up our push for conservation measures in these areas. To realise any further benefits from boat-based surveys, we would need to alter the current format to capitalise on what we had learned so far, spend more time remaining with the shoals, and concentrate on photo-identification. This wouldn't fit in well with the project obligations that we currently had, which included regular media commitments and the restricting regime that a 20-week volunteer-based schedule demanded. For a variety of reasons, it felt like our current chapter of work would have to come to its natural end after the 2007 season. It was a sad decision to have to make, but an unavoidable one.

As if to confirm that the right decision had been made, the phone in my home now began to ring at odd hours of the night and when I finally answered, it was to be regaled by some drunken hero making threats on my life. So I had the phone removed and switched off my mobile phone before going to bed. That diverted attention to my office phone, and on a Monday morning the answerphone would sometimes yield a graphic and painful description of my forthcoming demise at the hands of the brave soul at the other end of the line, out hiding in the ether. So I switched off the answerphone and forgot all about it.

But when we returned to Scotland at the beginning of April 2007 to bring the boat south for the start of the survey season, we could have done without the unwelcome news that awaited us there. Somehow our radar scanner, which was mounted on *Forever Changes*'s mast and stored off the boat in a separate shed, had been smashed. Nor did this seem to have been an accident, as the perpetrator had to have climbed over at least six other masts to get to it. This was a bad blow, not only due to the expense of replacing it, but also because at this time of year there would always be the likelihood of poor visibility. As we'd likely have to sail overnight at least once on the forthcoming delivery voyage back to Cornwall, we certainly would have liked to have the radar in working condition. But as it was too short notice to get it repaired or replaced, we would have to go without it.

After the launch and a thorough check of all systems, we set off for Dunstaffnage to fuel up on the first evening, arriving in the pitch darkness of an icy night that chilled the bones to the marrow. We were up at first light, and as soon as the fuel nozzle was out of the final spare jerrycan, we set off. On a glorious, clear, sharp spring morning Louise and I bade farewell to the magnificent snow-capped hills of Scotland and sped south on our way to Cornwall.

Inevitably with our broken radar, the weather forecast spoke of fog and partway down the Sound of Luing we ran into a thick bank of it. With the visibility down to less than 20 metres I decided that the best thing would be to keep out of the middle of the channel to avoid other boats taking the direct route, only to very nearly run into another yacht doing exactly the same thing! Rattled by this, I then set to work our way into the bay outside the Crinan Canal to anchor and wait for the fog to lift – we couldn't see a thing. After a few hours with no improvement, I tried to raise the canal lockkeepers via VHF radio; we could at least keep going in safety in the canal, even in thick fog. No answer. But luckily, as the afternoon wore on, the fog lifted enough for us to continue on to Craighouse Bay, Jura for the night, where we could mull over the history of the renowned distillery whilst huddled around the little heater on *Forever Changes*.

Fortunately for us, this was the last we saw of the fog for a few days, and we made our way down through the Irish Sea with all haste. The daylight hours were short, the cold intense and with no autopilot to take the load off us, one poor soul had to be at the wheel the whole time. No further mishaps occurred with the boat, but I managed to make a very hard landing on the pontoon at Arklow in the early hours of one morning, stepping off the deck onto a sheet of ice that sent me flying. I spent the rest of the voyage sitting on a very thick cushion.

As we approached the open water leg across the Celtic Sea to Land's End, we were confronted with a big dilemma. The high pressure system that had given us light east and north-east winds all the way so far was set to intensify, bringing freshening winds and perhaps more fog. If our luck held, we had a sufficient window to make the crossing, but by the time we reached Land's End we would have to be safely around the corner and heading for Newlyn before the wind really got up. The other option was to hole up somewhere in Ireland and wait for the weather to go through. But the worry so early in the season is always: what will come next? We could be stuck there as gale after gale went through, getting ever more desperate to leave. It wasn't much of a choice, but we decided that fortune favours the brave – we'd go for it.

We cleared the dreaded passage between Carnsore Point and Tuskar Rock off Rosslare in perfect conditions as the day began, wisps of mist and sea smoke hiding the land. Ploughing on through a chilly day, we swapped watches every two hours to keep our energy and spirits up, and occasional encounters with Common Dolphins and a Minke Whale cheered us on – we were not alone out there. By nightfall the wind was steadily freshening and we were making good speed with our wind vane finally steering. The forecast now spoke of north-east winds reaching force 6 the following day, so the window still looked good, even though the mention of fog was now more insistent. Rounding Land's End would put us into far more shipping and fishing boat traffic, so I decided to head instead to the Isles of Scilly, and wait out the weather there.

Dawn brought drizzle and diminishing visibility. Finally, we spotted Round Island lighthouse and altered course for the northern entrance to New Grimsby Sound, between Tresco and Bryher. These were my home waters and I knew the Isles of Scilly well; I reckoned that rather than go all the way around the islands in thickening weather it made sense to save time by threading our way through the islands on a rising tide to reach a secure anchorage. Entering New Grimsby Sound in early April was a world apart from high summer, when the moorings would be jammed with boats, their crews ashore enjoying this wonderful place. But on a grey, wet, windy morning like this there was little glory on display, and not a soul nor a boat was on the move.

With the tide near high water we crossed the Tresco Flats and set our course for the Spanish Ledge Buoy off Bryher. With the wind set to be strong from the north-east for several days I wanted to make for the most sheltered anchorage in the islands for that wind direction, the Cove, between St Agnes and Gugh. It was mid-morning by the time we made our way into the Cove and dropped our anchor. As expected, there was not another boat to be seen, so we had the pick of the best shelter and could let out as much anchor chain as we liked. By now the wind was up to a good force 6, with the rain much more persistent; conditions got steadily worse over the next two days. It was only after three days at anchor that we felt we could safely launch our dinghy and venture ashore for a well-earned pint at the Turks Head.

Two days later, on a lovely spring morning, we set sail for the mainland, and by nightfall were home in Falmouth. It had been a great voyage, and both Louise and *Forever Changes* had once again proved their mettle.

Now the real work would begin.

The 2007 season

We had hoped that a year of absence from the waters of Cornwall and Devon might have brought us better fortune in terms of weather and sharks. The memories of 2005 were still with us, and we desperately hoped the weather would give us a break. But it was not to be, and from the moment the season began the weather was anything but benign. We struggled desperately to make the best of it, shifting the boat constantly to seek sheltered conditions to continue the surveys, but for the most part to no avail. Easterly winds in the English Channel leave you few hiding places. Slowly we made progress, but it was hard going.

Conditions improved a little as the weeks went by, and slowly but surely the tally of transects began to mount up. But Basking Sharks were few and far between, and apart from three solitary animals encountered at the beginning of June we drew a blank. As conditions only remained settled for a few days at a time, it was hardly surprising. We managed two or three voyages west to the Isles of Scilly that were productive for cetacean sightings, but even down around the Basking Shark hotspots of the Lizard and Land's End, not a fin was to be seen.

By the middle of June we were running out of apologies for the lack of Basking Sharks to our ever-willing volunteers. But with a week ahead when the weather finally looked like it would improve, we carried a moderate easterly wind west to the Isles of Scilly, ready to carry a light westerly airflow back in what should be ideal conditions. It was as good a forecast as we'd had all season.

On the return leg, back along the mainland, we were sailing up the coast towards Polperro in a light south-easterly breeze. All seemed well, so leaving Lissa in charge on deck I went below to put the kettle on. Then I felt the boat gently round up to the wind for no apparent reason. I could hear Lissa telling the helmsman how to bring the boat back on course, but nothing appeared to be happening, so I climbed up into the cockpit to have a look. Taking hold of the wheel it felt very light, and I guessed that maybe a steering cable had snapped. But on looking at the top of the rudder stock below the helmsman's seat I could see that the shaft was turning with the movement of the wheel. Nothing I did with the wheel made any difference – the boat wouldn't answer the helm. In an awful moment it struck me. We had lost the rudder.

No warning, no noise – the rudder was gone. I looked around. We were nearly half a mile offshore, with a light onshore breeze. What to do? Fortunately, we had trained for such an eventuality, but as you never think these things will happen, it took a few moments to come up with a plan of action. Shifting the crew weight to the port side of the boat, we gradually got her moving, the only problem being we were going in the wrong direction – we needed to tack to make progress away from the shore. I rigged up a bucket on a long line and used it in an attempt to get the boat through the wind, but we couldn't quite make it, and she fell off onto the wrong tack again and again. There was nothing for it but to gybe the boat round and get her on the other tack that way. I started the engine and put it into gear at tick-over speed to provide just a little extra forward motion. Shifting the sails around and then moving the crew over to use their weight, we finally settled onto the right tack. Now if we could keep her there we should soon be clear of the land.

The bow kept rounding up, but by backing the staysail we were able to maintain very slow forward progress out to sea, albeit swinging around between 90° to the wind and 30° when she would almost stall and then pay off eventually, back on the right tack, continuing on her way to safety. It was time to get on the radio and let the coastguard know of our predicament. While we could just about manage this course, I knew that it would be a real problem sailing the boat into the narrow entrance at Fowey, where the wind would inevitably die. And as we were a commercial boat, it had to be safety first.

The coastguard were sympathetic and asked us if there were any other boats in sight that might be able to give us a tow, but it was a grey midweek day, so the sea was empty. Finally, they spoke to the Fowey lifeboat coxswain who said they would come out, even though we had explained this was not a life or death emergency. An hour or so later we had made good around three miles out to sea and were enjoying our long-delayed cup of tea. We saw the lifeboat roaring up the coast, and so called them up to pinpoint our position to them. They were surprised to see we had made progress out to sea, and when they ranged alongside asked why we hadn't used the emergency tiller, thinking we had only snapped a steering cable. I pointed out that as far as I could deduce, and as incredible as it sounded, we'd lost the whole rudder. Taking up station on our stern quarter and peering into the sea as our stern lifted to the swell, they confirmed my suspicions – the rudder was completely gone.

As always, the lifeboat crew were amazingly efficient and passed us a tow rope that I made fast to the bow cleats, followed by a drogue to be towed astern of us to keep us going in the right direction when the tow began. It was a long, nerve-wracking few hours before we were finally alongside at Polruan opposite Fowey, ready to be lifted out the next morning to inspect the damage. We all thanked the wonderful guys in the lifeboat crew as profusely as we could. That week's volunteers had to pack up prematurely, and although disappointed, were really gracious and generous in their wishes of good luck. Now that we were safely tied up I felt like I'd been run over by a truck. In the heat of the moment all my energy had been focused on positive action. Now there was nothing but exhaustion and concern for the future.

Once the boat was out of the water on the quayside we could see what had happened. The rudder shaft had sheared off completely around one foot below the hull (at a point that would have been invisible to any inspection as it was actually inside the rudder shell). The boat design meant the stock had been sleeved at that spot to allow the rudder to be tapered, and this had made a weak spot that had finally failed and the whole rudder had gone to the bottom. There was no way we could have known this would happen and a succession of surveys had not picked it up either – how could they? What was clear, though, was that we were out of action for some weeks at least. By now I had been on to our insurers who promised prompt action, and we had begun ringing around to try and find a boat to charter so that we could fulfil our survey commitments. Pausing only to say goodbye to our great volunteer crew, who had remained helpful and cheerful throughout, Lissa and I set about packing up all of our essential gear to enable us to swap it to the charter boat. That is, if we could find one at all.

We had two weeks of surveys left to complete in the south-west before we were due to leave for Scotland for the second half of the season. I knew there was no way we could get a new rudder made in such a short time, but was hopeful that something might be achieved in a month. But we'd need everything to go in our favour to do so, including a quick and positive decision from the insurers. I'd also have to plead with all of the craftsmen required to get the work done, to drop everything to help us out. It was an awfully long shot.

Finding a charter yacht available at such short notice would be the first major difficulty; paying for it would be the next. There is no consequential loss in marine insurance so I'd have to find the money out of my own pocket, but with people already commencing their travel to join us I had no honourable option but to pay up and hope we could get *Forever Changes* back in action. We'd been in constant communication with Earthwatch and The Wildlife Trusts that we might lose some surveys down the line, but there was at least time for the volunteers involved to make alternative arrangements. It was numbingly frustrating, but there was no other way forward.

At the last minute we found a boat of about the same size as *Forever Changes*, fully coded for charter work. It was nothing like the tough warbird we were used to, nor did it have many of the modifications that she had to make her fit for our purpose, but in the circumstances it was a godsend. The forecast was diabolical once again though, and first mate Larry Williams and I knew it wasn't going to be an easy couple of weeks.

It wasn't. The weather was terrible, with a strong wind warning out virtually every day. And our luck, which had always been so good, seemed to have finally deserted us, along with the rudder: when we were on a mooring the charter boat's overly-light dinghy was hit by a fierce gust and capsized, drowning the outboard engine; we caught a stray rope around the rudder where it became jammed and we had to get some passing divers to free it – just two examples of an endless litany of small disasters that hit us on a daily basis. At least one was amusing, though. We were about to manoeuvre the boat out of a tight marina and had set up the warps on to the pontoons to make it easy. The wind was blowing hard on our beam and this boat was very light and blew around like a toy, so we needed to be smart about getting her moving. Down to the last few warps and a quick nod from me to Larry to cast off the final line and I gunned the engine astern to back her out – and stopped dead. We had left a line attached to the pontoon. It gave the whole marina a good laugh…

But that was about all we had to smile about. The weather never gave us a break, there were no Basking Sharks about and it was with some relief that we tied up at the end of the second week. Larry and I were both all in. The moment I walked through my front door, the phone rang. It was my insurance broker. The insurers had refused our claim on the grounds that the failure had occurred due to metal fatigue of the rudderstock. That was it. The end came with a whimper, not a bang. There would be no voyage to Scotland, no surveys in the Sea of the Hebrides, and all of the volunteers would have to be informed and their money returned. I'd lost the money we'd paid out for the charter and now would have to find a far greater sum to build a new rudder and pay all of the yard fees. Like the hunters before me, I now learned the hard way that working with sharks could land you in deep waters.

And I thought back to my prophetic words at the end of the 2006 season. I should have quit while I was ahead. Those niggling feelings are seldom wrong.

You pick yourself up

For the first few days it felt like a bereavement in the family had taken place. The project had filled each and every waking hour so completely for so many years that I was almost at a loss as to what to do with myself. But we could afford no time for self-pity, so I quickly snapped out of it, and threw myself back into a long-forgotten way of life ashore during the summer, that actually involved days off. I'd almost forgotten what they were.

There was some consolation: it gave us time to pick up the data we'd been gathering assiduously for the last few years, and see exactly what we had got. My main contact at The Wildlife Trusts, Joan Edwards, was a tower of strength at this time, and confirmed that the Trusts would help us see the project through. Our good friend Roger Covey at Natural England offered further support. Dr Suzanne Henderson at Scottish Natural Heritage commissioned us to write a report on our work there, which was really welcome news. We might have lost the boat, but the project was still alive and well.

In common with many field projects, we had been gathering data at a prodigious rate without really considering the consequences of eventually handling and analysing it. So the first thing we had to do was get the data into shape and do a full quality assurance. Louise is a dab hand with a spreadsheet, but this needed additional volunteer support and we were very lucky to recruit an avowed data lover within the college where we were based. Jackie Pearson was an absolute wonder and a delight to work with, and I know that without her hard work we would have struggled desperately to fulfil our promises and meet our deadlines. The report on our survey of the Devon and Cornwall coastline for Natural England had to come first as we had committed to it before the report for Scotland, but it would form a template for the latter as well. The smaller number of sightings in the database meant less work on that front, but that was largely cancelled out by the greater number of transects owing to the longer duration of the survey.

Once the data was in a presentable format, the report could be built around it. Looking through the findings provided a colourful reminder of days at sea, and the memories flooded back. The extraordinary 1999 season stood out above all others, not only due to the high level of sightings (102 Basking Sharks) but also the regular breaching and courtship-like behaviour recorded. After that early peak, it had been all downhill for the south-west. But following analysis of all transects and shark observations, and some rudimentary Geographic Information System (GIS) outputs, a clear pattern had emerged overall, with the vast majority of sightings clustered around three distinct areas: in Falmouth Bay, between Lizard point and the Runnelstone buoy at the south-western point of Land's End, and out west of the Runnelstone buoy in the Atlantic. Apart from a very few sightings outside those areas, that was it. We had identified our surface sighting Basking Shark hotspots for south west England.

When we later came to examine the overall dataset including Scotland, one feature of the data stood out, highlighting the dramatic way that sightings had declined in the south-west by 2002 and the corresponding explosion of sightings in Scotland from 2003 onwards. It was almost as if someone had flicked a switch, that directed the entire stock of sharks farther north. We were really looking forward to working on the Scotland report, as we knew we had such strong data.

Louise and I had worked up the south-west England report between us, but we were both aware that we could have done more with the data in terms of spatial and temporal analysis and presentation. What was needed was a co-author with real expertise in analysing and expressing the data using the latest GIS technology, and then creating the graphic presentations that would really bring the report to life. Just up the hill from our home in Falmouth was the relatively new University of Exeter in Cornwall, and a young scientist who had formerly worked with David Sims had recently moved there to take up a new post. I had seen Dr Matthew Witt's work, and been hugely impressed, so I contacted him through our friend Dr Brendan Godley to arrange a meeting. I was a little nervous that Matthew would be reluctant to pool his considerable skills with us, but after half an hour and a brief run-through of our data, he seemed hooked and pledged support to help produce the report. Quite how much of an asset he would be was apparent from the very outset, and we knew we were immensely lucky to have him in the team.

Anyone who has read a report of this nature will recognise there is such a mass of data that it is almost impossible to take it all in. This is where visual presentation of spatial and temporal data is so effective and powerful; in addition to offering additional analysis and quality assurance of the data, that was exactly what Matthew was able to do. Watching him convert our spreadsheets of hard-won data into simple, accurate, digestible GIS graphics was like watching alchemy in action as far as we were concerned, and gave us something to aspire to with the accompanying text.

In order to argue the case for hotspot status we knew we had to build strong arguments that would stand up to scrutiny. If we only focused on numbers or Sharks Per Unit Effort (SPUE) values, we thought that these alone would not carry enough weight. What would be required were consistently high SPUE values over time. To back that up, we needed to be able to show which areas consistently hosted groups of sharks, including the biggest shoals, that themselves contained the greatest percentage of large, mature adult fish. Behaviour was important, too, particularly social behaviour associated with groups, such as courtship-like behaviour and breaching. The data we had gathered ticked all of those boxes unequivocally in our designated Areas 15 (Coll) and 17 (Hyskeir and Canna).

The average SPUE value for the Coll site was 1.74h^{-1} and over 11% of the sharks sighted were involved in courtship. At Hyskeir and Canna, the average SPUE value was 2.82h^{-1}. These figures greatly exceeded the proposed average SPUE value of 0.6h^{-1} for frontal areas established in the English Channel. All groups of more than 10 surface-sighted sharks occurred at these two sites, and the majority of sharks recorded (65%) were in the size categories of mature animals. All instances of courtship and breaching also came from Coll and Hyskeir and Canna. This was the sort of evidence we needed to justify the case for hotspot status. Writing the report, although arduous and intense, was greatly assisted by such solid data. Now all we could hope for was that publication of the report would lead to positive action.

Happily, that was immediately forthcoming. Scottish Natural Heritage supported our arguments for hotspot status for the two sites, and worked with us to develop simple educational materials targeted at pleasure craft owners visiting or passing through them. The aim was to come up with map-based guides that delineated the most sensitive zones at the two sites, and outlined sensible practices for boats passing through those zones. These included actions such as reduced speed (maximum six knots) in high awareness areas to avoid the risks of collision with Basking Sharks at, or just below, the surface. Additional information on safe boat handling when in the company of sharks, including the best angles of approach, safe observation distances and maximum duration of any encounter were also included. All of this information was delivered in leaflets and waterproof paper guides for retention and use on board boats and even to allow use on deck. The idea was not to stop people from enjoying a life-enhancing chance encounter with such an amazing creature, but to encourage best behaviour and reduce disturbance and injury.

The recognition that these areas were considered worthy of special conservation and protection measures more than made all those long days of rain-lashed survey work worthwhile. But even with the report conclusions accepted, the two hotspot sites successfully established and conservation measures being widely discussed, we couldn't help but admit that many questions remained unanswered. Neither site figured large in the records left by the last generation of hunters post-World War II, even though one of the hunters, Patrick Fitzgerald O'Connor, actually came from Coll, but spent the vast majority of his time hunting in the Outer Herbides. Gavin Maxwell mentioned hunting around Canna occasionally, but never Coll. Anthony Watkins never mentioned either site. More intriguingly still, the sites that all of the hunters agreed were the very best hunting grounds – Moonen Bay, Uishenish, Barra – although surveyed (whenever we got the chance), never figured at all as sites of any importance. In fact, we never recorded a single sighting at either Uishenish or Moonen Bay, and even around Barra we only rarely recorded sharks. How, we asked ourselves, with such extraordinarily consistent sightings around Coll and Hyskeir, could we have totally failed to record sharks at sites no distance away that had once been the cream of the hunting grounds?

Our photo-ID work had gone extremely well, and we had built up a substantial database of images. It was obvious from even a brief scan of the database that the bigger, older sharks had accrued more markings over their lifetime than the smaller individuals. This meant that we had more animals with distinctive features, which should have resulted in far more matches over the years. But the percentage of re-sightings of recognisable sharks remained disappointingly small. The obvious conclusion from this finding seemed to be that the stock was much bigger than we had originally thought, but we needed far more matches to test that theory. Another discovery had been made though; that it was possible to examine and categorise the markings to some degree, into natural or human-related injuries, with a view to identifying their provenance. This would in time give us useful insights, for example into the key areas of conflict with fisheries.

It was clear that what was badly needed was a substantial satellite tagging programme in the west of Scotland that would confirm whether the sharks around the hotspot sites remained there for considerable periods, or whether they simply visited for short periods and moved on. Such a programme might also address the number of sharks returning to the same sites on an annual basis, as the battery life of the tags was getting better all the time, allowing for longer-term deployments, even over consecutive seasons. The potential insights to be gained were massive – but so would be the cost. It was way beyond our budget, that was for sure.

Clearly there was still unfinished business to be addressed in the Hebrides. It was disappointing not to be able to continue our study, but we felt pleased with what we had achieved so far. Maybe one day we'd be able to pick up the baton once more. On a very local scale we had made a difference. Others were doing so on a global scale, and it was fascinating to see the progress that was being made.

Catching up with the research of others

Meanwhile, Basking Shark research in the waters around western Europe had moved on significantly. There were now active projects in many areas, and new approaches were being developed that had exciting potential to provide answers to many of the outstanding questions. And there was a new, more collegiate approach between the various research teams that was both welcome and timely, to spur the process on.

Scientists from different disciplines were actively pursuing a variety of interesting possibilities, such as genetic studies. Dr Les Noble of the University of Aberdeen had offered to collate and analyse tissue samples taken from Basking Sharks for their DNA. To date this had been a somewhat challenging task where live sharks were concerned, necessitating the use of either a pole or crossbow to take a small biopsy sample, and as a result, few samples had been obtained. But after a Basking Shark slapped its tail against the bow of an Irish tagging boat leaving a sample of body slime, a useful step forward was made. The slime was scraped off the bow and later successfully analysed for DNA. Henceforth, the field teams in Ireland and the Isle of Man collected samples of slime with the aid of scrubbing pads mounted on poles, and the number of samples obtained increased dramatically. PhD student Lilian Lieber would later validate this simple, rustic approach to gathering DNA material as an effective, low cost, non-destructive, non-invasive technique of at least comparable efficacy to far more intrusive methods.

The potential benefits of using DNA sampling as a research tool for the species had already been demonstrated, when a team led by Professor Rus Hoelzel at Durham University examined Basking Shark tissue samples from the North Atlantic, Eastern North Atlantic, Mediterranean, Indian Ocean and Western Pacific in an effort to establish an estimate of abundance and genetic diversity. What they found was surprising, to say the least: there were no significant genetic differences across the ocean basins, and there was no difference between the Pacific and Atlantic populations, as had previously been proposed. Many shark species have low genetic diversity, but the study found that the Basking Shark has the lowest genetic diversity of any shark yet studied, a factor that was put down to a population bottleneck (such as a near extinction event) during the Holocene period, or perhaps female-mediated gene flow. As a result, the proposed population estimate as a result of the study was very low, at just 8,200 individuals. Although the authors pointed out that this was a very rough approximation, such a low population estimate combined with very low genetic diversity had significant implications for the species in conservation terms. Further surprises were soon to follow.

In the Isle of Man, researchers Jackie and Graham Hall, working with the Manx Wildlife Trust, had developed the successful Manx Basking Shark Watch project. In June 2007 they had collaborated with Dr Mauvis Gore on a satellite tagging programme in Manx waters, attaching Pop-up Satellite Archival Tags (PSAT tags) to two individual Basking Sharks, one small animal of indeterminate sex and a large female. The former animal moved northwards and stayed within continental shelf waters at depths of up to 200 metres. The latter animal moved south-west down through the Irish Sea towards the Land's End peninsula before turning more to the west and moving out beyond the Continental Shelf into deep water. From that point, she made her way steadily westwards across the Atlantic Basin, diving to a new record depth of 1,264m along the way.

When the tag popped off after 82 days she had almost reached Newfoundland, having covered a horizontal distance of 9,589 kilometres, three times the previously recorded distance for the species. The fact that this was a large (presumably mature) female that had made a trans-ocean basin migration, also gave potential support to the hypothesis previously put forward by Professor Hoelzel that female-mediated gene flow might occur, as highlighted above. The Basking Shark had joined the elite band of sharks that make these extraordinary migrations, covering almost as much distance as the current record holder, a female white shark that had covered 11,000km in the Southern Ocean.

And more was to come. In the USA Greg Skomal and his team had been back in action tagging Basking Sharks off southern New England, attaching 25 PSAT tags to individuals. Of the 25 tags, 18 functioned as designed, transmitting data after detachment following deployment periods ranging between 12 and 423 days' duration. Eight of the tagged sharks remained within the previously described range of the species from their earlier studies, but the rest of the tagged animals significantly expanded the range of the shark to the sub-tropical and tropical western Atlantic, including the Caribbean Sea and down as far south as Brazil. This provided the first evidence that a Basking Shark could make a trans-equatorial journey. Given the recent discovery that a Basking Shark could cross an ocean basin, this new information suggested that there might be global migratory connectivity for the species.

In further tagging developments, in 2009 a French shark research team from the Association pour l'Étude et la Conservation des Sélaciens led by Eric Stéphan and Armelle Jung, attached two tags to Basking Sharks in the Iroise Sea, and a further eight tags in the Isle of Man, working in collaboration with Jackie and Graham Hall. The Irish Basking Shark Project were actively tagging sharks with both coloured Floy tags and satellite tags (two) around the Irish coastline, as part of a project led by Dr Simon Berrow and Emmett Johnston.

Taking stock, and getting back into the game

With the publication of our report in Scotland in 2009, our direct involvement in the field undertaking summer-long surveys was at an end and we had fulfilled all of our obligations to funding organisations and our supporters. We had a new rudder made for *Forever Changes*, and she was sold to a good new home. Of course I missed the work, but at the same time, new doors were opening, and there was much to look forward to. Louise and I had a boat built, named *Pèlerin* (the French name for our favourite fish) to become our future home, but with no intention of using her for work, just pleasure. It seemed like the end of an era.

But we couldn't keep away. With *Pèlerin* launched, we decided to fulfil that last dream to sail to Scotland, to spend the summer of 2009 cruising the Hebrides with just family and friends. It was wonderful to have the luxury of getting up when we felt like it, and if the weather was bad, just switching on the heater, having a cup of tea then going hiking instead. The sailing was just fantastic and we re-visited many of our favourite anchorages. It was wonderful to be back in these waters, seeing dolphins and whales alongside. As we'd hoped, the Basking Sharks wouldn't leave us alone either, and we optimistically spent time scouring the historic hotspots that had failed to deliver during the surveys. And as before, we didn't make a single sighting in those locations, although we did (as always) around Canna, Hyskeir and Coll. It was baffling! And I knew then that I was still not quite over the Shark Fever, which, like malaria, kept recurring – someday it would draw me back, I knew. Not if, but how and when.

We didn't have long to wait. In 2011 our project received some very welcome news in the form of a grant from our long-term supporters the Swiss Shark Foundation. This grant would enable us to conduct two weeks of boat-based surveys in Scotland each year, to focus on the remotest corners of our previous survey area. This would allow us to more properly cover the hunters' old stamping grounds – Moonen Bay, Shepherds Bight, Barra and the southern Outer Hebrides – in a far more concentrated manner. And we would also use the opportunity to continue to monitor our established hotspots of Coll and Hyskeir and Canna. All being well, we hoped to be able to continue this study for a number of years.

With *Pèlerin* now slowly *en route* down to West Africa and beyond, and our new life afloat, we would need to charter a boat from within Scotland. In order to best position ourselves in terms of the weather and the likelihood of sightings, we selected the last two weeks of July as our survey period, and chartered a 12m yacht from the Isle of Skye as our research vessel. Having read this saga so far, you will no doubt marvel at our supreme confidence in selecting dates, but knowing that bad weather could occur at any time, it was simply a case of picking what we believed to be the least worst option and hoping for the best! As it turned out, we had the usual mix of good and bad weather with a couple of gales, but also some almost ideal conditions. It was great to be back in action after such a long layoff, even if positive results at the old hunters' sites still remained elusive. We did encounter a few sharks during both weeks, mainly around Coll, which (given the weather) was a successful outcome.

For some time, Scottish Natural Heritage had been developing the idea of fielding a comprehensive satellite tagging programme taking in the three hotspot sites of Tiree, Coll and the Hyskeir and Canna. Basking Sharks would be tagged at the three hotspot sites with the aim of examining how long the sharks remained within those sites and how they used the habitat. This had the potential to confirm the importance of the hotspots for the species beyond any doubt. It would also provide much-needed support for a potential Marine Protected Area (MPA) development for Basking Sharks and Minke Whales within the Skye to Mull area, which encompasses all three hotspot sites.

This was a substantial and critical piece of work that would require a team with considerable expertise in the field of satellite tracking of marine life, together with the modelling and analysis capacity to interpret and deliver the findings. It was no surprise to us then that the work was awarded to our friend and co-author Dr Matthew Witt and the team at the University of Exeter in Falmouth. Louise and I were invited to provide advice on Basking Shark distribution and behaviour in the region, along with strategic advice on weather and tidal factors that might impact on the tagging programme. Our friends Jackie and Graham Hall from the Isle of Man, contributory as always, were on hand to provide technical advice and support for the team on the practical elements of tagging the Basking Shark. We were all delighted to be invited to contribute to this major study.

Matthew had definitely caught Shark Fever too, and was working on a paper that examined the distribution of Basking Sharks around the UK, merging data collected from the likes of our boat-based surveys and the MCS public sightings network, a substantial long-term dataset. This paper highlighted the seasonal nature of shark sightings, confirming (for example) the peak period for surface sightings in Scotland as August, derived from the data. It also further identified a close relationship between the duration of a sighting season and the North Atlantic Oscillation. As we saw earlier, this climatic feature is recognised as a driver of marine ecosystems and affects the Basking Shark's favourite prey, calanoid copepods. When the NAO is in a positive phase, production of the richer *Calanus Finmarchicus* increases and *Calanus helgolandicus* decreases, and when the NAO is in a negative phase the opposite is the case, therefore positive phase NAO years should favour Basking Shark sightings.

Most interesting of all from a conservation perspective was the analysis of the shark size data, using both public sightings and our boat-based data, between 1998 and 2008. These showed there had been a decrease in the number of recorded sharks of less than four metres, an increase in the number of four- to six-metre sharks, and the number of sharks above six metres in length had remained constant. This was encouraging, as it suggested that a slow recovery in the population might be taking place. Further evidence to suggest that such a recovery was indeed underway came from an analysis of photo-ID images from Scotland and Ireland between 2004 and 2010 by Dr Gore and others. These authors estimated a regional population of 19,151 (+/− 10,629) animals (citing evidence from some of the most recent mass sightings in support of this proposal). Bearing in mind that the theoretical population doubling rate for the species is estimated to be in the order of 14 years and this apparent recovery has taken around 60 years from the peak days of hunting in Scotland, it can be understood why this putative recovery may have been such a long time coming. Small glimmers of hope, perhaps, that prove that the conservation measures now put in place are delivering positive benefits.

The tagging programme in Scotland

The next few years were to see the most complete data collection exercise undertaken to date. The tagging team of Matthew Witt and Philip Doherty arrived in the Hebrides in July 2012 and settled into the picturesque port of Tobermory on the Island of Mull, close to all three hotspot locations where the tagging was scheduled to take place. This was to be the base for the survey programme. Matthew had chartered a motorboat from Sea Life Surveys, based in the port, to act as the main tagging vessel. As we were already actively surveying the area at the same time, we planned to maintain contact with the team by mobile phone or VHF radio to advise them on whether we were seeing sharks and, if so, where they were to be found.

Two different types of tags were being deployed, in order to provide insights into different aspects of Basking Shark behaviour. The first question the team sought to answer was how long the sharks stayed in the hotspot areas, so Smart Position Only Tags (SPOT) were employed. This was the first time that these new generation real-time tags had been attached on a Basking Shark. SPOT tags are generally deployed in more inshore locations where they can transmit accurate (to 350 metres) geolocation data over a period of weeks to months (up to 250 transmissions per day). As such, they are ideal for examining whether tagged animals display site fidelity, how long they remain within such sites and whether there are particular locations within those areas that they prefer.

The second type of tag was an updated and modified version of the classic Pop-up Archival Transmitting (PAT) tags that had previously been deployed on the species. These were PAT tags with Fastloc™ GPS, or PAT-F tags, which contain an integral GPS unit. They can transmit highly accurate geolocation data (to within approximately 100m) when at the surface and record water temperature and pressure (to give depth information) at 10-second intervals to provide information on habitat usage. With sufficient battery capacity to permit attachment for up to 280 days they are highly suitable for longer-term studies that could give insights into shark movements beyond the seasonal study in the Sea of the Hebrides. These tags can summarise the data they collect over four-hour periods for transmission via satellite uplink but require the use of a tag detachment device to release the entire tag at a pre-set date for future recovery (if possible), whereupon the full data set can be retrieved for a far more detailed analysis.

In order to provide as complete a picture as possible, the team were also equipped with a pole-operated camera system to inspect the pelvic area of each tagged animal and assess its sex. The total length of the animal was determined in the usual manner with reference to the length of the tagging vessel. A GPS waypoint was taken at the position of the tag attachment, together with a vertical plankton sample and a Conductivity, Temperature and Depth (CTD) assessment. Slime samples were also taken using a pole-mounted scrubber to be stored in alcohol for later use in the genetic study.

It was decided to attach the SPOT tags first, to enable the team to confirm that they were working correctly. With weather conditions ideal and Basking Sharks widely distributed throughout the area, this proved straightforward. The tags were successfully attached in a short period, with three sharks tagged at each of the three sites of Gunna Sound (Coll), Tiree and Hyskeir. As eight of the nine SPOT tags were apparently working correctly after three days, the team moved on to the PAT-F tags, attaching four tags at each of the same locations over a 10-day period. It was a textbook operation.

New technology doesn't only provide scientists with novel ways to answer complicated questions, it can also enable them to give the world an insight into what they are seeking to achieve and how they are going about it. In order to do this, Scottish Natural Heritage offered the public the chance to name the Basking Sharks tagged with the SPOT tags, and a near real-time display of the movements of these sharks was made available on the Wildlife Tracking website. This proved to be a highly attractive development with over 60,000 views of the website in the first season alone, further demonstrating the level of interest that the Basking Shark now engenders.

Initial results were positive, with SPOT location data providing considerable evidence of residency in the hotspot areas, where most of the sharks remained within those areas for periods ranging from weeks to months. This was especially true around Coll and Tiree and to a lesser extent, Gunna Sound, where the sharks remained for periods of up to several months. The shorter duration PAT-F attachments supported this, with most of the animals remaining within the area in which they were tagged, before they were presumed to move into deeper water offshore. Location data from both SPOT and PAT-F tags showed that offshore movement of some animals occurred in early autumn, between September and October. Two animals took off on long-distance migrations south towards Madeira and the Canary Islands, adding weight to the previous discoveries from the Isle of Man and New England, that some Basking Sharks can and will undertake such extended voyages.

The intention for 2013 was to increase the sample size and use updated technology to expand on the results gained so far. The PAT-F tags used previously had now been superseded by an improved tag, the Mini-PAT, with enhanced data compression and a longer battery life allowing attachment to a shark for up to one year. SPLASH-F tags that transmitted data in real time and also had an integral Fastloc™ GPS system supplemented the SPOT tags that had been used in the previous season for short-term attachments. To take even greater advantage of the new range of tags, the number of tags available for attachment had also been increased, with 4 SPLASH-F tags, 12 Mini-PAT tags and 15 SPOT tags. Through a combination of a larger sample size and an extended period of attachment, there was every reason to hope that the most broad-based data set could be achieved. As this was the most comprehensive satellite-tracking study for the Basking Shark anywhere in the world to date, it certainly had the potential to do so. When combined with the data from 2012, this would produce an authoritative picture of the movements and habitat use by Basking Sharks in the Sea of the Hebrides and beyond.

As we have seen throughout the history of the Basking Shark in Scotland, the weather will always play a major part in all dealings with the species, and success or failure in one year is no guarantee of the same result a year later. So it was in 2013, when weather conditions were more challenging; despite three exploratory voyages out into the exposed waters around the Hyskeir no sharks were to be found there. As a result, all of the tags deployed were attached to sharks at either Tiree or Coll, leaving an apparent gap around Hyskeir and Canna, although a number of sharks tagged off Coll and Tiree did make their way north to that site.

Once the data had been gathered and merged with 2012's findings, the results became more solid. High levels of seasonal residency were demonstrated around the hotspot sites of Coll, Tiree and Hyskeir during the months of July to September, once again supporting the previous findings from public sightings and our own boat-based surveys. Approximately 85% of the geolocations came from within the Skye to Mull search area, further highlighting the area as being of high importance for the Basking Shark. All the evidence pointed to the sharks staying in shallow water during the summer months, before moving offshore into deeper water in the late summer and autumn, varying their daily diving behaviour according to habitat. Merging Scottish data with that from sharks tagged at the Isle of Man helped to identify a potential migratory corridor between the Sea of the Hebrides, the Celtic Sea and south-west England. The data proved much that researchers had long suspected, and more. When combined with the high levels of putative courtship and other social behaviour observed during this and the previous studies, the Skye to Mull area was confirmed as a likely stronghold for the species in the North-east Atlantic.

Meanwhile, away from the technology

In the interim, our surveys of the old hunting hotspots continued. In 2012 we had mostly excellent weather with only a couple of windy spells to hamper our efforts, enabling us to explore the absolute extremities of the hunter's range. Still we found no sharks at those historic locations, but many substantial groups at both of our established hotspot sites of Coll and Hyskeir, with much breaching and courtship-like activity as usual. Once again, a few sharks were encountered south of Barra, offering a glimpse of success there that failed to materialise into anything more concrete. A total of 80 sharks recorded over the two weeks showed that the animals were still around in appreciable numbers.

2013 offered us almost perfect weather apart from a little fog, and we covered more distance and hours on survey than we had ever managed before. But the fickle nature of sightings reminded us not to take things for granted; we drew a complete blank during the first week, and only nine sharks during the second, all around Coll. Multiple breaches towards the end of the second week told us that a shoal was arriving in the area, but just not up at the surface – yet. It was a successful yet frustrating fortnight, ending in a mad dash back to the charter base on the final morning in 35 knots of wind, with gale warnings ringing alarm bells all around the Sea Areas adjacent to the Hebrides. A reminder (as if we needed one) that up in the Hebrides the weather will always have the final say in all matters.

The final year of our survey of the historic hunting sites in 2014 continued the pattern. We had better weather than we had any right to expect throughout the fortnight, enabling us to cover the ground effectively, and we explored the farthest flung corners of the assigned area with ease. Matthew Witt and Philip Doherty were up once again to attach a final batch of satellite tags, but didn't have much luck finding sharks, despite the good conditions. After a very short burst of sightings over the preceding days, the ocean telegraph had gone quiet and the ecotourism boats were not reporting any sightings. Not the best of news, but the weather forecast was good, so there was still everything to hope for – maybe they might arrive in numbers yet?

In almost perfect weather we headed south towards Coll, before turning west at the Cairns of Coll to go out to Barra. The fog came down when we were halfway across, but we still found a few sharks out in the murk, so we knew there was some life about. We headed up the west coast of Skye yet again to Gavin Maxwell's inevitable first port of call, Moonen Bay. No ghostly boom from 'Sugan' shattered the stillness of that flat calm day, and the surface remained undisturbed by fins, apart from a Minke Whale that was feeding along a well-defined front well right under the cliffs in the Bay. We altered course west across the Little Minch to the lonely lighthouse at Uishenish Point, where Maxwell and others had reported hunting for 'days without end', but as always we went away empty handed. As evening came, we made a last tour of Shepherd's Bight, the bay just to the north of Uishenish Point, but to no avail, so we quietly slipped into the lovely little anchorage in Wizard Pool, beneath the looming mass of Hecla. With not a soul in sight, we drank in the uncanny peace, as the light over the hills slowly faded.

We were away down the Outer Hebridean chain the next morning for a perfect voyage south to Barra for the night and an excuse to visit the superb little curry house – as we were finding so few sharks, we had to get our pleasures wherever we could. As we followed the channel out the next morning a solitary Bottlenose Dolphin took up station beneath our bow, seemingly leading us through the reefs out into clear water. Down at Barra Head the sea was swirling with tidal eddies, and auks scurried purposefully about on the surface, but still – no fins.

Hyskeir, then Canna followed in ideal conditions, but our luck was seemingly out. With a perfect forecast and workable tides, we decided to make for Soay, and spend a night in the harbour. I hadn't been there for 10 years; it was too good an opportunity to miss, visiting once more the epicentre of hunting that had drawn me here in the first place. Nothing had changed: the peace and solitude still intact, the decaying buildings and rusting machinery being slowly reclaimed by time, wind and rain. But there was none of the sense of unease I'd felt the last time I was here. I now felt I understood the men who had worked here a little better, and no longer felt like an interloper, surveying the remnants of their disaster. As we slipped out over the bar the next morning, with glorious sunshine lighting up the Cuillins on Skye, I felt at perfect peace with their world.

Stopping only to swap crews at our old haunt of Arisaig, we set off once more on our lonely circuit: Moonen Bay, Uishenish, Barra, all followed in excellent conditions, but still with nothing to show for it so far. These were the conditions that hunter and researcher alike pray for, but seldom get up here: warm, calm, high-pressure weather. But no sharks yet. For us this was merely disappointing, as we weren't worrying about money haemorrhaging out of a failing business, with ruin staring us in the face. Though I'll always wish they had never hunted the Basking Shark, I couldn't help but feel some sympathy for the hunters; I now understood the challenges and frustrations they had endured only too well. It must have been heartbreaking for them at times.

An old friend appeared one day out west of Canna. On a dreich and dreary morning, I spotted a large splash around a half-mile away, followed by a quick glimpse of a fin. Thinking it might have been a breaching shark I altered course in that general direction and we got the cameras ready. Then a huge fin broke the surface, unmistakably an Orca even at such a distance. Viewing through the binoculars we could identify three, maybe four animals, but all eyes were on the dominant bull with his two-metre tall dorsal fin. And then I saw the notch low down on the posterior edge of the fin – it was John Coe, who I'd last seen over 20 years before, no more than five miles from this same spot. Both of us older and more careworn perhaps, but both still defiantly resisting the ravages of time, we passed like two wandering ghosts of these wild and lonely waters.

By now we were becoming more than a little frustrated. Everything was perfect, the weather was calm, life was everywhere, Gannets were diving, dolphins were chasing shoals of fish and Puffins with sandeel-filled beaks mocked us as we gazed out upon fin-free waters. As we completed our transect circuit at Barra, we were feeling thoroughly disheartened – this was not doing our SPUE scores any good at all (all effort, no sharks). A quick phone call to Matthew and Philip in Tobermory confirmed that they were having no better luck than ourselves and were running out of time to get their tags attached. There was a lot riding on the next couple of days, and we all knew we had to find sharks soon.

As anyone familiar with those 'how we made the film' snippets at the end of wildlife documentaries, this is when the very patient cameraman who has sat still in the same puddle for a full six months without a sighting, manages to get 'the shot of a lifetime' during the final hour before packing up and going home … Didn't the sharks realise this could very well be our last season? Crossing the Sea of the Hebrides towards the Cairns of Coll we saw the first signs that our luck might change. Over a period of only a few hours we saw numerous breaches, always at a distance, the first characteristic signs that forewarned us that a shoal was arriving. As we rounded the Cairns we had our first quality Basking Shark sighting of the season, followed by several more – at last! Courtship, breaching, big animals, all of the signs were there that we were at the visible tip of a big shoal, with many more animals below us like the body of an iceberg. I called Matthew and gave him the position of these animals, knowing just how welcome that news would be.

The following morning, we were up and out bright and early, and had one of our best days for a long time; 50 sharks recorded through the day, finally into a big shoal and an incredible amount of breaching, all around us. Our photo-ID folder was finally filling up again. Matthew and Philip were able to get their tags attached, so we were all very relieved. Coll had proved as always to be a happy hunting ground for us, as it had now become for Matthew and the tagging team. And we had all shared one of the most amazing wildlife encounters of all, being out in a boat amongst a big shoal of Basking Sharks. Our research has always been about simply trying to ensure that others may have the opportunity to witness such a unique, life-enhancing experience. If that can be achieved, it will be reward enough.

Chapter 18
Where Do We Go from Here?

At the time that we produced our report for Scottish Natural Heritage we were aware that there were plans afoot to develop a major offshore wind installation to the south-west of Tiree, just outside our survey area. Like most people within the environmental movement we were favourably inclined towards the development of offshore wind installations as an alternative to fossil fuel power generation, but naturally we had concerns about the potential impacts they might have on marine life, including Basking Sharks. As a result, we included a short section in our report on the likely impacts on the species related to offshore energy installations, including wind farms.

Those impacts included the likelihood of collision and injury to Basking Sharks from support boats, especially during the construction phase of any installation. As the majority of the work would almost certainly be undertaken in summer, and most especially during calmer weather when Basking Sharks were highly likely to be present, this could prove to be a major concern. Other potentially harmful impacts included an increase in suspended sediments during the installation phase that might adversely affect zooplankton levels and the risk of disturbance or injury from the high levels of noise generated by pile driving. A final concern was the potential effect of electromagnetic field emissions generated by underwater cabling between individual turbines and from turbines to the shore. These might cause disruption for elasmobranchs (sharks, skates and rays) that rely on electroreception to find their prey and may use magnetic fields for navigation.

While on the face of it, some of these potential threats appeared to be at the lower end of the risk scale, nonetheless they needed thorough assessment and mitigation. The major threat (it seemed to me) was from collision with support boats. We believed we had a viable measure to mitigate this impact, recommending that all boat operators should undergo suitable wildlife awareness training on how to manoeuvre boats in wildlife-rich areas. If that training was combined with slow-speed operations (six knots maximum boat speed) and a careful visual watch regime kept throughout the development area, the threat of collision with Basking Sharks could be reduced. These measures would also deliver notable benefits for other species such as the Harbour Porpoise.

Now the question was whether the area off Tiree designated for the wind farm was a seasonal home to the Basking Shark. On the face of it, the answer almost certainly had to be yes, as it was almost contiguous with the Coll site we had already identified as a hotspot, but we had no data to support that hypothesis as it was just outside our survey area. Therefore, it was to everyone's good fortune that our friends and colleagues at the Hebridean Whale and Dolphin Trust (HWDT) had conducted extensive cetacean (whale and dolphin) surveys within the area in question over many years in their sailing research vessel *Silurian*, and had recorded all Basking Shark sightings alongside their cetacean sightings. Perhaps unsurprisingly, their data confirmed that the site was indeed a hotspot of at least equal importance to the two we had identified. As Scottish Natural Heritage wanted to examine the greater picture, we agreed to pool our data with those gathered by HWDT to create Sharks Per Unit Effort (SPUE) values across the larger region. These combined data would provide vital baseline information from which the significance of potential impacts associated with the proposed wind farm could be assessed.

The developer commissioned a matrix of additional line transect surveys through the area of the proposed wind farm between 2009 and 2012, with 33 surveys conducted within that period. The results were, as predicted, significant, and the overall average SPUE value turned out to be $1.00h^{-1}$, well within hotspot criteria as we had assessed them. Breaking the site down into smaller sectors was even more impressive, with some truly astonishing SPUE values in the sectors between the south-west and south-east of Tiree, close inshore, with a peak value of $21.26h^{-1}$. On the final survey on 5 August 2012, no fewer than 918 Basking Sharks were recorded, an astonishing number that put all of our own 'best days' firmly into the shade. There was now no question that the proposed location was deserving of hotspot status. The general view from those concerned with the conservation of the Basking Shark was that, with such elevated surface sighting values and the acknowledged potential for disturbance and injury to the species, the proposed wind farm should not go ahead. When the developer later shelved the application there was a heartfelt sigh of relief in the world of Basking Shark conservation.

Is the Basking Shark out of the woods yet?

Hunting has now been banned in many important areas like the North-east Atlantic, and after many years of conservation campaigns, scientific research and protective international treaties, the Basking Shark is now arguably the most highly protected shark species in the world. From a conservationist's perspective, these campaigns have been very successful. We have seen the species go from a lucrative prey and a pariah to a much-loved marine icon, with a chance to recover from the depredations of the past in some highly significant areas around the world. Yet still, in huge swathes of the world's oceans within the known habitat of the species, the Basking Shark has no official protection at all. But as there is currently no known targeted fishery within any of those areas, surely the future survival of the species must now be reasonably secure. Or are there still other, less obvious threats that could affect the shark's future, even in areas where it already enjoys a high degree of protection?

As history has shown repeatedly, when liver oil prices hit all-time highs, fisheries can spring up where they had not existed within living memory, such as on the west coast of Canada, in Scotland and in Ireland. This could easily happen again in sea areas where the species is not protected, driven (for example) by the staggering price of fins on the global market. With the volume of Basking Shark fins reaching the market currently at an all-time low, perhaps, like ivory, that scarcity might drive prices to more stratospheric levels and might encourage the re-opening of targeted fisheries, such as happened in Newfoundland in the 1980s.

The inclusion of the Basking Shark in Appendix II of CITES is intended to ensure that international trade in their fins and other products will only take place if the exporting country has ensured that the sharks were caught legally (e.g. not in protected areas or when the species is itself protected or under a zero catch quota), and that catches are not detrimental to the survival of the population in the wild. New Zealand, for example, has in the past allowed some exports of fins from bycaught Basking Sharks in their waters, on the basis that only a very small proportion of the Southern Ocean's stock is exposed to fisheries. Given the law of unintended consequences, perhaps (again, like ivory) even modest amounts of legally available fins might help to keep the demand alive, and resurrect hunting in countries where it has long been moribund, or has never even existed before.

Tidal turbines

In addition to the potential threats posed by offshore wind farm developments, other offshore renewable energy devices (OREDs) arguably pose a far more direct threat. Tidal turbines, for example, are recognised to pose a risk to sharks feeding around them. This is particularly the case in areas that support regular aggregations of Basking Sharks. Not all of the sharks will be at the surface all of the time, and so the likelihood of injury and mortality through collision with underwater turbines cannot be ruled out.

Many key hotspots for Basking Sharks are characterised by the presence of strong tidal streams that might also make them attractive sites for the installation of tidally driven OREDs. This creates a grim dilemma for conservationists and policy makers. Generating renewable 'green' energy, independent of fossil fuel, is a key cornerstone of securing the planet's future, yet in this case it would be almost certain to cause major impacts on highly important local populations of a much-loved conservation icon. It would be nice to think that we can have both renewable energy and Basking Sharks, but that may be hard to achieve in general, and all but impossible at the key sites.

In order to understand these risks more thoroughly, a number of existing ORED installations are undertaking impact-monitoring studies, such as in Strangford Lough in Northern Ireland and at the proposed Wave Hub development off St Ives in Cornwall. A potential significant conflict with Basking Sharks has been identified at both of these sites, albeit at a far lower level than would exist at any of the key hotspot sites identified off the west coast of Scotland. At the Strangford Lough tidal development, Appropriate Assessment investigations were required under the Habitats Directive, and the authorities had the power to revoke the consent in the event that significant adverse effects occurred. This recognised that potential problems could exist, and so was a responsible way forward, but it may still be difficult to extrapolate the lessons learned at single sites for more widespread application.

Nonetheless, it is true that the more we learn from the small-scale pilot sites, the better prepared we will be when the almost inevitable conflict of conservation interests falls into our lap. Much of the technology for tidal power generation is still in the early stages, and devices that do not depend on potentially injurious blades to generate power are under development. As conservationists, that might one day allow us to have our cake and eat it.

Small sailing vessels are often used for simple 'citizen science' type research projects concerning marine life such as Basking Sharks. Safe, seaworthy vessels with a low environmental impact, capable of operating independently in remote areas, they are ideal platforms for this type of work. Our current yacht Pèlerin is a good example of just such a vessel. © Colin Speedie

Soay was Gavin Maxwell's 'Island valley of Avalon', but siting his shark factory there turned it into a charnel house. The scattered remnants of his abortive enterprise still litter the shore of the tiny harbour on the north shore of the island, mute testament to his folly. © Colin Speedie

This male neonate Basking Shark that stranded at Millisle, County Down in 2014 has the slender form and pointed nose typical of the youngest sharks. At just over 2.6m this rare stranding might indicate that the new born pups may be larger than had previously been suggested. © Gary Burrows/DAERA Marine & Fisheries Division

The second largest known fish in the world, an encounter with a Basking Shark makes for an unforgettable experience. Sinuous swimmers, beautifully marked, primeval and of colossal size and girth they are charismatic ambassadors of the marine world, well worthy of their protected status. © *Colin Speedie*

Our work established a number of important 'hotspot' sites around Britain, such as Lands End, Lizard Point, Coll (here) and Canna and the uninhabited Hyskeir islet. Other established hotspots include the Isle of Man and Cork and Donegal in Ireland. These are all areas where large shoals of Basking Sharks have been regularly recorded, sometimes running into hundreds of animals. © Colin Speedie

These big shoals tend to be made up of large mature animals, and are where what may be courtship behaviour is most regularly recorded. The animals here are engaged in 'parallel swimming', one form of such behaviour, another classic form being the nose-to-tail strings of multiple sharks that Patrick Fitzgerald O'Connor called the 'Love Dance'. © E. Johnston - Irish Basking Shark Study Group

The skin of the Basking Shark is covered in small placoid scales called 'dermal denticles', which are highly abrasive to the touch. Sharks engaged in parallel swimming and nose-to-tail following are often seen to touch each other which results in abraded noses and fins. © Colin Speedie

Breaching is another form of activity we came to associate with the big shoals at the hotspot sites. Early researchers believed that breaching was associated with the sharks attempting to dislodge ectoparasites such as Sea Lampreys, but today this is believed to be another form of social behaviour linked to courtsthip. © Youen Jacob

One of the most striking and obvious features associated with the Basking Shark is the colossal first dorsal fin. Marks, scars and injuries to this fin can allow individuals to be identified using a simple technique known as photo-identification. This can allow insights into how often, or for how long individuals inhabit or return to key sites, sometimes over years. Or, as here, to assess how often human impacts such as collisions with boats may occur. © Colin Speedie

There are only a few documented predators of Basking Sharks: Humans, Sperm Whales and Orcas. This large shark has lost a substantial section of its first dorsal fin to what appears to be a bite – perhaps from an Orca? Slow and docile, a Basking Shark is an easy target for any predator, as the hunters of old proved all too effectively when conditions were in their favour. © Colin Speedie

Basking Shark fins vary greatly in their height, shape and chord, some being tall and with a distinctly pointed tip, others with a long base and rounded tip. The fins in the older sharks are often ragged along the trailing edge and due to their size and weight may flop over to one side. With some of the real leviathans (such as this 10m. animal) the fin lies completely over and drags alongside in the water. © Colin Speedie

Basking Sharks act as hosts to a variety of external parasites. Some of these are large such as the Sea Lamprey, Petromyzon marinus, that leaves a raw, white, over lesion on the skin after it has been dislodged, which may then be colonized by smaller ectoparasites such as the copepod Dinematura producta. © Colin Speedie

A Basking Shark feeding on a dense patch of plankton, most likely its favourite prey in British waters, the reddish calanoid copepod Calanus. Far from blindly chancing on their prey, Basking Sharks are now believed to be highly efficient foragers capable of finding their way to the densest patches of plankton whether at the surface (as here) or at great depth. © Alex Mustard/2020VISION/NPL

When feeding the mouth is wide open and the radial gill arches expand to allow the water that enters the mouth to pass through filamentous gill rakers in the arches that filter out their tiny copepod prey. It has been estimated that an average sized shark (5–7m) may consume over 30kg of food in a day when plankton is abundant. Note also the Sea Lampreys attached to the shark in the genital area © Alex Mustard/NPL

Bycatch

Incidental bycatch in a range of fisheries has long been recognised as a cause of mortality for Basking Sharks. A 1986 report from eastern Canada recorded 410 Basking Sharks bycaught in salmon gill nets and cod traps off Newfoundland between 1980 and 1983. A study off Ireland in 1994 estimated an annual bycatch of between 77 and 120 Basking Sharks a year in bottom-set gill nets, and recorded 287 incidences of shark entanglement in 1993 alone, mainly in surface-set gill nets. Scientists in New Zealand recorded an incidental bycatch of 203 Basking Sharks in deepwater trawls for Hoki (*Macruronus novaezelandiae*) on their spawning grounds (where the sharks were perhaps feeding on Hoki eggs).

Some of the fisheries implicated in incidental bycatch of the Basking Shark have themselves all but died out due to the collapse of their target fish stocks (salmon nets off Ireland, cod traps off Newfoundland). But others still exist, such as the setting of gill nets in tidal waters inshore around hotspot sites, where seasonal conflict with visiting Basking Sharks is all but inevitable.

The greatest number of bycatch reports from around England and Scotland have identified Basking Sharks being caught in fixed fishing gear, especially in gill nets in the south-west; a smaller number of animals becoming entangled in lobster, crab and prawn creel ropes, and just a very few animals being caught in trawls throughout the year. In years when Basking Sharks have been abundant around Cornwall, bycatch in gill nets in some sheltered bays has been regularly reported, although this remains difficult to quantify due to a lack of reliable information. For fishermen, the occasional catch of a Basking Shark in a net is distressing, costly and time consuming; between those elements and the public opprobrium they fear as a result of adverse publicity, there is little to encourage them to report bycatch. This is a pity, as bycatch is accidental, they are doing nothing illegal and most of us still enjoy eating fish. The lack of open and honest dialogue doesn't help the development of a better understanding between all parties, or allow strategies to minimise bycatch to be mutually developed, for the benefit of all concerned.

Most witnesses agree that bycatch numbers around the British Isles remain at a low level at this time. Yet it is hard to argue that if the use of these nets were to become more widespread, and most especially if they were allowed to proliferate around Basking Shark hotspot sites, bycatch would inevitably rise dramatically. As we know from the devastating efficiency of set net fisheries at Keem Bay on Achill Island, such developments on a larger scale could hardly fail to have a significant impact.

Human disturbance

The waters around the British Isles have seen a dramatic increase in the level of pleasure craft activity in recent years, as boat ownership has become more affordable and water sports activities have blossomed. Whilst it is marvellous that people are out enjoying the sea, this new enthusiasm can present problems for marine life that now has to coexist with the much higher levels of boat traffic. The question is how to minimise disturbance of species such as Basking Sharks whilst still allowing a wildlife-friendly public to view these magnificent creatures without restriction. And it is true that some species need more protection from human activities than others.

Basking Sharks do not always respond negatively to the approach of vessels, perhaps because they have historically lacked predators. Indeed, this was clearly their Achilles heel in the days of the hunters. But as we saw earlier in this book, Basking Sharks *can* be disturbed by boats that approach too closely, or are carelessly or recklessly handled around them. In a study I was involved in, we found that engine noise has an effect, and repeated approaches have a cumulative impact, eventually resulting in the shark diving and not returning to the surface. Disturbance of this nature on a regular basis may not be as sudden or damaging as a collision, but nonetheless has the potential to disrupt feeding or courtship behaviour and so should be avoided at all costs.

As both the hunters of old and the new generation of tagging teams are aware, a slow, quiet approach from behind a Basking Shark very seldom generates any kind of reaction from the target animal, until the vessel becomes apparent visually to the shark as it comes alongside – the eyes are well forward on the head and are largely focused frontwards.

In my experience, big, mature animals are generally less susceptible to disturbance, while small, younger animals are more easily spooked, perhaps because they are more vulnerable to large natural predators such as Orcas. In both cases, though, this may have more to do with their relative ability to locate the densest patches of zooplankton; when plankton is dense, the sharks seem to simply concentrate on feeding, and the adults appear to be far more efficient at finding those rich patches of food. Juvenile sharks, generally solitary, tend to swim fast and in a far more erratic manner. Sharks engaged in what we believe to be courtship show minimal reaction to disturbance unless approached at close range, seemingly fixated on the all-important 'Love Dance'. This is a major concern at all of the hotspot sites, as disturbance of animals engaged in possible reproductive behaviour could lead to very serious consequences for the maintenance or recovery of stocks.

Close approach by boats doesn't just have the potential to disturb, however – the risk of collision also increases dramatically. Basking Shark movements are unpredictable, can be sudden, and directional changes are often extremely hard to anticipate. Not all sharks present are easily detectable, as others may be just out of sight below the surface. As sharks do not always take evasive action, collisions can and do occur. Injuries can be severe and potentially fatal to Basking Sharks (and potentially humans, too), as we have seen ourselves and heard of on occasion.

For all of the above reasons, careful boat handling around Basking Sharks is critical, and very close approaches (closer than 100 metres) are not recommended. While several of the hotspot sites identified around Britain are remote and seldom visited by surface craft, some – Lizard Point and Land's End, for example – are busy turning points, the 'roundabouts' of the sea-lanes where small craft congregate to move from one sea area to another. At certain times in the calendar, these shark-rich spots are busy with boats, as when the famous Fastnet Race for yachts or the Round Britain Powerboat Race pass through. Working with my friend Sarah Brown (then at the Royal Yachting Association) some years ago, we drew up advisory 'no go' areas around both of these sites for the latter race to use, based on the sightings data we had gathered. Our data was put to good use: as the tiny, fragile, high-speed boats were endangered by the possibility of collision with a giant shark, the race organisers sensibly agreed to route the race boats outside the protection zones. Working in an open and honest partnership can yield real conservation benefits, and help others secure benefits, too.

For some of the sites that see less traffic, but where shark density is high during the busy summer period (Coll and Canna, for example), our project worked with Scottish Natural Heritage and others to raise awareness amongst the boat-owning public of the seasonal presence of Basking Sharks. Slow speeds and high awareness zones may not sound like dramatic conservation measures, but if they have the desired effect in reducing collision and disturbance, then they may be all that is needed. If boat owners follow the simple guidelines that have been developed, the experience and interaction doesn't need to be to the detriment of the shark. Failing that, in UK waters, the Law is there as the ultimate backstop to protect the Basking Shark from reckless or intentional disturbance – may it never be necessary to use it.

That is, of course, as long as people are aware of the laws that exist. A 2004 study into the extent of reported wildlife disturbance in south-west England assessed people's perceptions of current levels of disturbance and awareness of existing legislation. The results revealed a very low level of reported incidents, along with a significant lack of awareness of marine wildlife protection legislation. Given the lack of support given to important players like Police Wildlife Liaison Officers by cash-strapped constabularies, this was hardly surprising, and it is my belief that the situation is not greatly improved all these years later. This is not the case everywhere, in fairness, as the government in Scotland places a far higher importance on education about wildlife protection, partly due to national pride in its rich wildlife heritage, but also because of the economic value of wildlife in tourism terms. But it should be acknowledged that there is still a major task to be undertaken in regard to educating boat owners in ways to help minimise disturbance of marine wildlife, and to explain wildlife law and how it applies to them. This is being addressed in a number of ways at all levels of boat activity, from ecotourism boats to the pleasure craft-owning public.

On the commercial level, marine ecotourism still remains one of the fastest growing tourism sectors globally, and the Basking Shark is fast rising up the rankings of 'must-see' species. The sea around the British Isles is our Serengeti, and the Basking Shark is our elephant. While Gavin Maxwell saw the liver oil as the elephant's ivory to be exploited for gain, nowadays the financial gain comes from taking ecotourists out to see our elephant intact in its natural habitat. Given that there are currently very few places in the world that the species can regularly be seen at the surface, the British Isles have become something of a Mecca for ecotourists, especially in places like west Cornwall, the Isle of Man and the west coast of Scotland, where many of the people who make the pilgrimage do so in the hope that they may encounter a Basking Shark (amongst other wonders).

Marine ecotourism has generated a significant financial benefit for these places, and has played a major role in creating and sustaining jobs. In 2010 the Scottish Government commissioned a report into the economic benefits of ecotourism in Scotland which established that net spending through wildlife tourism amounted to £63 million. Net spending for marine wildlife watching was estimated at £15 million and supported 633 full-time equivalent jobs. In a remote area such as the western Highlands and Islands, those are not negligible figures. The Basking Shark once more has a price on its head – but now in a way that will only be enhanced by its ongoing presence.

That access to wildlife needs to be managed is not in doubt, but that needn't necessarily imply the need for draconian action. Access to well-presented education at all levels is the best way to explain the responsibilities we must all accept if we want to continue to enjoy the benefits of our wonderful marine life. This was the driver for the development of the Wildlife Safe (WiSe) training course for commercial boat operators, which contains a specific module on the Basking Shark. Since we launched the project in 2003 we have trained thousands of operators of small commercial craft in the best ways to handle their vessels to watch marine wildlife safely and sustainably. This has included include all manner of craft including yachts, kayaks, dive boats and workboats that support offshore energy installations. Most commercial wildlife-watching boat operators in the British Isles have attended a standard WiSe course and many have also attended the advanced WiSe Master course specifically created for ecotourism operators. At the pleasure craft level, WiSe has also created a simplified version of the course to be delivered to the boat-owning public, recognising that it's in everyone's interest to safeguard our precious living marine heritage.

Although there have been very few confirmed reports of disturbance of Basking Sharks in recent years, I have personally witnessed a small number of incidents that would definitely fall within that category. Whilst reducing the likelihood of disturbance without spoiling one of the greatest wildlife watching experiences in the world is undoubtedly a challenge, compared with the battles that were fought to halt the hunting of the Basking Shark, it pales into insignificance. At current levels of boat activity, the problem should be manageable – if we all play our part in minimising the risks.

Climate change

As we have already seen, a number of studies have demonstrated that Basking Sharks migrate into coastal waters around the UK to feed on seasonally abundant aggregations of zooplankton, principally copepods such as *Calanus*. It has also been shown that surface-feeding Basking Sharks actively forage along thermal fronts seeking the richest and densest patches of large zooplankton. Foraging behaviour might therefore be regarded as an indicator of plankton abundance, especially in relation to oceanographic and climatic fluctuations (notably the state of the North Atlantic Oscillation) that affect the abundance of *Calanus*.

Research from the Sir Alistair Hardy Foundation for Ocean Science, which operates the Continuous Plankton Recorder survey, has shown a progressive shift in distribution of the cold temperate *Calanus finmarchicus* and the warm temperate *C. helgolandicus*, in response to rising sea surface temperature. This mass biogeographical displacement northward of the two plankton species has been estimated to represent a movement in the order of 10 degrees of latitude over the 50 years from 1958, and this trend appears to have accelerated in recent years. However, the trend in copepod abundance has been declining, by 70% in the North Sea for example. This pattern has been recognised on the west coast of Scotland, with substantial reductions in *Calanus* being recorded, with levels that are now below the long-term mean.

If *C. helgolandicus* is replacing *C. finmarchicus* as the dominant calanoid species off western Scotland, it could be argued that this might not pose a major problem for the Basking Shark, as long as the overall *Calanus* levels do not continue to decline. *Calanus finmarchicus* has a single seasonal maximum in May, whereas *C. helgolandicus* has two maxima, the first and lesser one in May–June and the second and major one from September to October, thus potentially offering the Basking Shark a longer feeding season in that location. This change in prey species and timing of abundance might explain the temporal change in seasonal abundance that we observed during our surveys. The post-war hunters reckoned on a season that started in late April and was over by July. From our surveys and the public sightings data, the regime today is very different, with the season starting in June, at its peak in July, and lasting well into September in many years. This is a dramatic change, and the only likely cause must be a change in seasonal abundance of suitable prey. We know that the Basking Shark feeds selectively on *C. helgolandicus* in the English Channel, so it might equally do so in the waters off Scotland, especially because the two *Calanus* species are so closely related.

That the Basking Shark is an efficient forager was determined by Dr Peter Cotton in a study from the English Channel. His research demonstrated that long-term patterns in relative abundance of Basking Sharks within a region are integrally linked with climate-driven changes in sea surface temperature, and to a lesser extent, *C. helgolandicus* density. The same study also postulated that Basking Sharks use thermal fronts as 'foraging or migration corridors', and that this may be one means by which they orientate themselves towards the most abundant prey. If, however, the strength or persistence of these foraging corridors are affected by climate change, there may be significant effects on spatial and temporal distribution of the species. A case in point is the regional stronghold for the Basking Shark in western Scotland, where the sharks are thought to use the Malin Head and Islay fronts as corridors to orientate themselves to the hotspots.

A study by Professor David Sims and Professor Philip Reid suggested that the decline (and eventual cessation) of the Achill Island shark fishery in the West of Ireland described in Chapter 11 might not have been solely due to overfishing of a discrete local population, but might also have been caused by a parallel decline in *Calanus* spp. within the area over the same period. The study also suggested there had been a considerable increase in the Norwegian Basking Shark fishery during that period – this might in turn indicate that a spatial shift took place northwards in response to the movement of zooplankton prey. Recent research shows that *C. finmarchicus* is able to respond to changes in habitat availability and maintain stable population sizes, suggesting that the 70% decline in biomass in the North Sea may be viewed as a range shift, as opposed to a straightforward decline in abundance. Anecdotal reports also suggest that commercial whale-watching vessels in Iceland have observed significant numbers of Basking Sharks in recent years. Perhaps this may be linked to the multi-decadal shift of *C. finmarchicus*, with sharks tracking the increase in available prey further north, or perhaps a population recovery from the decline and eventual cessation of the North-east Atlantic fishery. And yet there is still no sign of any major recovery in numbers at Achill Island.

Our current understanding is that Basking Sharks utilise thermal cues (temperature) as a means of orienting themselves to high levels of available prey. Add to that the ability of individual sharks to make ocean basin-scale migrations, and it may be reasonable to conclude that the species will continue to track changes in the availability and distribution of zooplankton efficiently, even over long distances. It also seems reasonable to conclude that surface sightings of sharks in hotspot areas will be sustained as long as the thermal regime is acceptable and sufficient quantities of food in the form of their favoured zooplankton prey persist. Let us hope so. Otherwise, like the hunters before them, the shark ecotourism ventures may lose out.

Ocean acidification

The chemistry of the marine environment is changing rapidly, as absorption of carbon dioxide (CO_2) into the marine environment has lowered the average pH of the oceans by about 0.1 units from pre-industrial (1750) levels (IPCC 2007). The Royal Society warns that the pH of our oceans is reducing in a manner that is unprecedented, foreseeing a drop of a further 0.5 units by 2100 – this is a level lower than has been experienced for hundreds of thousands of years, decreasing at a rate 100 times faster than at any time within that period. It has been predicted that even current levels of acidification are irreversible within our lifetime. This may have severe consequences for the marine environment, not least for the survival of carbonate-shelled organisms, as acid dissolves calcium carbonate. Examples would include coccolithophore plankton, molluscs, echinoderms and coldwater corals. In addition, the sublethal effects on copepods (in particular egg production rate and early development) are a concern and might have direct implications on available feedstock for the Basking Shark.

As ocean acidification is only a recently observed phenomenon, scientists suggest that there are few outcomes that can be accurately predicted. As the scale of anticipated disruption to normal ecosystem functioning has not yet been experienced, it may therefore be impossible to predict how successfully the Basking Shark will be able to acclimatise to the resulting changes in food distribution and composition. Even OSPAR, the organisation by which 15 governments and the EU cooperate to protect the North-east Atlantic, recognises that understanding the likely effects of ocean acidification, and what to do about it, will be a major challenge. The only clear mitigating action identified so far appears to be to reduce the amount of CO_2 emissions released into the atmosphere, even acknowledging that there will be a considerable time lag before atmospheric CO_2 levels might stabilise, let alone start to reduce.

Marine pollution

Perhaps unsurprisingly, given that an estimated eight million tonnes of plastic are dumped in the sea each year, there may be harmful impacts upon the Basking Shark from the plastic waste that is now found throughout the world's oceans. The surface fronts that aggregate the plankton that the sharks prefer also concentrate plastic debris, some of which is accidentally ingested by the sharks along with their plankton prey. Some (such as packaging straps) has even been observed entangled upon a shark's rostrum.

And that is just the obvious, visible, plastic material. Filter-feeding Basking Sharks also directly ingest microplastics, the tiny particles of larger items that break down and persist in the marine environment. And, horrifyingly, the zooplankton that the sharks eat have now also been shown to ingest microplastics, which may affect their own ability to feed. Direct and indirect ingestion of microplastics may therefore be linked to the recent discovery of elevated levels of the plastic softener MEHP in the muscle tissue of Basking Sharks sampled in the Mediterranean. MEHP may be viewed as an indicator of elevated levels of microplastic ingestion, and thus also contaminants that are known to be persistent, bioaccumulative and toxic (PBT) such as PCBs, DDT and mercury that can leach out of or be carried by microplastics. These can become concentrated in large predators such as Basking Sharks with as yet unknown consequences, although their effects are believed to include reduced life expectancy and reproductive ability in marine mammals.

What we still don't know

My old friend Kelvin Boot once remarked to me that although he thought it wonderful that new technology was allowing us to make so many amazing discoveries about the Basking Shark, he 'hoped there would some mysteries left'. He needn't have worried – there is much that remains unknown, or at the very least, unclear, about the species.

One fascinating aspect that still needs further elucidation is just what importance site fidelity (termed philopatry) might mean for the Basking Shark. Philopatry may be where a species remains in an area full-time, or where it returns on a regular basis to a home site or a natal site, rather as salmon do when they return to their natal rivers to spawn. Philopatry has been recorded in a wide range of shark species, such as the White Shark, Lemon Shark (*Negaprion brevirostris*) and Blacktip Shark (*Carcharhinus limbatus*). Philopatry in these cases is probably linked to environmental factors such as the location of safe pupping and nursery areas (female sharks often show stronger philopatric behaviour than males), food availability and finding potential mating partners.

Photo-identification studies and evidence from the Scottish Basking Shark tagging programme both suggest that the species exhibits varying degrees of philopatry, possibly related to high levels of zooplankton abundance and the potential to find mating partners. Given the high levels of putative courtship behaviour that we recorded during our surveys, the latter may well prove to be a significant factor.

This has been both seasonal, when individual sharks remain within the same local area for months at a time, and annual when animals have returned to the same geographic area in consecutive seasons or consecutive breeding cycles. In fact, in a few instances, photo-ID has shown that in one case a large female was re-sighted after a period of just over three years in the same region, and that in other cases, individual animals have returned to the same region on multiple occasions.

But just how 'local' is that site fidelity? Tex Geddes certainly thought there were distinct groups that favoured certain sites:

> As we drew nearer the splashes grew less frequent and we were about a hundred yards off him when he stopped his antics and began to behave more normally, swimming lazily on the surface, away from us. On closer inspection he appeared to be lighter in colour than the fish we had caught in Moonen Bay. I have often noticed this, not that there were always lighter-coloured fish in Canna, but there seemed to be distinct families which we thought we could recognize. Certainly, for instance some 'families' had more white on their bellies than had others.

Certainly pigmentation varies considerably, and of the (well over) 1,000 Basking Sharks I have personally observed, the diversity of coloration has been from almost black through brown to sand colour, where the elegant python-like stripes are very noticeable. However, I could not honestly say that I ever considered pigmentation to have a site-specific dimension. But given the discoveries being made that suggest fine-scale site fidelity in other related shark species, perhaps in time we shall find that some Basking Sharks display similar behaviour. Although it may seem far-fetched, site fidelity on the micro-scale could be one possible explanation for the total lack of sightings we have recorded over many years of surveys at what were the most historically important hunting sites in the Hebrides.

In a fascinating study of the implications of shark philopatry in regard to fisheries management, Dr Robert Hueter explains:

> If sharks are philopatric for specific parts of their ranges, be they nursery areas, feeding grounds, mating areas or other locations, fishing within those areas can remove individual animals that depend on and, in a sense, 'belong to' those localities rather than are part of a larger, fully mixed stock. This would be true regardless of the highly migratory nature of shark species. In this case, fishery removals can have a more dramatic effect on the relative abundance of species in localised areas, with the appearance that species density has been 'hole-punched' in a specific part of its range.

Known as localised stock depletion, Dr Hueter reports that this effect has been noticed in both commercial fisheries in Australia and in South Africa, where shark meshing has been employed to protect bathers. The drastic reduction in the number of Basking Sharks recorded at former strongholds like Barkley Sound, Achill Island and Nakiri suggest that localised hunting and control measures, taken to extreme levels, had absolutely devastating effects on the Basking Shark populations at those sites. This certainly looks like localised stock depletion.

Perhaps the same may be true on a finer scale for the historic hunting sites in Scotland. Maxwell, Watkins, Geddes, O'Connor and the Norwegians all hunted, as Maxwell put it, for 'days without end' at sites such as Moonen Bay and Shepherd's Bight, where we have never recorded a single shark at the surface despite many visits over a decade. If that were to be the case then there is all the more reason to put in place measures that safeguard the species at the local hotspot sites where the Basking Shark is still seen today – Tiree, Coll, Hyskeir and Canna – to help the stock recover.

That the decline in the number of Basking Sharks recorded around Scotland, Ireland and Norway may not be entirely due to hunting is not under dispute, as shifts in their zooplankton prey may also have had an effect. But the removal of up to 106,000 animals from the North-east Atlantic stock in the last 50 years of the twentieth century cannot have failed to have a major impact. During that period, there was a 90% decline in catches from the late 1960s to the 1990s in the North-east Atlantic. Even during a period when the price paid for Basking Shark oil rose (between the early 1970s and mid-1980s) and the hunt expanded to take advantage of that opportunity, shark landings fell, even though the Norwegian fleet travelled far and wide to try and find Basking Sharks. This looks like a mopping-up operation targeting the remnants of the big shoals wherever they could be found, and suggests that the decline in catch was caused by stocks falling across the whole region rather than any reduction in the number of vessels engaged in the hunt at that time. It could likely be as a result of the much higher numbers taken across the North-east Atlantic in the preceding decades; this resembles the effects on populations seen elsewhere with fisheries for large sharks, where bust has inevitably followed boom.

Currently (2017), Scotland continues to record high levels of surface sightings in the summer season, in a relatively unbroken upswing since 2003. This appears to mirror the long spell of abundance in the region between the mid-1930s until the late 1950s that encouraged the hunters of that era. But as we have seen throughout the history of the species, these periodic abundances may not last, and may be followed by equally long periods of almost complete absence. Until more is known about the effects of climate change on plankton abundance and distribution, and until the status of the Basking Shark population post-hunting is better understood, we cannot say for certain that any putative recovery is secure. Therefore, it seems fair to argue that we would do well to safeguard and monitor important local populations as a matter of urgency.

This goes to the heart of the debate about the inclusion of a mobile species (i.e. those species that do not permanently inhabit a site) in a Marine Protected Area (MPA). It has long been argued by some parties that it is impossible to effect protection of animals at such sites given that they spend parts of their lives outside the MPA. However, given the growing body of evidence that Basking Sharks inhabit hotspots for extended periods seasonally, probably for reproductive purposes, and are at their most vulnerable when 'basking' at the surface (as they do at hotspot sites), then surely it is imperative to do everything necessary to protect them while they are in residence. And with new, statistical ways to predict potential shark hotspot areas in development, perhaps new, so far unidentified areas may be discovered, or historic areas further confirmed in their importance.

At the moment, the Scottish Government is considering a proposal to create an MPA in the Sea of the Hebrides. It would include the Basking Shark as a mobile species and would encompass all of the hotspot sites highlighted by the boat-based surveys and confirmed by the satellite tagging programme. The final report for the joint Exeter University/Scottish Natural Heritage satellite-tagging project (2012–2014) has just been published, and the results appear to conclusively support the value of MPA establishment there. In the three successive years of the tagging programme, the tagged Basking Sharks spent 78%, 89% and 80% of their time within the boundaries of the proposed MPA during the months of July to the end of September. Factor in the associated and repeatedly recorded social behaviour amongst large groups of mature animals recorded during our boat-based survey, and the case for an MPA for the purpose of maintaining or even generating a local Basking Shark population recovery, surely becomes indisputable.

Another, perhaps more difficult, question not yet answered is: where are the pregnant females? The extreme sex segregation recorded by the hunters suggests that the ratio of females to males at the surface in summer is very much in favour of the females (around 18:1), and is possibly one of the reasons why their hunting had such devastating effects on local populations. Furthermore, in many shark species, females are more philopatric than males. Kill all the females and it is hardly surprising that local populations collapse. But segregation by sex in sub-surface gill net bycatch off Newfoundland showed twice as many males as females, and a similar ratio was found in New Zealand. Whatever the cause, we still have only one observed birth on record. And despite what is believed to be courtship being recorded regularly, there has never been a single verified sighting of copulation. Yes, much remains unknown.

Over the centuries

The Basking Shark is a survivor. Genetic studies have suggested that the species survived a near extinction event at some stage in the distant past, probably during the Holocene period. But it has endured.

As we have seen in this book, during a period of over 250 years man has hunted the Basking Shark, in some cases to apparent near extinction on a local level. In times of historic hardship, it might have been possible to understand the rationale for such hunting, but in an era when synthetic oils have altogether replaced the need for shark liver oil, those days are surely over. The only market that is left is for the fins, which now attract 'trophy' prices, their value spiralling endlessly upward as they become ever more scarce. In 2006 the *Guardian* newspaper reported a single Basking Shark tail fin fetching US$10,000, and 10 years later, reports suggest that a single pectoral fin from the species can fetch as much as US$50,000. There is something obscene about an ongoing trade that kills an iconic creature simply for its most decorative parts; after all, no one is going to starve because shark's fin soup is no longer on the menu. But like the rhino poached for its horn or the elephant for its ivory, the Basking Shark tragically suffers from the same disease of human affluence. Surely man has done enough harm to the Basking Shark, as he has to the rhino and the elephant. Let it stop. Enough is enough.

The campaign by conservationists and scientists to secure recognition and protection for the Basking Shark since the 1970s was fought against just such a background of stupendous financial reward against drastically declining populations. That the species is now one of the world's most widely protected sharks is a testament to the success of that campaign, and it can be argued that the lessons learned in the battle to save the Basking Shark have been of benefit in helping to achieve similar levels of protection for other large sharks. Being one of the huge 'charismatic megafauna' species has undoubtedly helped the Basking Shark to survive, and it is now a much more valuable commodity as a living creature on the global ecotourism circuit. Like the great whales that draw millions of people worldwide to view them, the

Basking Shark still has a price on its head, but now it is no longer a one-off payment for a carcass, but a long-term return on a living creature. The Basking Shark has, in that sense, become an 'honorary cetacean', attracting top billing with the great whales wherever it is found.

It is, of course, a good thing that creatures like the Minke Whale and the Basking Shark now have a solid financial value that derives from more peaceable activities than hunting. If the ecotourism jobs that bring much-needed revenue to remote coastal communities ultimately depend on the continued presence of such creatures in our waters, then that is a powerful argument for their protection at every level. As one of the key weapons in the battle to save the Basking Shark, those of us who were involved from the earliest days wielded that argument at every opportunity. But the financial argument isn't the only valid one. Even the most beautiful places without their wild inhabitants are simply barren, devoid of a crucial part of their appeal, the very magic of life. Sometimes we should agree to conserve what we have for no other reason than that it is the right thing to do. In doing so, we sustain its intrinsic value for future generations to share. That has certainly been the underlying driving force for most of the people I know who have been directly involved in the campaign to protect the Basking Shark.

One perfect day in the Hebrides we were gliding along the north coast of Canna, one of the loveliest of all the islands. Away to the east, the Cuillins of Skye stood out stark and rugged against a cornflower-blue sky dotted with soft puffs of cloud. Below the mountains I could make out the low outline of Soay, the beautiful little island that I knew to have been such a key site in the post-war saga of the shark hunt. And then ahead of us we saw the tiny tip of a fin breaking the surface, artfully carving a neat 'V' in the oily calm of the sea's surface. Slowly it rose to become a magnificent, tall symbol of power and grace, silhouetted against the backdrop of the hunting ground that had taken so many of its forebears. A single, huge Basking Shark, one of the quintessential elements that make the Western Isles one of the most magnificent wild places on earth. In that single moment, all of the time, toil and trouble that so many of us in the conservation movement have expended in the fight to protect the Basking Shark seemed utterly worthwhile. Never did a wild creature seem so perfectly in its place.

Long may it remain so.

Sources

Prologue

COMPAGNO, L.J.V. 2002. Sharks of the World. FAO Species Catalogue No. 1, Vol. 2. Food and Agriculture Association of the United Nations, Rome.

Chapter 1

BLAND, K.P. & SWINNEY, G.N. 1978. Basking Shark: Genera *Halsydrus* Neill and *Scapasaurus* Marwick as synonyms of *Cetorhinus* Blainville. *Journal of Natural History* 12: 133–135.

COUCH, J. 1862. *A History of the Fishes of the British Islands* Vol. 1. Groombridge & Sons, London.

COUCH, J. 1825. Some Particulars of the Natural History of Fishes Found in Cornwall. *Transactions of the Linnaean Society of London* 14 (3): 67–92.

De BLAINVILLE, H. 1816. Prodrome d'une Nouvelle Distribution Systematique du Regne Animal. *Bulletin des Sciences par La Société Philomatique de Paris.* 113–124.

FAIRFAX, D. 1998. *The Basking Shark in Scotland   natural history, fishery and conservation.* Tuckwell Press, East Linton.

Falmouth's myths and legends. Available at: www.falmouthport.co.uk/commercial/html/documents/Mythsandlegends.pdf

GUNNERUS, J.E. 1765. Brugden (*Squalus maximus*). *Det Tiondhiemske Selskabs Skrifter* 3: 33–49.

HOME, E. 1809. An Anatomical Account of the Anatomy of the *Squalus maximus* (of Linnaeus) Which in the Structure of its Stomach Forms an Intermediate Link in the Gradation of Animals Between the Whale Tribe and Cartilaginous Fishes. *Philosophical Transactions of the Royal Society of London* pt 2: 227–241.

LEBOND, P.H. & BOUSFIELD, E.L. 2000. *Cadborosaurus – survivor from the deep.* Heritage House Publishing, Victoria B.C.

LINNAEUS, C. 1766. *Systema Naturae.* Editio 12, Tomus 1, Regnum Animale. Catalogue of the Books, Manuscripts, Maps and Drawings in the British Museum, London.

Memoirs of the Wernerian Natural History Society, 1818 Vol. 2: 638–640.

PENNANT, T. 1769. *British Zoology* Vol. 3. Printed by Eliz. Adams for Benjamin White, Chester.

PENNANT, T. 1774. *A Tour in Scotland and Voyage to the Hebrides 1772.* Printed by John Monk, Chester.

PENNANT, T. 1776. *British Zoology* Vol. 3. Printed for Benjamin White, London.

PONTOPPIDAN, E., 1755. *The Natural History of Norway* (Translation). Printed for A. Linde, London.

SWINNEY, G. 1983. The Stronsay Monster: A case of mistaken identity. *Journal of Marine Education* 4 (2): 15–17.

WALLACE, S. & GISBORNE, B. 2006. *Basking Sharks: the slaughter of B.C.'s gentle giants.* Transmontanus, Vancouver B.C.

Chapter 2

BALDWIN, J. 2008. Subsistence whaling in the Western and Northern Isles of Scotland. In:

Whaling and the Hebrides. Island Book Trust, Isle of Lewis.

BRABAZON, W. 1848. *The Deep Sea and Coast Fisheries of Ireland.*

BRADY, T.F. 1873. *Distress on the islands of Bofin and Shark, county Galway, 1873.* Dublin.

DUTTON, H. 1824. *A Statistical and Agricultural Survey of the County of Galway.* Dublin.

DRUCKER, P. 1951. *The Northern and Central Nootkan Tribes.* Smithsonian Institution Bureau of American Ethnology. *Bulletin* 144. Washington: U.S. Government Printing Office.

FAIRFAX, D. 1998. *The Basking Shark in Scotland – natural history, fishery and conservation.* Tuckwell Press, East Linton.

FAIRLEY, J.S. 1981. *Irish Whales and Whaling,* pp.119–120. First report of the Commissioners of Inquiry into the state of the Irish Fisheries, Dublin, 1836.

GEORGE, E.M. 2003. *Living on the Edge: Nuu-chah-nulth History from an Ahousat Chief's Perspective.* Winslow, B.C.

GUNNERUS, J.E. 1765. Brugden (*Squalus maximus*). Det Trondhiemske Selskabs Skrifter 3: 33–49.

HARDIMAN, J. 1820. *The History of the Town and County of Galway. From the Earliest Period to the Present Times.*

McGONIGLE, M. 2008. Oil and Water – 18th Century Whale and Basking Shark Fisheries of Donegal Bay, Ireland. *The International Journal of Nautical Archaeology* 37.2: 302–312.

McNALLY, K. 1974. *The Sun-Fish Hunt.* Blackstaff Press, Belfast.

Memorial of Donald MacLeod, Canna, 1766. In FAIRFAX, D. 1998. *The Basking Shark in Scotland – natural history, fishery and conservation,* Appendix B, pp.183–185.

MENZIES, A. 1768. *Voyage through the Western Isles and the and west coast of Scotland.*

NEW STATISTICAL ACCOUNT OF SCOTLAND 1845. Blackwood & Sons, Edinburgh, 15 volumes.

PENNANT, T. 1769. *British Zoology* Vol. 3.

PENNANT, T. 1774. *A Tour in Scotland and Voyage to the Hebrides 1772,* pp.169–170 and plate 13.

SINCLAIR, J. (ed.) 1799. *The Statistical Account of Scotland 1791–1799,* pp.340–341. Also known as the OLD STATISTICAL ACCOUNT.

SMITH, C. 1750. *The ancient and present state of the county and city of Cork.* Dublin.

SUCKLEY, G. 1860. All fish exclusive of the salmonidae. *Reports of Explorations and Surveys to Ascertain the Most Practicable and Economical Route for a Rail Road from the Mississippi River to the Atlantic Ocean made under the direction of the Secretary of State of War in 1853–5.*

WALLACE, S. & GISBORNE, B. 2006. *Basking Sharks: the slaughter of B.C.'s gentle giants.* Transmontanus, Vancouver B.C.

YOUNG, A. 1780. *A Tour in Ireland* Vol. 1, pp.1776–79. New York.

Chapter 3

BREEN, J, 2010. Pers. comm.

FAIRFAX, D. 1998. *The Basking Shark in Scotland – natural history, fishery and conservation.* Tuckwell Press, East Linton.

FLAHERTY, R. 1934. *Man of Aran.* Gainsborough Films, London.

McNALLY, K. 1974. *The Sun-Fish Hunt.* Blackstaff Press, Belfast.

WALLACE, S. & GISBORNE, B. 2006. *Basking Sharks: the slaughter of B.C.'s gentle giants.* Transmontanus, Vancouver B.C.

WILSON, S.G. 2004. Basking sharks (*Cetorhinus maximus*) schooling in the southern Gulf of Maine. *Fisheries Oceanography* 13 (4): 283–286.

Chapter 4

WATKINS, A. 1958. *The Sea My Hunting Ground.* William Heinemann, London.

Chapter 5

BOTTING, D. 1993. *Gavin Maxwell – A life.* Harper Collins, London.

MAXWELL, G. 1952. *Harpoon at a Venture.* Rupert Hart-Davis, London.

WATKINS, A. 1958. *The Sea My Hunting Ground.* William Heinemann, London.

Chapter 6

CAMPBELL, J.L. 1984. *Canna – The Story of a Hebridean Island.* Oxford University Press.

MAXWELL, G. 1952. *Harpoon at a Venture.* Rupert Hart-Davis, London.

Chapter 7

BOTTING, D. 1993. *Gavin Maxwell – A Life.* Harper Collins, London.

DYER, R. 2015. Pers. comm.

MAXWELL, G. 1952. *Harpoon at a Venture.* Rupert Hart-Davis, London.

WATKINS, A. 1958. *The Sea my Hunting Ground.* William Heinemann, London.

Chapter 8

BOTTING, D. 1993. *Gavin Maxwell – A Life.* Harper Collins, London.

FAIRFAX, D. 1998. *The Basking Shark in Scotland – natural history, fishery and conservation.* Tuckwell Press, East Linton.

MAXWELL, G. 1952. *Harpoon at a Venture.* Rupert Hart-Davis, London.

WATKINS, A. 1958. *The Sea My Hunting Ground.* William Heinemann, London.

Chapter 9

BOTTING, D. 1993. *Gavin Maxwell – A Life.* Harper Collins, London.

COCKER, M. 1992. *Loneliness and Time.* Pantheon Books, New York,

FAIRFAX, D. 1998. *The Basking Shark in Scotland – natural history, fishery and conservation.* Tuckwell Press, East Linton.

FRERE, R. 1976. *Maxwell's Ghost.* Victor Gollancz, London.

MATTHEWS, L.H. 1950. Reproduction in the Basking Shark. *Philosophical transactions of the Royal Society of London* 234B: 247–31.

MATTHEWS, L.H. & PARKER, H.W. 1950. Notes on the anatomy and biology of the Basking Shark *Cetorhinus maximus* (Gunnerus). *Proceedings of the Zoological Society of London* 120 (3): 535–576.

MAXWELL, G. 1952. *Harpoon at a Venture.* Rupert Hart-Davis, London.

MITCHELL, I. 1999. *Isles of the West.* Canongate Books, Edinburgh.

RAINE, K. 1976. *The Lions Mouth.* Hamish Hamilton, London.

WATKINS, A. 1958. *The Sea My Hunting Ground.* William Heinemann, London.

Chapter 10

BOTTING, D. 1993. *Gavin Maxwell – A Life*. Harper Collins, London.

FAIRFAX, D. 1998. *The Basking Shark in Scotland – natural history, fishery and conservation*. Tuckwell Press, East Linton.

FITZGERALD O'CONNOR, P. 1953. *Shark-O!* Secker & Warburg, London.

GEDDES, J.T. 1960. *Hebridean Sharker*. Herbert Jenkins, London.

KUNZLIK, P. 1988. *The Basking Shark*. Scottish Fisheries Information Pamphlet No. 14. Department of Agriculture and Fisheries for Scotland, Aberdeen.

MAXWELL, G. 1952. *Harpoon at a Venture*. Rupert Hart-Davis, London.

MAXWELL, G. 1960. *Ring of Bright Water*. Longmans, London.

MITCHELL, I. 1999. *Isles of the West*. Canongate Books, Edinburgh.

Chapter 11

BERROW, S.D. 1994. Incidental capture of elasmobranchs in the bottom set gill-net fishery off the south coast of Ireland. *Journal of the Marine Biological Association of the UK* 74: 837–847.

BERROW, S. & HEARDMAN, C. 1994. The basking shark *Cetorhinus maximus* (Gunnerus) in Irish waters – patterns of distribution and abundance. *Proceedings of the Royal Irish Academy* 94B (2): 101–107.

CONVENTION ON INTERNATIONAL TRADE IN ENDANGERED SPECIES (CITES) 2000. Consideration of proposals for amendment of Appendices I and II. London.

DARLING, J.D. & KEOGH, K.E. 1994. Observations of basking sharks *Cetorhinus maximus* in Clayoquot Sound, British Columbia. *Canadian Field Naturalist* 108: 199–210.

EARLL, R. C. 1990. The Basking Shark – its fishery and conservation. *British Wildlife* No. 3: 121–129.

EARLL, R.C. 2015. Pers. comm.

FLAHERTY, R. 1934. *Man of Aran*. Gainsborough Films, London.

FALKUS, H. 1951. *Shark Island*. Anglo American Film Company, London.

HORSMAN, P.V. 1987. *The Basking Shark – Cetorhinus maximus – A species under threat?* Marine Conservation Society, Ross-on-Wye.

IUCN 2004. *2004 IUCN Red List of Threatened Species*. IUCN, Gland, Switzerland and Cambridge, UK.

KUNZLIK, P. 1988. *The Basking Shark*. Scottish Fisheries Information Pamphlet No. 14. Department of Agriculture and Fisheries for Scotland, Aberdeen.

MAXWELL, G. 1952. *Harpoon at a Venture*. Rupert Hart-Davis, London.

McNALLY, K. 1974. *The Sun-Fish Hunt*. Blackstaff Press, Belfast.

MYKLEVOLL, S. 1968. Basking Shark Fishery. *Commercial Fisheries Review*, July, 59–63.

SPRINGER, S. & GILBERT, P.W. 1976. The basking shark *Cetorhinus maximus* from Florida and California, with comments on its biology and systematics. *Copeia*: 47–54.

SQUIRE, J.L. 1967. Observations of basking sharks and great white sharks in Monterey Bay 1948–1950. *Copeia* 1:247–250.

SQUIRE, J.L. 1990. Distribution and apparent abundance of the basking shark *Cetorhinus maximus* off the central and southern California coast 1962–85. *Marine Fisheries Review* 52(2): 8–11.

WALLACE, S. & GISBORNE, B. 2006. *Basking Sharks: the slaughter of B.C.'s gentle giants*. Transmontanus, Vancouver B.C.

WATKINS, A. 1958. *The Sea my Hunting Ground*. William Heinemann, London.

WENT, A.E.J. & Ó Suilleabháin, S. 1967. Fishing for the sun-fish or basking shark in Irish waters. *Proceedings of the Royal Irish Academy* 65C: 91–115.

Chapter 12

EARLL, R.C. 1990. The Basking Shark – its fishery and conservation. *British Wildlife* No. 3: 121–129.

EARLL, R.C. 2015. Pers. comm.

EARLL, R.C. & TURNER, J.R. 1992. *A review of methods and results from a sightings scheme and field research on the Basking Shark in 1987–92.* Marine Conservation Society, Ross-on-Wye.

EVANS, P.G.H. 1987. *The Natural History of Whales and Dolphins.* Christopher Helm Mammal Series, Random House, New Zealand.

HORSMAN, P.V. 1987. *The Basking Shark* Cetorhinus maximus *– A species under threat?* Marine Conservation Society, Ross-on-Wye.

KUNZLIK, P. 1988. *The Basking Shark.* Scottish Fisheries Information Pamphlet.

MATTHEWS, L.H. 1950. Reproduction in the Basking Shark. *Philosophical transactions of the Royal Society of London* 234B: 247–31.

MATTHEWS, L.H. & PARKER, H.W. 1950. Notes on the anatomy and biology of the Basking Shark *Cetorhinus maximus* (Gunnerus). *Proceedings of the Zoological Society of London* 120 (3): 535–576.

MAXWELL, G. 1952. *Harpoon at a Venture.* Rupert Hart-Davis, London.

MAXWELL, G. 1960. *Ring of Bright Water.* Longmans, London.

NICOLSON, A. 2001. *Sea Room – An Island Life.* Harper Collins, London.

ROBERTS, C.M. 2007. *The Unnatural History of the Sea.* Island Press. Washington D.C.

ROBINSON, W. 1985. Pers. comm.

STOTT, F.C. 1974. Why ignore the basking shark? *Fishing News.*

WATKINS, A. 1958. *The Sea My Hunting Ground.* William Heinemann, London.

Chapter 13

Basking Shark Photo-Identification Project. Available at: www.sharktrust.org/en/basking_shark_photo-identification. The Shark Trust, UK.

Blue Water White Death, 1971. Cinema Center Films, USA.

CONVENTION ON INTERNATIONAL TRADE IN ENDANGERED SPECIES (CITES) 2000. *Consideration of proposals for amendment of Appendices I and II.* London.

EARLL, R. C. 1990. The Basking Shark – its fishery and conservation. *British Wildlife* No. 3: 121–129.

EARLL, R.C. 2015. Pers. comm.

FAIRFAX, D. 1998. *The Basking Shark in Scotland – natural history, fishery and conservation.* Tuckwell Press, East Linton.

FARMERY, J.P. 1992. Zooplankton of fronts associated with feeding behaviour of *Cetorhinus maximus* off the coast of the Isle of Man. Unpublished BSc thesis, Univ. of Wales Bangor.

FITZGERALD O'CONNOR, P. 1953. *Shark-O!* Secker & Warburg, London.

FOWLER, S.L. 1996. Status of the basking shark *Cetorhinus maximus* (Gunnerus). *Shark News* 6: 4–5. Newsletter of the IUCN Shark Specialist Group.

HORSMAN, P.V. 1987. *The Basking Shark – Cetorhinus maximus – A species under threat?* Marine Conservation Society.

IUCN 2004. *2004 IUCN Red List of Threatened Species*. IUCN, Gland, Switzerland and Cambridge, UK.

MATTHEWS, L.H. 1950. Reproduction in the Basking Shark. *Philosophical Transactions of the Royal Society of London* 234B: 247–31.

MATTHEWS, L.H. 1962. The shark that hibernates. *New Scientist* No. 280.

MATTHEWS, L.H. & PARKER, H.W. 1950. Notes on the anatomy and biology of the Basking Shark *Cetorhinus maximus* (Gunnerus). *Proceedings of the Zoological Society of London* 120 (3): 535–576.

MAXWELL, G. 1952. *Harpoon at a Venture*. Rupert Hart-Davis, London.

McLACHLAN, H. 1991. An analysis of sightings of the basking shark *Cetorhinus maximus* in British coastal waters from 1987–1990, and behavioural observations of sharks in the waters around the Isle of Man in 1990. *Unpublished MSc thesis, University of Wales*, Bangor.

OKE, J.E. 2000. An examination of the feasibility of using photo-identification techniques for the Basking Shark (*Cetorhinus maximus*) studies. Unpublished MSc thesis, University of Plymouth, UK.

PARKER, H. & BOESEMAN, M. 1954. The Basking Shark (*Cetorhinus maximus*) in winter. *Proceedings of the Zoological Society of London* 124: 185–194.

PARKER, H.W. & STOTT, F.C. 1965. Age, size and vertebral calcification in the Basking Shark *Cetorhinus maximus* (*Gunnerus*). *Zoologische Mededelingen* 40: 305–319.

PRIEDE, I.G. 1984. A Basking Shark (*Cetorhinus maximus*) tracked by satellite together with simultaneous remote sensing. *Fisheries Research* 2: 201–216.

SIMS, D.W. 1997. *Distribution and behaviour of the Basking Shark,* Cetorhinus maximus *(Gunnerus) off southwest England: Implications for perception of "endangered" status*. The Nature Conservancy Council for England (English Nature).

SIMS, D.W. 1999. Threshold foraging behaviour of Basking Sharks on zooplankton: life on an energetic knife edge? *Proceedings of the Royal Society of London* B, 266: 1437–1443.

SIMS, D.W. 2000. Filter feeding and cruising swimming speeds of Basking Sharks compared with optimal models: they filter-feed slower than predicted for their size. *Journal of Marine Biology and Ecology* 249 (2000): 65–76.

SIMS, D.W., FOX, A.M. & MERRETT, D.A. 1997. Basking shark occurrence off south-west England in relation to zooplankton abundance. *Journal of Fishery Biology* 51: 436–440.

SIMS, D.W. & MERRETT, D.A. 1997. Determination of zooplankton characteristics in the presence of surface feeding Basking Sharks *Cetorhinus maximus*. *Marine Ecological Progress Series* 158: 297–302.

SIMS, D.W. & QUAYLE, V.A. 1998. Selective foraging behaviour of Basking Sharks in a small-scale front. *Nature* 393: 460–464.

SPEEDIE, C.D. 1996. *The Basking Shark in the waters of Devon and Cornwall – A review of sightings in 1995 and 1996*. Cornwall Wildlife Trust, Truro.

SPEEDIE, C.D. 2000. The European Basking Shark Photo-identification Project. In: VACCHI, M., La MESA, G., SERENA, F. and SERET, B. (eds) *Proceedings of the 4th European Elasmobranch Meeting*, Livorno (Italy), 2000. ICRAM, ARPAAT & SFI, 2002: 157–160.

STAGG, A. 1990. A study of the behaviour and migration of the basking shark, *Cetorhinus maximus* (Gunnerus), in British and Irish waters. Unpublished MSc thesis, University of Wales, Bangor.

STRAWBRIDGE, J. 1992. The behavioural ecology of the basking shark *Cetorhinus maximus* in inshore waters off the Isle of Man. Unpublished MSC thesis, University of Wales, Bangor.

STRONG, P. 1991. Local oceanographic parameters associated with the distribution and feeding behaviour of *Cetorhinus maximus* off the Isle of Man. Unpublished MSc thesis, University of Wales, Bangor.

THOM, T., O'CONNELL, M. & LUCAS, M. 1999. A satellite tagging study of basking sharks (*Cetorhinus maximus*) and investigations of associated plankton distribution in the Firth of Clyde. *Scottish Natural Heritage Research Report* No. 117.

TREGENZA, N.J.C. 1992. Fifty years of cetacean sightings from the Cornish coast, SW England. *Biological Conservation* 59: 65–70.

Chapter 14

CAMPBELL, J.L. 1984. *Canna – The Story of a Hebridean Island*. Oxford University Press, Oxford.

CONVENTION ON INTERNATIONAL TRADE IN ENDANGERED SPECIES (CITES) 2000. Consideration of proposals for amendment of Appendices I and II. London.

FAIRFAX, D. 1998. *The Basking Shark in Scotland – natural history, fishery and conservation*. Tuckwell Press, East Linton.

FITZGERALD O'CONNOR, P. 1953. *Shark-O!* Secker & Warburg, London.

GEDDES, J.T. 1960. *Hebridean Sharker*. Herbert Jenkins, London.

HARVEY-CLARK, C.W., STOBO, W.T., HELLE, E. & MATTSON, M. 1999. Putative mating behaviour in Basking Sharks off the Nova Scotia coast. *Copeia* 1999 (3): 780–782.

HILL, P. 2003. *Stargazing – Memoirs of a Young Lighthouse Keeper*. Canongate, Edinburgh.

KUNZLIK, P. 1988. *The Basking Shark*. Scottish Fisheries Information Pamphlet.

MATTHEWS, L.H. & PARKER, H.W. 1951. Basking shark leaping. *Proceedings of the Zoological Society of London* 121: 461–462.

MAXWELL, G. 1952. *Harpoon at a Venture*. Rupert Hart-Davis, London.

PYLE, P., ANDERSON, S.D., KLIMLEY, A.P. & HENDERSON, R.P. 1996. Environmental factors affecting the occurrence and behavior of white sharks at the Farallon Islands, California. In A.P. KLIMLEY & D.G. AINLEY (eds) *Great White Sharks: The Biology of Carcharodon carcharias*., Academic Press, San Diego, USA.

SIMS, D.W., SOUTHALL, E.J., QUAYLE, V.A. & FOX, A.M. 2000. Annual social behaviour of Basking Sharks associated with coastal front areas. *Proceedings of the Royal Society of London B* (2000), 267: 1897–1904.

SIMS, D.W, SPEEDIE, C.D. & FOX, A.M. 2000. Movements and growth of female

Authors?? Shark re-sighted after a three year period. *Journal of the Marine Biological Association of the U.K.* (2000), 80: 1141–1142.

SPEEDIE, C. 2006. *The Ulster Wildlife Trust Basking Shark Project. Environment and Heritage Service Research and Development Series* No. 06/16.

WALLACE, S. & GISBORNE, B. 2006. *Basking Sharks: the slaughter of B.C.'s gentle giants*. Transmontanus, Vancouver B.C.

WATKINS, A. 1958. *The Sea My Hunting Ground*. William Heinemann, London.

WHITEHEAD, H. 1985. Why whales leap. *Scientific American* 252: 70–75.

Chapter 15

CONVENTION ON INTERNATIONAL TRADE IN ENDANGERED SPECIES (CITES) 2000. Consideration of proposals for amendment of Appendices I and II. London.

COMPAGNO, L.J.V. 2002. Sharks of the World. *FAO Species Catalogue* No. 1, Vol. 2. Food and Agriculture Association of the United Nations, Rome.

COTTON, P.A., SIMS, D.W., FANSHAWE, S. & CHADWICK, M. 2005. The effects of climate variability on zooplankton and Basking Shark relative abundance off southwest Britain. *Fisheries Oceanography* 14: 151–155.

DOYLE, J.I., SOLANDT, J-L, FANSHAWE, S., RICHARDSON, P., DUNCAN, C. 2005. *Marine Conservation Society Basking Shark Watch 1987–2004*. Marien Conservation Society, Ross-on-Wye.

FAIRFAX, D. 1998. *The Basking Shark in Scotland – natural history, fishery and conservation*. Tuckwell Press, East Linton.

FOWLER, S.L. 2007. IUCN 2007. *IUCN Red List of Threatened Species*. Available at: www.iucnredlist.org/details/4292/0

FOWLER, S.L. 1996. Status of the basking shark *Cetorhinus maximus* (Gunnerus). *Shark News* 6: 4–5. Newsletter of the IUCN Shark Specialist Group.

FOWLER, S.L. 2005. Status of the Basking Shark *Cetorhinus maximus* (*Gunnerus*). In: FOWLER, S.L., CAMHI, M., BURGESS, G., FORDHAM, S. & MUSICK, J. *Sharks, Rays and Chimaeras – The status of Chondrichthyan fishes*. IUCN Species Survival Commission Shark Specialist Group. IUCN, Gland, Switzerland and Cambridge.

FOWLER, S.L., REED, T.M. & DIPPER, F.A. (eds) 2002. *Elasmobranch Biodiversity, Conservation and Management*. Proceedings of the International Seminar & Workshop, Sabah, Malaysia, July 1997. (258pp.). Occasional paper of the Species Survival Commission No. 25.

FRANCIS, M.P. & DUFFY, C. 2002. Distribution, seasonal abundance and bycatch of Basking Sharks (*Cetorhinus maximus*) in New Zealand, with observations on their winter habitat. *Marine Biology* 140: 831–842.

FRANCIS, M.P. & SMITH, M.H. 2010. Basking Shark (*Cetorhinus maximus*) bycatch in New Zealand Fisheries, 1994–1995 to 2007–2008. *New Zealand Aquatic Environment and Biodiversity Report* No. 449. Ministry of Fisheries, Wellington.

GEDDES, J.T. 1960. *Hebridean Sharker*. Herbert Jenkins, London.

KUNZLIK, P. 1988. *The Basking Shark*. Scottish Fisheries Information Pamphlet.

LIEN, J. & FAWCETT, L. 1986. Distribution of basking sharks *Cetorhinus maximus* incidentally caught in inshore fishing gear in Newfoundland. *Canadian Field Naturalist* 100: 246–252.

MATTHEWS, L.H. 1962. The shark that hibernates. *New Scientist* No. 280.

MAXWELL, G. 1952. *Harpoon at a Venture*. Rupert Hart-Davis, London.

McNALLY, K. 1974. *The Sun-Fish Hunt*. Blackstaff Press, Belfast.

MILLER, P. 2004. Multispectral front maps for automatic detection of ocean colour features from SeaWiFS. . *International Journal of Remote Sensing* 25(7–8): 1437–1442.

PAULY, D. 1976. *A critique of some literature data on the growth, reproduction and mortality of the lamnid shark Cetorhinus maximus (Gunnerus)*. International Council for the Exploration of the Sea. Council meeting 1978/H:17. Pelagic Fish Committee. 10pp.

PAULY, D. 1997. Growth and mortality of the Basking Shark *Cetorhinus maximus* and their implications for the management of Whale Sharks *Rhincodon typus*. In:

LIEN, J. & ALDRICH, D. 1982. *The basking shark Cetorhinus maximus in Newfoundland*. Report to the Department of Fisheries, Government of Newfoundland and Labrador. 186pp.

PRIEDE, I.G. 1984. A Basking Shark (*Cetorhinus maximus*) tracked by satellite together with simultaneous remote sensing. *Fisheries Research* 2: 201–216.

SIMS, D.W., SOUTHALL, E.J., RICHARDSON, A.J., REID, P.C. & METCALFE, J.D. 2003. Seasonal movements and behaviour of Basking Sharks from archival tagging: no evidence of winter hibernation. *Marine Ecological Progress Series* 248: 187–196.

SIMS, D.W., SOUTHALL, E.J., METCALFE, J.D. & PAWSON, M.G. 2005. *Basking Shark population assessment*. Report for Global Wildlife Division of DEFRA.

SIMS, D.W., SOUTHALL, E.J., TARLING, G.A., METCALFE, J.D. 2005. Habitat specific normal and reverse diel vertical migration in the plankton-feeding Basking Shark. *Journal of Animal Ecology* 74: 755–761.

SIMS, D.W. & REID, P.C. 2002. Congruent trends in long-term zooplankton decline in the northeast Atlantic and Basking Shark (*Cetorhinus maximus*) fishery catches off west Ireland. *Fisheries Oceanography* 11 (1): 59–63.

SKOMAL, G.B., WOOD, G. & CALOYIANIS, N. 2004. Archival tagging of a Basking Shark, *Cetorhinus maximus*, in the western North Atlantic. *Journal of the Marine Biological Association of the UK* 84: 795–799.

SKOMAL, G.B., ZEEMAN, S.I., CHISHOLM, J.H., SUMMERS, E.L., WALSH, H.J, McMAHON, K.W. & THORROLD, S.R. 2009. Transequatorial migrations by Basking Sharks in the western Atlantic Ocean. *Current Biology* 19.

SOUTHALL, E.J., SIMS, D.W., METCALFE, J.D., DOYLE, J.I., FANSHAWE, S., LACEY, C., SHRIMPTON, J., SOLANDT, J-L. & SPEEDIE, C.D. 2005. Spatial distribution patterns of Basking Sharks on the European shelf: preliminary comparison of satellite-tag geolocation, survey and public sightings data. *Journal of the Marine Biological Association of the U.K.* (2005) 85: 1083–1088.

SOUTHALL, E.J., SIMS, D.W., WITT, M.J. & METCALFE, J.D. 2006. Seasonal space use estimates of Basking Shark in relation to protection and political economic zones in the northeast Atlantic. *Biological Conservation* 132: 33–39.

SUND, O. 1943. Et brugdebarsel. *Naturen* 67: 285–286.

WALLACE, S. & GISBORNE, B. 2006. *Basking Sharks: the slaughter of B.C.'s gentle giants.* Transmontanus, Vancouver B.C.

WATKINS, A. 1958. *The Sea My Hunting Ground.* William Heinemann, London.

Chapter 16

FOWLER, S.L. 2016. Pers. comm.

FITZGERALD O'CONNOR, P. 1953. *Shark-O!* Secker & Warburg, London.

GEDDES, J.T. 1960. *Hebridean Sharker.* Herbert Jenkins, London.

HOARE, P. 2008. *Leviathan, or the Whale.* Fourth Estate, London

MAXWELL, G. 1952. *Harpoon at a Venture.* Rupert Hart-Davis, London.

SPEEDIE, C.D. 2003. The value of public sightings schemes in relation to the Basking Shark in U.K. waters. *Cybium* 2003, 27(4): 255–259.

SPEEDIE, C.D. & JOHNSON, L.A. 2008. The Basking Shark (*Cetorhinus maximus*) in West Cornwall. Key sites, anthropogenic threats and their implications for conservation of the species. *Natural England Research Report* NERR 0018.

SPEEDIE, C.D., JOHNSON, L.A. &d WITT, M.J. 2009. Basking Shark hotspots on the west coast of Scotland: Key sites, anthropogenic threats and implications for conservation of the species. *Scottish Natural Heritage Commissioned Report* No. 339.

WATKINS, A. 1958. *The Sea My Hunting Ground.* William Heinemann, London.

WILSON, E. 2000. Determination of boat disturbance on the surface feeding behaviour of Basking Sharks *Cetorhinus maximus*. Unpublished MSc thesis. University of Plymouth, UK.

Chapter 17

BERROW, S. & JOHNSTON, E. 2009. *Basking Shark Survey: tagging and tracking.* Final Report to the Heritage Council, Ireland.

BLOOMFIELD, A. & SOLANDT, J-L. 2010. *The Marine Conservation Society Basking Shark Watch 20-year report (1987-2006).* Marine Conservation Society, Ross-on-Wye.

COTTON, P.A., SIMS, D.W., FANSHAWE, S. & CHADWICK, M. 2005. The effects of climate variability on zooplankton and Basking Shark relative abundance off southwest Britain. *Fisheries Oceanography* 14: 151–155.

FITZGERALD O'CONNOR, P. 1953. *Shark-O!* Secker & Warburg, London.

GEDDES, J.T. 1960. *Hebridean Sharker*. Herbert Jenkins, London.

GORE, M.A., ROWAT, D., HALL, J., GELL, F.R. & ORMOND, R.F. 2008. Transatlantic Migration and Deep Ocean diving by Basking Shark. *Biological Letters* doi:10.1098/rsbl.2008.0147.

GORE, M.A., ALLAN, H., BERROW, S., CORREALE, V., DICK, C., FREY, P.H., GILKES, G., ORMOND, R.F. & SPEEDIE, C.D. 2013. *Just how many Basking Sharks are there? Using photo-identification to assess population abundance*. Poster, 17th European Elasmobranch Association Conference.

HOELZEL, A.R., SHIVJI, M.S., MAGNUSSEN, J. & FRANCIS, M.P. 2006. Low worldwide genetic diversity in the Basking Shark (*Cetorhinus maximus*). *Biological Letters* 2: 639–642.

JOHNSTON, E. 2010. Surveying the Basking Shark (*An Liamhán Gréine, Cetorhinus maximus*) Investigating a robust methodology. Unpublished MSc thesis, University of Ulster.

LIEBER, L., BERROW, S., JOHNSTON, E., HALL, G., HALL, J., GUBILI, C., SIMS, D.W., JONES, C.S. & NOBLE, L.R. 2013. Mucus: aiding elasmobranch conservation through non-invasive genetic sampling. *Endangered Species Research* 21: 215–222.

MAXWELL, G. 1952. *Harpoon at a Venture*. Rupert Hart-Davis, London.

MUSICK, J.A., HARBIN, H.M., BERKELEY, S.A., BURGESS, G.H., EKLUND, A.M., FINDLEY, L., GILMORE, R.G., GOLDEN, J.T., HA, D.S., HUNTSMAN, G.R., McGOVERN, J.C., PARKER, S.J., POSS, S.G., SALA, E., SCHMIDT, T.W., SEDBERRY, G.R., WEEKS, H. & WRIGHT, S.G. 2000. Marine, estuarine and diadromous fish at risk of extinction in North America (exclusive of Pacific salmonids). *Fisheries* 25 (11): 6–30.

REEVE, A., SMALL, D., SPEEDIE, C. 2008. *Characteristic markings of British Basking Sharks*. Poster, 12th European Elasmobranch Association Conference.

SKOMAL, G.B., ZEEMAN, S.I., CHISHOLM, J.H., SUMMERS, E.L., WALSH, H.J, McMAHON, K.W. & THORROLD, S.R. 2009. Transequatorial migrations by Basking Sharks in the western Atlantic Ocean. *Current Biology* 19.

SPEEDIE, C.D. & JOHNSON, L.A. 2008. The Basking Shark (*Cetorhinus maximus*) in West Cornwall. Key sites, anthropogenic threats and their implications for conservation of the species. *Natural England Research Report* NERR 0018.

SPEEDIE, C.D., JOHNSON, L.A. & WITT, M.J. 2009. Basking Shark hotspots on the west coast of Scotland: Key sites, anthropogenic threats and implications for conservation of the species. *Scottish Natural Heritage Commissioned Report* No. 339.

STÉPHAN, E., GADENNE, H. & JUNG, A. 2011. *Sur les traces du Requin Pèlerin – satellite tracking of Basking Sharks in the northeast Atlantic*. Association Pour L'Etude et la Conservation des Selaciens, Brest.

WILSON, S.G. 2004. Basking sharks (*Cetorhinus maximus*) schooling in the southern Gulf of Maine. *Fisheries Oceanography* 13 (4): 283–286.

WITT, M.J., HARDY, T., JOHNSON, L.A., McLELLAN, C.M., PIKESLEY, S.K., RANGER, S., RICHARDSON, P.B., SOLANDT, J.L., SPEEDIE, C.D., WILLIAMS, R. & GODLEY, B.J. 2012. Basking Sharks in the northeast Atlantic: Spatio-temporal trends from sightings in U.K. waters. *Marine Ecology Progress Series* 457: 121–134.

WITT, M.J., DOHERTY, P.D., HAWKES, L.A., GODLEY, B.J., GRAHAM, R.T. & HENDERSON, S.M. 2013. Basking Shark satellite tagging project: post-fieldwork report. *Scottish Natural Heritage Commissioned Report* No. 555.

WITT, M.J., DOHERTY, P.D., GODLEY, B.J., GRAHAM, R.T., HAWKES, L.A. & HENDERSON, S.M. 2013. Basking Shark satellite tagging project: insights into Basking Shark (*Cetorhinus maximus*) movement and distribution using satellite telemetry (Interim report November 2013). *Scottish Natural Heritage Commissioned Report* No. 700.

WITT, M.J., DOHERTY, P.D., GODLEY, B.J., GRAHAM, R.T., HAWKES, L.A. &

HENDERSON, S.M. 2014. Basking Shark satellite tagging project: insights into Basking Shark (*Cetorhinus maximus*) movement and distribution using satellite telemetry (Phase 1, July 2014). *Scottish Natural Heritage Commissioned Report* No. 752.

WITT, M.J., DOHERTY, P.D., GODLEY, B.J., GRAHAM, R.T., HAWKES, L.A. & HENDERSON, S.M. 2016. Basking Shark satellite tagging project: insights into Basking Shark (*Cetorhinus maximus*) movement and distribution using satellite telemetry (Final Report). *Scottish Natural Heritage Report* No. 908.

Chapter 18

ANDERSON, S.D., CHAPPLE, T.K., JORGENSEN, S.J., KLIMLEY, A.P. & BLOCK, B.A. 2011. Long-term individual identification and site fidelity of White Sharks *Carcharinus carcharias*, off California using dorsal fins. *Marine Biology*, 158 (6), 1233–37.

AUSTIN, R.A., WITT, M.J., PIKESLEY, S.K., JOHNSON, L.A.. & SPEEDIE, C. 2016 (In Press).*Ecological niche modelling for basking sharks in the U.K. and Ireland: an ensemble approach.*

BEAUGRAND, G., EDWARDS, M., BRANDER, K., LUCZAKI, C. & IBENEZ, F. 2008. Causes and projections of abrupt climate-driven ecosystem shifts in the North Atlantic. *Ecology Letters* 11: 1157–1168.

BEAUGRAND, G., REID, P.C., IBANEZ, F., ALISTER LINDLEY, J. & EDWARDS, M. 2002. Reorganization of North Atlantic Marine Copepod Biodiversity and Climate. *Science* 296: 1692–1694.

BENHAM, D. 2010. Pers. comm.

BERROW, S.D. 1994. Incidental capture of elasmobranchs in the bottom set gill-net fishery off the south coast of Ireland. *Journal of the Marine Biological Association of the UK* 74: 837–847.

BESSELING, E., FOEKEMA, E.M., VAN FRANEKER, J.A., LEOPOLD, M.F., KÜHN, BRAVO REBOLLEDO, E.L., HEBE, E., MIELKE, L., IJZER, J., KAMMINGA, P & KOELMANS, A.A. 2015. *Marine Pollution Bulletin* Available at: www.sciencedirect.com/science/article/pii/S0025326X15001952

BINDOFF, N.L., J. WILLEBRAND, V. ARTALE, A, CAZENAVE, J. GREGORY, S. GULEV, K. HANAWA, C. Le Quéré, S. LEVITUS, Y. NOJIRI, C.K. SHUM, L.D. TALLEY & UNNIKRISHNAN, A. 2007. Observations: Oceanic Climate Change and Sea Level. In: SOLOMON, S., D. QIN, M. MANNING, Z. CHEN, M. MARQUIS, K.B. AVERYT, M. TIGNOR & H.L. MILLER (eds) *Climate Change 2007: The Physical Science Basis. Contribution of Working Group I to the Fourth Assessment Report of the Intergovernmental Panel on Climate Change*. Cambridge University Press, Cambridge, UK and New York, USA.

BONFIL, R., MEYER, M, SCHOLL, M.C., JOHNSON, R., O'BRIEN, S., OOSTHIZEN, H., SWANSON, S., KOTZE, D. & PATERSON, M. 2005. Transoceanic migration, spatial dynamics, and population linkages of White Sharks. *Science* 310: 100–103.

BOOTH, C.G., KING, S.L. & LACEY, C. (2013). 'Argyll Array Wind farm Basking Draft Chapter for Environmental Statement'. SMRU Ltd. Unpublished report no. SMRUL-WSP – 2013 – 001. January 2013.

BOUSTANY, A.M., DAVIS, S.F., PYLE, P., ANDERSON, S.D., LE BOEUF, B. & BLOCK, B.A. 2002. Satellite tagging: expanded niche for White Sharks. *Nature* 415: 35–36.

BRITISH MARINE FEDERATION 2006. *Watersports and leisure participation survey 2006*. Arkenford Market Modelling and Research.

COLE, M., LINDEQUE, P., FILEMAN, E., HALSBAND, C., GOODHEAD, R., MOGER, J. & GALLOWAY, S. 2013. Microplastic Ingestion by Zooplankton. *Environmental Science & Technology* 18, 47(12): 6645–6656.

CONVENTION ON INTERNATIONAL TRADE IN ENDANGERED SPECIES (CITES) 2000. Consideration of proposals for amendment of Appendices I and II. London.

COTTON, P.A., SIMS, D.W., FANSHAWE, S. and CHADWICK, M. 2005. The effects of climate variability on zooplankton and Basking Shark relative abundance off southwest Britain. *Fisheries Oceanography* 14: 151–155.

DOHERTY, P.D., BAXTER, J.M., GELL, F.R., GODLEY, B.J., GRAHAM, R.T., HALL,G.,HALL, J.,HAWKES, L.A., HENDERSON, S.M, JOHNSON, L., SPEEDIE, C., & WITT, M.J. *In Press.* Long-term satellite tracking reveals variable seasonal migration strategies of basking sharks in the north-east Atlantic.

DOHERTY, P.D., BAXTER, J.M., GODLEY, B.J., GRAHAM, R.T., HALL,G.,HALL, J.,HAWKES, L.A., HENDERSON, S.M, JOHNSON, L., SPEEDIE, C., & WITT, M.J. *In Press.* Testing the boundaries: Seasonal recidency and inter-annual site fidelity of basking sharks in a proposed marine protected area.

DREWERY, H.J., 2012. Basking Shark (*Cetorhinus maximus*) literature review, current research and new research ideas. *Marine Scotland Science Report 2012.*

EDWARDS, M., BRESNAN, E., COOK, K., HEATH, M., HELAOUET, P., LYNAM, C., RAINE, R. & WIDDICOMBE, C. 2013. Impacts of climate change on plankton. *MCCIP Science Review 2013*: 98–112.

EDWARDS, M. & JOHNS, D. 2005. *Monitoring the ecosystem response to climate change in the North-East Atlantic.* Sir Alistair Hardy Foundation for Ocean Science (SAHFOS).

EDWARDS, M. & RICHARDSON, A.J. 2002. *Ecological Status Report 2001/2002.* SAHFOS Technical Report.

EDWARDS, M. & RICHARDSON, A.J. 2004. Impact of climate change on marine phenology and trophic mismatch. *Nature* 430: 881.

EUROPEAN COMMISSION, 1992. *EC Habitats Directive* (Council Directive 92/43/EEC on the Conservation of natural habitats and of wild fauna and flora).

FELDHEIM, K.A., GRUBER, S.H., DIBATTISTA, J.D., BABCOCK, E.A., KESSEL, S.T., HENDRY, A.P., PIKITCH, E.K., ASHLEY, M.V. & CHAPMAN, D.D. 2013. Two decades of genetic profiling yields first evidence of natal philopatry and long-term fidelity to parturition sites in sharks. *Molecular Ecology*, doi: 10.1111/mec.12583.

FITZGERALD O'CONNOR, P. 1953. *Shark-O!* Secker & Warburg, London.

FOSSI, M.C., COPPOLA, D., BAINI, M., GIANNETTI, M., GUERRANTI, C., MARSILI, L., PANTI, C., de SABATA, E. & CLO, S. 2014. Large filter feeding marine organisms as indicators of microplastics in the pelagic environment: the case studies of the Mediterranean Basking Shark (*Cetorhinus maximus*) and the Fin Whale (*Balaenoptera physalus*). *Marine Environmental Research* doi: 10.1016/j.marenvres.2014.02.002.

FOWLER, S.L. 1996. Status of the basking shark *Cetorhinus maximus* (Gunnerus). *Shark News* 6: 4–5. Newsletter of the IUCN Shark Specialist Group.

FOWLER, S.L. 2005. Status of the Basking Shark *Cetorhinus maximus* (*Gunnerus*). In: FOWLER, S.L., CAMHI, M., BURGESS, G., FORDHAM, S. & MUSICK, J. *Sharks, Rays and Chimaeras – The status of Chondrichthyan fishes.* IUCN Species Survival Commission Shark Specialist Group. IUCN, Gland, Switzerland and Cambridge, UK.

FRANCIS, M.P. & DUFFY, C. 2002. Distribution, seasonal abundance and bycatch of Basking Sharks (*Cetorhinus maximus*) in New Zealand, with observations on their winter habitat. *Marine Biology* 140: 831–842.

FRANCIS, M.P. & SMITH, M.H. 2010. Basking Shark (*Cetorhinus maximus*) bycatch in New Zealand Fisheries, 1994–1995 to 2007–2008. *New Zealand Aquatic Environment and Biodiversity Report* No. 449. Ministry of Fisheries, Wellington.

FROMENTIN, J.M. & PLANQUE, B. 1996. Calanus and environment in the eastern North Atlantic. II. Influence of the North Atlantic Oscillation on *C. finmarchicus* and *C. helgolandicus*. *Marine Ecological Progress Series* 134: 111–118.

GEDDES, J.T. 1960. *Hebridean Sharker.* Herbert Jenkins, London.

GILL, A.B. 2005. Offshore renewable energy – ecological implications of generating electricity in the coastal zone. *Journal of Applied Ecology* 42: 605–615.

GILL, A.B. & KIMBER, J.A. 2005. The potential for cooperative management of elasmobranches and offshore renewable energy development in UK waters. *Journal of the Marine Biological Society of the UK* 85: 1075–1081.

GORE, M.A., ALLAN, H., BERROW, S., CORREALE,V., DICK, C., FREY, P.H., GILKES, G., ORMOND, R.F. & SPEEDIE, C.D. 2013. *Just how many Basking Sharks are there? Using photo-identification to assess population abundance.* Poster, 17th European Elasmobranch Association Conference.

HOELZEL, A.R., SHIVJI, M.S., MAGNUSSEN, J. & FRANCIS, M.P. 2006. Low worldwide genetic diversity in the Basking Shark (*Cetorhinus maximus*). *Biological Letters* 2: 639–642.

HUETER, R.E. & HEUPEL, M.R. 2004. Evidence of Philopatry in Sharks and Implications for the Management of Shark Fisheries. *Journal of Northwest Atlantic Fishery Science* 35: 239–247.

JORGENSEN, S.J., CHAPPLE, T.K., ANDERSON, S., HOYOS, M., REEB, M.C. & BLOCK, B.A. 2012. Connectivity among White Shark Coastal Aggregation Areas in the Northeast Pacific. In: M.L. DOMEIER (ed) *Global perspectives on the biology and life history of the White Shark.* CRC Press, Boca Raton, 13: 159–167.

KELLY, C., GLEGG, G.A. & SPEEDIE, C.D. 2004. Management of marine wildlife disturbance. *Ocean and Coastal management* 47: 1–19.

LIEN, J. & ALDRICH, D. 1982. The basking shark *Cetorhinus maximus* in Newfoundland. Report to the Department of Fisheries, Government of Newfoundland and Labrador. 186pp.

LIEN, J. & FAWCETT, L. 1986. Distribution of basking sharks *Cetorhinus maximus* incidentally caught in inshore fishing gear in Newfoundland. *Canadian Field Naturalist* 100: 246–252.

LYONS, K., CARLISLE, A., PRETI, A., MULL. C., BLASIUS, M. O'SULLIVAN, J., WINKLER, C. & LOWE, C.G. 2013. Effects of trophic ecology and habitat use on maternal transfer of contaminants in five species of young of the year lamniform sharks. *Marine Environmental Research* 90, 72–38.

MAXWELL, G. 1960. *Ring of Bright Water.* Longmans, London.

NASH, R.D.M. & GEFFEN, A.J. 2004. Seasonal and interannual variation in abundance of *Calanus finmarchicus* (*Gunnerus*) and *Calanus helgolandicus* (*Claus*) in inshore waters (west coast of the Isle of Man) in the central Irish Sea. *Journal of Plankton Research* 26 (3): 265–273.

OSPAR 2006. OSPAR Commission Biodiversity Series. Effects on the marine environment of ocean acidification resulting from elevated levels of CO_2 in the atmosphere.

PRIEDE, I.G. & Miller, P.I. 2009. A Basking Shark (*Cetorhinus maximus*) tracked by satellite together with simultaneous remote sensing II: New analysis reveals orientation to a thermal front. *Fisheries Research* 95: 370–372.

PYLE, P., ANDERSON, S.D., KLIMLEY, A.P. & HENDERSON, R.P. 1996. Environmental factors affecting the occurrence and behaviour of White Sharks at the Farallon Islands, California. In: KLIMLEY, A.P. & AINLEY, D.G. (eds) *Great white sharks: the biology of Carcharodon carcharias*, pp.281–291. Academic Press, San Diego.

REID, P.C., EDWARDS, M., BEAUGRAND, G., SKOGEN, M. & STEVENS, D. 2003. Periodic changes in the zooplankton of the North Sea during the twentieth century linked to oceanic inflow. *Fisheries Oceanography* 12 (4/5): 260–269.

REID, P.C., EDWARDS, M., HUNT, H. & WARNER, A.E. 1998. Phytoplankton change in the North Atlantic. *Nature* 391: 546.

REID, P.C., Planque, B. & EDWARDS M. 1998. Is observed variability in the long-term results of the Continuous Plankton Recorder survey a response to climate change? *Fisheries Oceanography* 7 (3/4): 282–288.

ROYAL SOCIETY 2005. Ocean acidification due to increasing atmospheric carbon dioxide, *Royal Society Policy Document* 12/05.

SAMPLE, I. 2006. Sharks pay high price as demand for fins soars. *The Guardian*. Available at: www.theguardian.com/environment/2006/aug/31/fish.frontpagenews

SCOTTISH EXECUTIVE, 2003. *Securing a renewable future: Scotland's Renewable Energy.*

SCOTTISH GOVERNMENT SOCIAL RESEARCH, 2010. *The Economic Impact of Wildlife Tourism in Scotland.* International Centre for Tourism and Hospitality Research, Bournemouth University. 51pp.

SCOTTISH RENEWABLES 2008. *Summary of Renewable Energy Projects in Scotland, Sept 2008.*

SHARK TRUTH, 2016. *Shark Fin Trade – Bowling for Truth.*

SIMS, D.W. 2008. Sieving a living: A review of the biology, ecology & Conservation status of the plankton-feeding Basking Shark *Cetorhinus maximus*. *Advances in Marine Biology*, 54.

SIMS, D.W. & REID, P.C. 2002. Congruent trends in long-term zooplankton decline in the northeast Atlantic and Basking Shark (*Cetorhinus maximus*) fishery catches off west Ireland. *Fisheries Oceanography* 11 (1): 59–63.

SIMS, D.W, SPEEDIE, C.D. & FOX, A.M. 2000. Movements and growth of female Basking Shark re-sighted after a three year period. *Journal of the Marine Biological Association of the U.K.* (2000) 80: 1141–1142.

SOUTH WEST OF ENGLAND REGIONAL DEVELOPMENT AGENCY Hayle. *Wave Hub Environmental Statement, June 2006.*

SOUTHALL, E.J., SIMS, D.W., METCALFE, J.D., DOYLE, J.I., FANSHAWE, S., LACEY, C., SHRIMPTON, J., SOLANDT, J.-L. & SPEEDIE, C.D. 2005. Spatial distribution patterns of Basking Sharks on the European shelf: preliminary comparison of satellite-tag geolocation, survey and public sightings data. *Journal of the Marine Biological Association of the U.K.* (2005) 85: 1083–1088.

SOUTHALL, E.J., SIMS, D.W., WITT, M.J. & METCALFE, J.D. 2006. Seasonal space use estimates of Basking Shark in relation to protection and political economic zones in the northeast Atlantic. *Biological Conservation* 132: 33–39.

Shark . Available at: www.sharktruth.com/learn/shark-finning/

SPEEDIE, C.D. 2001. *Marine ecotourism potential in the waters of south Devon and Cornwall.* In: GARROD, B. & WILSON, J.C. (eds) *Marine Ecotourism: Issues and Experiences*, pp.204–214. Channel View Publications, Clevedon.

SPEEDIE, C.D., JOHNSON, L.A. & WITT, M.J. 2009. Basking Shark hotspots on the west coast of Scotland: Key sites, anthropogenic threats and implications for conservation of the species. *Scottish Natural Heritage Commissioned Report* No. 339.

SUSTAINABLE DEVELOPMENT COMMISSION, 2007. *Turning the tide: tidal power in the UK.*

TIRARD, P., MANNING, M.J., JOLLITT, I., DUFFY, C. and BORSA, P. 2010. Records of Great White Sharks (*Carcharodon carcharias*) in New Caledonian waters. *Pacific Science* 64 (4): 567–576.

WARBURTON, C.A., PARSONS, E.C.M, WOODS-BALLARD, A., HUGHES, A. & JOHNSTON, P. 2001. *Whale watching in West Scotland.* Department for Environment, Food and Rural affairs, London.

WILSON, E. 2000. Determination of boat disturbance on the surface feeding behaviour of Basking Sharks *Cetorhinus maximus*. Unpublished MSc thesis. University of Plymouth, UK.

WiSe Scheme. *Wildlife Safe Training and Accreditation.* Available at: www.wisescheme.org

WITT, M.J., HARDY, T., JOHNSON, L.A., McLELLAN, C.M., PIKESLEY, S.K., RANGER, S., RICHARDSON, P.B., SOLANDT, J.L., SPEEDIE, C.D., WILLIAMS, R. & GODLEY, B.J. 2012. Basking Sharks in the northeast Atlantic: Spatio-temporal trends from sightings in U.K. waters. *Marine Ecology Progress Series* 457: 121–134.

WITT, M.J., DOHERTY, P.D., HAWKES, L.A., GODLEY, B.J., GRAHAM, R.T. & HENDERSON, S.M. 2013. Basking Shark satellite tagging project: post-fieldwork report. *Scottish Natural Heritage Commissioned Report* No. 555.

WITT, M.J., DOHERTY, P.D., GODLEY, B.J., GRAHAM, R.T., HAWKES, L.A. & HENDERSON, S.M. 2013. Basking Shark satellite tagging project: insights into Basking Shark (*Cetorhinus maximus*) movement and distribution using satellite telemetry (Interim report November 2013). *Scottish Natural Heritage Commissioned Report* No. 700.

WITT, M.J., DOHERTY, P.D., GODLEY, B.J., GRAHAM, R.T., HAWKES, L.A. & HENDERSON, S.M. 2014. Basking Shark satellite tagging project: insights into Basking Shark (*Cetorhinus maximus*) movement and distribution using satellite telemetry (Phase 1, July 2014). *Scottish Natural Heritage Commissioned Report* No. 752.

WITT, M.J., DOHERTY, P.D., GODLEY, B.J., GRAHAM, R.T., HAWKES, L.A. and HENDERSON, S.M. 2016. Basking Shark satellite tagging project: insights into Basking Shark (*Cetorhinus maximus*) movement and distribution using satellite telemetry (Final Report). *Scottish Natural Heritage Report* No. 908.

Acknowledgements

I owe an immense debt to my wife, Louise Johnson, for her unstinting support throughout the creation of this book, and for reading and commenting on the text as it evolved. Not only that, she tolerated my mental absence over a considerable period while I wrestled with the text. Such dedication went well beyond the call of duty.

I would like to thank Marc Dando and Julie Dando for their encouragement and support in bringing this book to fruition. Julie's many years of experience helped to guide me at key stages through the challenges of writing my first book and Marc worked his familiar magic via the creation of the wonderful illustrations that please the eye and help to bring the text to life.

The book has greatly benefitted from the wisdom and keen eye of a wonderful editor, Rowena Millar. Finding someone who combines not only the editing ability she possesses but also an encyclopaedic knowledge of wildlife and wild places is exceptionally rare and I was extraordinarily lucky to have those skills brought to bear on this book. I thank her warmly for all of her hard work.

Living aboard a voyaging boat is a wonderful life, but hardly conducive to writing. Nonetheless, most of this book was completed afloat, apart from some key stages where reliable internet access and a stable desk was essential. As a result I was entirely reliant on the kindness of friends who live more sedentary lifestyles, and who generously allowed me to share their space to work on the book during those periods, especially Guy Davies and Tracy Sherman in London, Dr Nick Tregenza in Cornwall, Mike Brown and Heather Johnson in Canada and Carol Hardman and Tim Williams in Kenya. I owe them all a heartfelt thank you.

Throughout my time at sea I was lucky enough to have the support of a wonderful cohort of first mates aboard a variety of vessels. All of them added something positive to the experience that benefited us all, volunteers and crew alike. As such I would like to offer my thanks to: Dr Lissa Batey, Leon Edwards, Louise Johnson, David Marshall, John Gyll-Murray, Carlos Ogier, Mick Roberts, Juliet Savage, Guy Speedie, Patrick Truscott and Larry Williams.

I would also like to thank Jackie Pearson for the astonishing level of commitment and time she expended in translating our paper-based data into digital format. Jackie made an immense contribution to the success of our project, and always in a spirit of good cheer.

Many hundreds of volunteers assisted us over the years, from Cornwall to the Outer Hebrides. All of them shared the adventure with us and whatever small successes we achieved are theirs too. Nothing would have been possible without their support, and we offer them all our sincere thanks.

I was lucky enough to work with some excellent people at a number of conservation NGOs over many years and during the period of the surveys we conducted around the British Isles. I would like to thank the following people:

At the Marine Conservation Society: Dr Bob Earll, Sam Famshawe, Dr Peter Richardson, Dr Jean-Luc Solandt

At the Shark Trust: Sarah Fowler OBE, Kelvin Boot, George Bowser, Clive James, Jeremy Stafford-Deitsch.

At the Wildlife Trusts: Joan Edwards, Dr Lissa Batey, Andrew Davis.

At WWF-UK: Dr Sian Pullen, Janet Powers, Sylvette Peplowski.

At Born Free: David Jay, Will Travers OBE.

At Earthwatch Institute (Europe): Jen Alger, Dr Michael Humphries, Nigel Winser, Nat Spring.

At Natural England: Roger Covey, Victoria Copley.

At Scottish Natural Heritage: Dr Suzanne Henderson, Dr Fiona Manson

I also benefited from the support and friendship of many wonderful people within the Basking Shark Community, including: Gary Burrows, Joe Breen, Graham and Jackie Hall, Dr Fiona Gell, Emmett Johnston, Armelle Jung and Sylvain Guérain.

And I owe a special word of thanks to Dr Alexander Godknecht of the Swiss Shark Foundation (Hai-Stiftung) for loyally supporting our project for so many years.

At various stages over the years I have benefited hugely from the advice, support and friendship of the following people: Matt Borne, Gary Burrows, Dr Johanna Bailey, Roger Covey, Mike Daines, Dr Peter Evans, Dr Bob Earll, Joan Edwards, Sarah Fowler OBE, Professor Brendan Godley, Dr Miles Hoskin, Ronnie Mackie, Dr Nick Tregenza, Stephen Westcott and Dr Matthew Witt.

I would also like to thank our supporters and volunteers from Cornwall Wildlife Trust and the Scottish Wildlife Trust, for their help and input throughout the survey.

On a practical note, I owe a sincere vote of thanks to the many people who helped to keep us going at sea and made us welcome at our home ports:

In Falmouth: Kevin Green of Marine Electrical Services, Chris Jones of Hardman Jones Marine Services.

In Arisaig: Ronnie Dyer and Martine Wagenaar, Graham and Susan MacLellan at Arisaig Marine, James MacLellan at Moidart Engineering.

In Mallaig: James McClean, Robert MacMillan and the staff at Mallaig Harbour Authority.

In Tobermory: Jim Traynor at Tobermory Harbour Authority.

I would also like to thank Mark and Charmian Entwistle of Isle of Skye Yacht Charters for providing us with excellent yachts during the latter years of the survey.

I owe a great debt to John Harries and Phyllis Nickel of Attainable Adventure Cruising for giving me a platform to write about sailing, marine life and wild places for many years. Without that training ground this book might never have happened.

Finally, I would like to pay tribute and thanks to all of the authors whose fantastic books provided me with inspiration and material for this book: Denis Fairfax, Joseph 'Tex' Geddes, Gavin Maxwell, Ken McNally, Patrick FitzGerald O'Connor, Scott Wallace and Brian Gisborne and Anthony Watkins.

Index

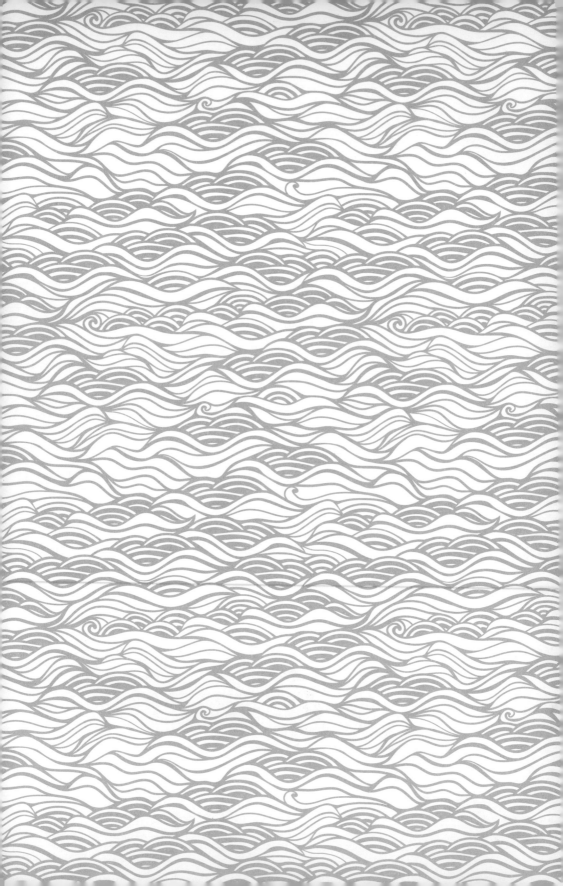